Secure Control of Networked Control Systems and Its Applications

Dong Yue · Songlin Hu · Zihao Cheng

Secure Control of Networked Control Systems and Its Applications

Springer

Dong Yue
Nanjing University of Posts
and Telecommunications
Nanjing, Jiangsu, China

Songlin Hu
Nanjing University of Posts
and Telecommunications
Nanjing, Jiangsu, China

Zihao Cheng
Henan University of Chinese Medicine
Zhengzhou, Henan, China

ISBN 978-981-33-6729-6 ISBN 978-981-33-6730-2 (eBook)
https://doi.org/10.1007/978-981-33-6730-2

This Springer imprint is published by the registered company Springer Nature Singapore Pte Ltd.
The registered company address is: 152 Beach Road, #21-01/04 Gateway East, Singapore 189721, Singapore

Preface

Networked control systems (NCSs) are a kind of closed-loop control systems where the sensors, the controllers and the actuators exchange data through a limited bandwidth communication network. With the rapid development of computer and communication technologies, NCSs have received increasing attention in the last two decades and have been widely applied in different fields including power system, transportation system and so on. Compared with traditional point-to-point control systems, NCSs offer many advantages, such as low cost, easy maintenance and increased system flexibility. Yet despite the many benefits of NCSs, they also bear some inherent challenges. Specifically, the use of band-limited communication network (especially wireless communication channel) increases the exposure of NCSs to vulnerabilities and cyberthreats, which leads to several challenges regarding cybersecurity and resilience while saving the limited communication resources.

Thus, secure control problems of NCSs have attracted much attentions from academia and industry in recent years. The investigation of secure control can be classified into twofold: resilient control and active defense control. Resilient control is currently a hot research topic in NCSs since it can provide the ability to maintain acceptable levels of operation of NCSs in the presence of abnormal conditions. Resilient control focuses on mitigating the negative effect of network attacks on system performance of NCSs. In general, the existing resilient control strategies guarantee the resilience of NCSs if attack signal satisfies a certain condition. In fact, it is not enough to guarantee the security of NCSs for system defenders, since the actual attack signal does not satisfy the designed resilient condition. Hence, active defense control methods need to be developed to enhance system security under arbitrary network attacks. To achieve active defense control, redundant resources like communication, computation and control are taken into consideration. The functionality of redundant resources is to preserve the secure operation of NCSs when the primary resources are exhausted by network attacks. The related problems of active defense control can be discussed in two steps. The first step is to construct primary-redundancy control structure by redundancy resource. The second step is to generate switching laws to manage redundancy resources from the perspective of system control. Within the framework of event-triggered stabilization, resilient control design problems of NCSs and multi-agent systems with various dynamics are investigated and elaborated in detail. An application of resilient control/consensus for a variety of power systems is introduced and some practical scenarios are provided to show the superiority of the resilient control methodology.

Structure and readership. This monograph is concerned with developing resilient control and estimation methods for NCSs under cyberattacks and its applications to power systems. In this book, Chap. 1 is a summary of security of NCSs, attack models, and security control methods, they are the premises of the following chapters. The rest of this monograph is structured into four parts, Part I is devoted to considering the resilient control synthesis of NCSs; Part II is devoted to designing resilient consensus control protocol of networked multi-agent systems; Part III is devoted to considering active defense control of NCSs; and Part IV is devoted to considering some applications of secure control methods on power system under network attacks.

Part I: The majority of the previous studies require that the feedback gain matrix of the resilient controller is known in advance, while few works consider the synthesis of feedback controller. In addition, most works on resilient control of NCSs so far are assumed that the sampled data are periodically transmitted by the sensors over the communication networks, i.e., in the time-triggered control framework. In general, a time-triggered control leads to inefficient utilization of limited network resources. Therefore, in Chap. 2, an event-triggered resilient state feedback controller synthesis method for NCSs under DoS attacks is proposed. In Chap. 3, observer-based resilient event-triggered state feedback controller design method is proposed for NCSs under DoS attacks. A observer-based resilient H_∞ state feedback controller design method is proposed for nonlinear cyber-physical systems under DoS attacks in Chap. 4.

Part II: Some control strategies that ensure normal operation of NCSs subject to disturbances and malicious DoS attacks have been established in Part I. They are called resilient control strategies in which the controller is designed to ensure system security. As a complement to the previous results shown in Part I, the objective of Part II is to introduce some resilient control strategies that ensure normal operation of a multi-agent system in the presence of cyberattacks. In Chap. 5, an observer-based distributed secure consensus control protocol is presented for a class of linear multi-agent systems subject to random DoS attacks. In Chap. 6, an event-based secure leader-following consensus control scheme is proposed for nonlinear multi-agent systems under both DoS and replay attacks.

Part III: Most works on resilient control of NCSs so far can be reviewed as passive defense control strategy. With the rapid development of computer network, the existing passive defensive technology cannot meet the increasing requirements about complex network security. Therefore, in Chap. 7, a model-based predictive control (MPC) framework is proposed by embedding predictors within the closed-loop NCSs, which can actively compensate the missing data caused by DoS attacks. In Chap. 8, bandwidth allocation-based switched dynamic triggering control strategy is proposed for NCSs under DoS attacks and time delay. To compensate the effect of the DoS attacks actively, a primary-redundancy (PR) communication structure on the basis of limited bandwidth allocation is also established.

Part IV: The main focus of the previous three parts is on developing resilient estimation and control methods for uncertain NCSs under cyberattacks. In this part, the developed methods are applied for a variety of power systems. Specifically, the resilient controller design method proposed in Chap. 2 is applied to examine the stabilization problem of DC microgrids with constant power load in Chap. 9, develop a resilient dynamic event-triggered controller design strategy for multi-area power systems with renewable energy penetration in Chap. 10, and propose a resilient load frequency controller for multi-area power systems with uncertainties and time delay in Chap. 11. In Chap. 12, secure distributed Optimal frequency regulation of power grid with time-varying voltages under false data injection attacks is discussed.

Nanjing, China Dong Yue
Nanjing, China Songlin Hu
Zhengzhou, China Zihao Cheng
August 2020

Acknowledgements

We would like to acknowledge the collaborations with Prof. Xiangpeng Xie and Dr. Hui Ge on the work of security control of cyber-physical system based on switching approach and security control of networked T-S fuzzy system, Prof. Yang Yang on the work of observer-based distributed secure consensus control, Prof. Jinliang Liu on the work of event-based secure leader-following consensus control and Prof. Chunxia Dou and Dr. Shengxuan Wen on the work of secure distributed optimal frequency regulation reported in the monograph. The supports from the Natural Science Foundation of Jiangsu Province of China under Grant (BK20202011 and BK20201377), the National Natural Science Foundation of China under Grant (61833008 and 61673223), the "Six Talent Peaks Project" of Jiangsu Province of China under Grant (RLD201810), the Qing Lan Project of Jiangsu Province of China under Grant (QL 04317006), and the 1311 Talent Project of Nanjing University of Posts and Telecommunications, are gratefully acknowledged.

Contents

Chapter 1
Introduction

In the past few decades, the development of new network technology has made it possible to apply to significantly different control systems. These applications have led to a series of new control architectures called Network Control Systems (NCSs). In NCSs, sensors, actuators and controllers are connected through a shared band-limited digital communication network. The advantages of using digital communication networks are very obvious, not only from an application perspective. However, the network also brings some challenges, such as bandwidth limitations, delays or packet loss. In order to solve the above problems, in the past two decades, a new theoretical basis has been established [1]. Recently, NCSs faces many potential threats from network attacks, which can invade system nodes or communication channels, as shown in Fig. 1.1. Therefore, the security of NCSs against malicious attackers has become the focus of attention of scholars [2]. Security in NCSs is a relatively new area of research, and it is the main theme of this book. We will discuss it in detail in the next section.

1.1 Security of Networked Control Systems

Wireless communication networks may make NCSs more vulnerable to attacks, because attackers can cause performance degradation or instability by intruding into the communication channel [3], thereby injecting malicious signals into the closed loop system. Network attacks on control systems have severely damaged many industrial systems [4, 5]. One of the most famous examples is the attack on Maroochy Shire Council's sewage control system in Queensland, Australia in January 2000 [6]. The attacker hacked into the controller and modified the electronic data of certain sewer pumping stations, causing one million liters of untreated sewage to enter the drain and flow from there to the local waterways. Another incident is the StuxNet

D. Yue et al., *Secure Control of Networked Control Systems and Its Applications*,
https://doi.org/10.1007/978-981-33-6730-2_1

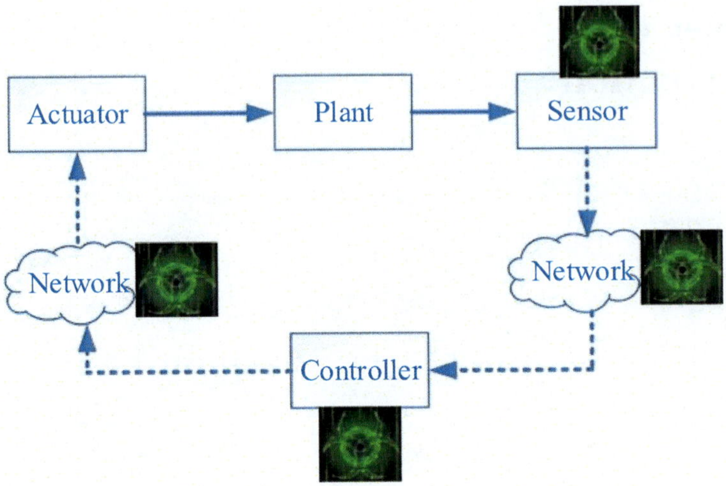

Fig. 1.1 NCS under network attacks

virus, which targets the Siemens supervisory control and data acquisition system used in many industrial systems [7, 8]. In 2014, the computer of a German steel plant was hacked through a small environmental control system, causing the steel plant's blast furnace to be destroyed on a large scale. In addition to these incidents, a report released by the U.S. Department of Homeland Security showed that from 2009 to 2015, the number of cyber attacks on critical U.S. infrastructure has increased significantly. Therefore, active defensive measures are needed to resist attacks against NCSs.

1.2 Network Attack Incidents in Power Systems

With the increased dependency on communication network, cyber attacks bring serious threats to security operation of power systems. For the production and living of modern society is closely relying on power systems, power outage would lead to disaster. The types of attacks on the power industry are divided into ransomware, DDoS attacks, APT attacks, vulnerabilities, and malware. In addition to ordinary power companies, nuclear power plants are also a key target of cyber attacks. Once a nuclear power plant is attacked, serious consequences such as loss of employees or trade secret information, reactor shutdown, or physical damage may occur.

In 2003, the Davis–Besse nuclear power plant in USA was attacked by the SQL Slammer worm. The network data transmission volume increased sharply. The computer processing speed of the nuclear power plant was slowed down. The process control computer stopped working for several hours. In 2006, the Browns Ferry nuclear power plant in Alabama, USA was attacked by a network attack. The reactor's recirculation pump and condensing demineralization controller malfunctioned and caused 3 unit to be shut down. In 2007, Aurora project of Idaho national laboratory proofed that network attacks can successfully destroy the controller. In the experiment, it was found that hackers could successfully open and close the circuit breaker of diesel generators through loopholes in the system. Further, it would cause power system explosion and shocks or even paralysis [9]. In 2010, Iron nuclear power system was intruded by Stuxnet. It leaded to one thousand centrifuges paralyzed by attacks [10]. In 2015 ukrainian power outage, Black Energy attacked the monitoring and management system. Three power areas were attacked. Seven 110 KV substations and 23 35 KV substations failed which affected 225,000 people [11, 12]. In 2019, the Venezuelan power system was attacked causing three times large-scale power outages. The reason for the power outage was the deliberate destruction of network attacks on the Guli hydropower station.

1.3 Network Attack Models

Note that the traditional IT community has been protecting the network components of NCSs to prevent cyber attacks. However, IT security tools cannot handle the dynamic processes of NCSs physical components. The physical layer of NCSs can be manipulated by attack without tampering with IT components. Therefore, in order to complement IT security methods, we need systems and control theory perspectives to protect NCSs. The control community mainly focuses on resilience to attacks, that is, tolerating a certain degree of confrontational attacks. The first step towards tolerance of attacks is to model cyber attacks. In the following sections, we will introduce classic attack models.

1.3.1 Denial-of-Service Attacks

Denial-of-Service (DoS) attack aims to damage availability of information. How does DoS attack intrude on information system? One way is to launch jamming signals in communication channels such that the system transmitted packets can not arrive in their destination. The other way is to paralyze communication system or exhaust communication resources such that the transmission service is broken or rejected for system lawful tasks. When DoS attacks occur, transmission packets will be lost or dropped. However, the time characteristics of packet drops are different from that of network induced. Because attackers could formulate various attack strategies with malicious attention while network induced packet loss often obey fixed probability distribution. Several attack model are given to represent various attack behaviors.

(1) Random Model

The attacks possess random nature since the successes of the attacks largely depends on the detection ability of protection equipment or software, the communication protocols and the network conditions (e.g., network load, network congestion, and network transmission rate) that are typically randomly fluctuated [13].

(1) Bernoulli distribution: Wang [5] and Yuan [14] described the randomness behavior of DoS attacks.

$$\bar{y}_k = \gamma_k y_k, \tag{1.1}$$

where γ_k subjected to Bernoulli distribution takes value 0 or 1.

(2) Markov model: Sun hongtao [15] proposed a Markov jump model to describe DoS attack duration.

$$p_{ij}(a_i, d_j) = Pr(\gamma_j(k+1) = j|\gamma_i(k) = i) \tag{1.2}$$

where the $p_{ij}(a_i, d_j)$ represents that packet dropouts caused by DoS attacks shift from i to j if attackers adopt attack strategy a_i and defenders adopt defend strategy d_j.

(2) Periodic DoS Jamming Attack

Foroush [3] studied periodic DoS jamming attacks with constrained energy.

$$u_{\text{jmd}}(t) = \begin{cases} 1, & (n-1)T \le t \le (n-1)T + T_{\text{off}}, \\ 0, & (n-1)T + T_{\text{off}} \le t \le nT, \end{cases} \tag{1.3}$$

where $n \in \mathbb{N}$ is the period number, T, T_{off} are period and the length of sleep interval, respectively.

(3) Average Dwell Time Model

Persis [16, 17] presented a general attack model by adopting average dwell time to constraint upper bound of attack duration and frequency.

(1) DoS frequency: There exist $\eta \in \mathbb{R}_{\ge 0}$ and $\tau_D \in \mathbb{R}_{\Delta}$ such that

$$n(\tau, t) \le \eta + \frac{t-\tau}{\tau_D} \tag{1.4}$$

for all $\tau, t \in \mathbb{R}_{\ge 0}$ with $t \ge \tau$.

(2) DoS duration: There exist $\kappa \in \mathbb{R}_{\ge 0}$ and $T \in \mathbb{R}_1$ such that

$$|\varXi(\tau, t)| \le \kappa + \frac{t-\tau}{T} \tag{1.5}$$

for all $\tau, t \in \mathbb{R}_{\ge 0}$ with $t \ge \tau$.

(4) Model-Based Average Dwell Time Model

Based on the model-based average dwell time concept [18], Cheng [19] proposed a more general model with concept of model-based average dwell time. This model described upper bound of average dwell time of attack interval and lower bound of average dwell time of sleep interval.

(1) ADT of sleep interval: Denote $N_F(t, t_0)$ the numbers of sleep intervals and $\Xi_F(t, t_0)$ the sum of sleep intervals length during $[t_0, t]$. There exist scalars $\zeta_F \in R_{\geq 0}$, $\tau_F \in R_{\geq h}$ satisfying

$$N_F(t, t_0) \leq \zeta_F + \frac{\Xi_F(t, t_0)}{\tau_F} \tag{1.6}$$

(2) ADT of attack interval: Denote $N_D(t, t_0)$ the numbers of attack intervals and $\Xi_D(t, t_0)$ the sum of attack intervals length during $[t_0, t]$. There exist scalars $\zeta_D \in R_{\geq 0}$, $\tau_D \in R_{\geq h}$ satisfying

$$N_D(t, t_0) \geq \zeta_D + \frac{\Xi_D(t, t_0)}{\tau_D} \tag{1.7}$$

1.3.2 Deception Attacks

Deception attack aims to damage integrity of information. For example, measurements or control commands can be tampered illegally by deception attack. Controller or actuator will operate by using false data caused by deception attack. Then, feedback control under deception attack cumulates control errors to lead to system instability. Specially, False data injection attack is one of typical deception attack, which injects false measurements and commands without being detected [20–22]. Other attack type like zero dynamic attack as one kind of false data injection attacks has been studied concretely [23, 24]. In following, several deception attack models are introduced.

(1) Bounded Deception Attack Model

Most of deception attack models are studied by treating deception attack as outer disturbance with bound. The occurrence of deception attack is in form of stochastic or deterministic feature. Ding [13] presented a stochastic deception attacks on measurements and controller outputs.

$$\begin{aligned} &u_k = \tilde{u}_k + \alpha_k r_k, \quad y_k = \tilde{y}_k + \sigma_k \nu_k, \\ &r_k = -\tilde{u}_k + \zeta_{1k}, \quad \nu_k = -\tilde{y}_k + \zeta_{2k}, \\ &\|\zeta_{1k}\| + \|\zeta_{2k}\| \leq \varrho. \end{aligned} \tag{1.8}$$

where $\tilde{u}_k$, $\tilde{y}_k$ are the controller outputs and the sensor measurements. The stochastic variables α_k, σ_k are subjected to independent Bernoulli distribution. r_k, ν_k represent the signals injected by deception attacker.

(2) Unbounded Deception Attack Model

Unknown attack strategy is the nature of deception attack. The study [25] relaxes the constraints on bounded deception attacks and investigates an unknown unbounded deception attack with the proposed distributed attack-resilient control framework.

$$\tilde{u} = u + \delta \tag{1.9}$$

where $\tilde{u}$ is the corrupted command and δ with bounded $\dot{\delta}$ represents the potential unbounded injections launched in the control communication channel.

(3) Stealthy Attack

Stealthy attack is implemented through false data injection attacks [26, 27]. Attackers formulate stealthy strategy without being detected by the existing algorithms with combining the known system configuration information. In [26], both online and offline cases of stealthy attacks have been studied in distributed energy management problem of smart grid.

(4) Linear Function Model

Guo [28] proposed linear function model for deception attack. This kind of model has its structure to describe more attack behaviors.

$$\tilde{y}_k = T_k y_k + b_k, \tag{1.10}$$

where T_k is an arbitrary matrix and $b_k \sim \mathcal{N}(0, \mathfrak{L})$ is an i.i.d. Gaussian random variable which is independent of y_k.

1.3.3 Replay Attacks

Replay attacks repeat the messages of the compromised controllers (sensors) to actuators (controllers). The implementation of replay attacks requires no knowledge of system dynamics. Meanwhile, the repeated messages are difficult to be detected. In practice, replay attacks have been successfully launched by Stuxnet virus. Replay attacks are executed with the following steps [29]: (i) erases the messages sent by the lawful transmission nodes; (ii) sends the previous data stored in adversary memory to system nodes to be compromised; (iii) repeat above steps;

1.3.4 Multiple Attacks

In practice attack scenario, attackers would take multiple attacks to achieve the maximum profits of attack. Multiple attacks are launched by adopted more than two kinds of attack simultaneously. For example, DoS attack cooperates with deception attack; replay attack cooperates with DoS attack. Ding [30] considered random DoS attacks and deception attacks.

$$y_k = \alpha_k(\tilde{y}_k + \beta_k \nu_k) + (1 - \alpha_k) y_{k-1}, \tag{1.11}$$

and

$$\begin{aligned} &\nu_k = -\tilde{y}_k + \xi_k, \ \|\xi_k\| \le \delta, \\ &\text{Prob}\{\alpha_k = 0\} = 1 - \bar{\alpha}, \text{Prob}\{\alpha_k = 1\} = \bar{\alpha}, \\ &\text{Prob}\{\beta_k = 0\} = 1 - \bar{\beta}, \text{Prob}\{\beta_k = 1\} = \bar{\beta}, \end{aligned}$$

where $\bar{\alpha} \in [0, 1)$ and $\bar{\beta} \in [0, 1)$.

1.4 Secure Control Methods

To preserve system performance under network attack, secure techniques based on control method are important and necessary. Secure control method focus on physical system performance ignored by network security techniques. Secure control method consists of two class of methodologies. One is resilient control method to guarantee system performance can tolerate some level of attack. The other one is active defense control method that improve the security of networked control systems to against arbitrary attack.

1.4.1 Resilient Control Methods

To explore the influence of attack on NCS performance, resilient control methods are provided with considering attack modeling, system modeling under attack and resilient analysis and synthesis. Most of researches have paid attention to resilient control methods, such as, stochastic control method, Lyapunov stability method, combined method LKF with switched system method.

(1) Event-Based Control

Event-based control methods combine the design of the feedback control and the transmission mechanism. Under the assumption of event-driven mode in controller and actuator, event-triggered mechanism regulates the control time to achieve system stability as well as resources reduction. In case of DoS attacks , however, event-triggered mechanism plays a new role in guaranteeing sufficient transmission for feedback control. In [16], Lyapunov function method is applied for input-to-state stability of networked control system under DoS attacks. Event-based sampling logics are proposed to achieve suitable transmission schedule. In [31], a resilient event-triggering communication scheme is proposed and a stabilization criteria is derived by using uniform Lyapunov–Krasovskii functional method. In [32], co-designed of event-triggered mechanism and resilient control is proposed by combing piecewise Lyapunov–Krasovskii functional method and switched system theory.

(2) Stochastic Control

Considering the randomness of attack behaviors, control systems under attacks are formulated by stochastic system. In [15], Markovian process is used to describe DoS attack behavior and a Markovian jump linear system is established for NCS under DoS attacks. Lyapunov method and LMIs technique are utilized to design the resilient control gain. In [13], discrete-time stochastic nonlinear systems under deception attacks is secure with probability $1 - \varepsilon$ and with an upper bounded for the given quadratic cost functional by employing the discrete time version of ISS in probability (ISSiP).

(3) Game-Theory Based Resilient Control

To model and analyze the interaction between defender and attacker, game theoretical framework are applied for security of NCSs. In [33], an optimal stationary cyber-policy is proposed to ensure system security through an infinite-horizon zero-sum stochastic game between a defender and an attacker. In [14], a unified game approach is applied to study the resilient control of NCS under DoS attacks. Optimal criteria for both cyber defenders and DoS attackers is obtained with the optimal control method.

1.4.2 Active Defense Control Methods

The desired resilient performance of NCS can be preserved within the tolerated attack level under resilient control methods. When strong attack breaks the resilient condition, however, the performance of NCS can not be guaranteed by resilient control methods. Hence, NCS with resilient control can only tolerate some kind of weak attack negatively. Thus, it is necessary to study active defense control method to enhance the security of NCS positively. Active defense control methods often make use of more communication or computation resources to support the security method such as, predictive control method, redundancy-based control method.

(1) Predictive Control Method

Predictive control is an effective method to compensate for network-induced delays and packet dropouts actively. For the influence of DoS attacks can be viewed as packet dropouts, predictive control scheme is still suitable. Further, false data detected can be replaced by the predictive commands generated by predictive control. In [34], predictive cloud control method is proposed to compensate packet dropouts caused by DoS attacks actively. In [35], a recursive networked predictive control (RNPC) method is proposed to compensate for the weak DoS attacks.

(2) Redundant Control Method

In practice, redundant channels are simultaneously equipped in networked control systems. Redundant controls are supported with the redundant channels for a same control task [36]. In case of network attacks, the primary channel would be compromised to lost its function of providing safe and reliable service for feedback control. Then, the directed and efficient way to eliminate the influence of attacks is to build a new secure control channels by starting up redundant channels.

In [35], the switched system theory is applied for the closed-loop stability. For the strong DoS attacks , a multi-controller switching scheme is presented. In [37], a primary-redundancy (PR) communica- tion structure is presented to prevent DoS attacks. The convex combination method is proposed to design switching law for PR channel with the function of keeping the concealment of redundancy channel.

1.5 Structure of This Book

The book is organized as follows: This chapter introduces NCS: background, security, overview; Part I consists of Chaps. 2–4, which is devoted to resilient control of networked control systems ; Part II consists of Chaps. 5 and 6, which refer to resilient consensus control of multi-agent systems; Part III includes Chaps. 7 and 8, devoting to active Defense Control of Networked Control System; Part IV includes Chaps. 9–12, devoting to the application of secure control method on power system.

In Chap. 2, the event-based controller synthesis for networked control systems (NCSs) under resilient event-triggering communication scheme (RETCS) and periodic denial-of-service (DoS) jamming attacks is proposed.

In Chap. 3, observer-based event-triggered control for a continuous networked linear system subjects to denial-of-service (DoS) attacks is proposed.

In Chap. 4, switching system approach is proposed to study the finite-time nonlinear system robust stability under intermittent DoS jamming attacks.

In Chap. 5, event-based secure leader-following consensus for multi-agent systems (MASs) with multiple cyber attacks is obtained by using Lyapunov stability theory.

In Chap. 6, an observer-based distributed secure consensus control strategy for a class of linear multi-agent systems (MASs) with random attacks is proposed.

In Chap. 7, a model-based predictive control (MPC) framework is proposed for networked T-S fuzzy system under intermittent DoS jamming attack.

In Chap. 8, a primary-redundancy (PR) communication structure is proposed to against DoS attacks.

In Chap. 9, the attack-resilient event-triggered controller of a DC microgrid under DoS attacks is designed by event-triggered control method.

In Chap. 10, resilient dynamic event-triggered control is proposed for frequency stabilization of multi-area power systems with renewable energy penetration under transmission delay and denial-of-service (DoS) attacks.

In Chap. 11, resilient additional control design of a multi-area power system with uncertainty and physical constraint under denial-of-service (DoS) attacks and transmission delay is proposed with delay system method and switched system method.

In Chap. 12, a novel type of secure distributed optimal frequency regulation is developed by introducing secure-based communication network to improve the resilience against cyber-attack.

References

1. J. Baillieul, P.J. Antsaklis, Control and communication challenges in networked real-time systems. Proc. IEEE **95**(1), 9–28 (2007)
2. S. Amin, G.A. Schwartz, S.S. Sastry, Security of interdependent and identical networked control systems. Automatica **49**(1), 186–192 (2013)
3. H.S. Foroush, S. Martinez, On event-triggered control of linear systems under periodic denial-of-service jamming attacks, in *Proceedings of the 51st Annual Conference on Decision and Control* (2012), pp. 2551–2556
4. A.A. Cardenas, S. Amin, S. Sastry, Secure control: Towards survivable cyber-physical systems, in *International Conference on Distributed Computing Systems Workshops* (2008), pp. 495–500
5. D. Wang, Z. Wang, B. Shen, F.E. Alsaadi, T. Hayat, Recent advances on filtering and control for cyber-physical systems under security and resource constraints. J. Franklin Inst. **353**(11), 2451–2466 (2016)
6. L. Bayou, D. Espes, N. Cuppensboulahia, F. Cuppens, *Security Issue of WirelessHART Based SCADA Systems* (Springer International Publishing, Berlin, 2015)
7. R.S. Smith, Covert misappropriation of networked control systems: presenting a feedback structure. IEEE Control Syst. **35**(1), 82–92 (2015)
8. E.D. Knapp, *Industrial Network Security: Securing Critical Infrastructure Networks for Smart Grid, SCADA, and Other Industrial Control Systems*. Syngress (2014)
9. M. Zeller, Myth or reality does the aurora vulnerability pose a risk to my generator?, in *2011 64th Annual Conference for Protective Relay Engineers* (2011), pp. 130–136
10. N. Falliere, L.O. Murchu, E. Chien, W32.stuxnet dossier, *White Paper* (2011)
11. G. Liang, S.R. Weller, J. Zhao, F. Luo, Z.Y. Dong, The 2015 ukraine blackout: implications for false data injection attacks. IEEE Trans. Power Syst. **32**(4), 3317–3318 (2017)
12. R.M. Lee, *Analysis of the cyber attack on the ukrainian power grid*. Technical Report, SANS Institute and Electricity Information Sharing and Analysis Center(E-ISAC) (2016)
13. D. Ding, Z. Wang, Q.-L. Han, G. Wei, Security control for discrete-time stochastic nonlinear systems subject to deception attacks. IEEE Trans. Syst. Man Cybern.: Syst. **48**(5), 779–789 (2018)
14. Y. Yuan, H. Yuan, L. Guo, H. Yang, S. Sun, Resilient control of networked control system under dos attacks: a unified game approach. IEEE Trans. Ind. Inf. **12**(5), 1786–1794 (2016)
15. H. Sun, C. Peng, T. Yang, H. Zhang, W. He, Resilient control of networked control systems with stochastic denial of service attacks. Neurocomputing **270**, 170–177 (2017)
16. C.D. Persis, P. Tesi, Input-to-state stabilizing control under denial-of-service. IEEE Trans. Autom. Control **60**(11), 2930–2944 (2015)
17. S. Feng, P. Tesi, Resilient control under denial-of-service: Robust design. Automatica **79**, 42–51 (2017)
18. X. Zhao, Y. Kao, B. Niu, T. Wu, *Control Synthesis of Switched Systems* (Springer, Cham, 2017)

19. Z. Cheng, D. Yue, S. Hu, C. Huang, C. Dou, X. Ding, Resilient dynamic event-triggered control for multi-area power systems with renewable energy penetration under dos attacks. IET Control Theory & Appl. (2020).https://doi.org/10.1049/iet-cta.2019.1478
20. Z. Pang, G. Liu, D. Zhou, F. Hou, D. Sun, Two-channel false data injection attacks against output tracking control of networked systems. IEEE Trans. Ind. Electr. **63**(5), 3242–3251 (2016)
21. X. Liu, Z. Li, Z. Li, Optimal protection strategy against false data injection attacks in power systems. IEEE Trans. Smart Grid **8**(4), 1802–1810 (2017)
22. K. Manandhar, X. Cao, F. Hu, Y. Liu, Detection of faults and attacks including false data injection attack in smart grid using kalman filter. IEEE Trans. Control Netw. Syst. **1**(4), 370–379 (2014)
23. A. Rasoolzadeh, F.R. Salmasi, Mitigating zero dynamic attack in communication link-enabled droop-controlled hybrid ac/dc microgrids. IET Cyber-Phys. Syst.: Theory Appl. **5**(2), 207–217 (2020)
24. A. Hoehn, P. Zhang, Detection of covert attacks and zero dynamics attacks in cyber-physical systems, in *American Control Conference (ACC) 2016* (2016), pp. 302–307
25. S. Zuo, T. Altun, F.L. Lewis, A. Davoudi, Distributed resilient secondary control of dc microgrids against unbounded attacks. IEEE Trans. Smart Grid **11**(5), 3850–3859 (2020)
26. C. Zhao, J. He, P. Cheng, J. Chen, Analysis of consensus-based distributed economic dispatch under stealthy attacks. IEEE Trans. Ind. Electr. **64**(6), 5107–5117 (2017)
27. M. Liu, C. Zhao, R. Deng, P. Cheng, W. Wang, J. Chen, Nonzero-dynamics stealthy attack and its impacts analysis in dc microgrids, in *American Control Conference (ACC) 2019* (2019), pp. 3922–3927
28. Z. Guo, D. Shi, K.H. Johansson, L. Shi, Optimal linear cyber-attack on remote state estimation. IEEE Trans. Control Netw. Syst. **4**(1), 4–13 (2016)
29. M. Zhu, S. Martinez, On the performance analysis of resilient networked control systems under replay attacks. IEEE Trans. Autom. Control **59**(3), 804–808 (2013)
30. D. Ding, Z. Wang, G. Wei, F.E. Alsaadi, Event-based security control for discrete-time stochastic systems. IET Control Theory & Appl. **10**(15), 1808–1815 (2016)
31. C. Peng, J. Li, M. Fei, Resilient event-triggering H_∞ load frequency control for multi-area power systems with energy-limited DoS attacks. IEEE Trans. Power Syst. **32**(5), 4110–4118 (2017)
32. S. Hu, D. Yue, X. Xie, X. Chen, X. Yin, Resilient event-triggered controller synthesis of networked control systems under periodic DoS jamming attacks. IEEE Trans. Cybern. **49**(12), 4271–4281 (2019)
33. Q. Zhu, T. Basar, Game-theoretic methods for robustness, security, and resilience of cyber-physical control systems: games-in-games principle for optimal cross-layer resilient control systems. IEEE Control Syst. **35**(1), 46–65 (2015)
34. H. Yang, S. Ju, Y. Xia, J. Zhang, Predictive cloud control for networked multiagent systems with quantized signals under DoS attacks. IEEE Trans. Syst. Man Cybern.: Syst. (2019). https://doi.org/10.1109/TSMC.2019.2896087
35. Z. Pang, G. Liu, Z. Dong, Secure networked control systems under denial of service attacks. IFAC Proceed. Vol. **44**(1), 8908–8913 (2011)
36. Y. Zhu, L. Zhang, W.X. Zheng, Distributed H_∞ filtering for a class of discrete-time markov jump lur systems with redundant channels. IEEE Trans. Ind. Electr. **63**(3), 1876–1885 (2015)
37. S. Hu, Z. Cheng, D. Yue, C. Dou, Y. Xue, Bandwidth allocation-based switched dynamic triggering control against DoS attacks. IEEE Trans. Syst. Man Cybern.: Syst. 1–12 (2019), Early Access. https://doi.org/10.1109/TSMC.2019.2956945

Part I
Resilient Control of Networked Control System

Chapter 2
Resilient Event-Triggered Controller Synthesis of NCSs Under Periodic DoS Attacks

This chapter addresses the event-based controller synthesis problem for networked control systems (NCSs) under resilient event-triggering communication scheme (RETCS) and periodic denial-of-service (DoS) jamming attacks is studied. Firstly, a new periodic RETCS is designed under the assumption that the DoS attacks imposed by power-constrained Pulse-Width Modulated (PWM) jammers is partially identified, i.e., the period of the jammer and a uniform lower bound on the jammer's sleeping periods are known. Secondly, a new state error dependent switched system model is constructed, including the impacts of the RETCS and DoS attacks. According to this new model, the exponential stability criteria are derived by using piecewise Lyapunov functional (PLF). In these criteria, the relationship among DoS parameters, the triggering parameters, the sampling period, and the decay rate are quantitatively characterized. Then, a criterion is also proposed to obtain the explicit expressions of the triggering parameter and event-based state feedback controller gain simultaneously. Finally, the obtained theoretical results are verified by a satellite yaw angles control system [1].

This chapter is organized as follows. The description of system security problem is given in Sect. 2.1. Exponential stability analysis and resilient control synthesis are provided in Sects. 2.2 and 2.3. A simulation example about a satellite system is given in Sect. 2.4. Finally, Sect. 2.5 concludes the chapter.

2.1 Problem Formulation

2.1.1 System Description

The configuration of a NCS under RETCS and periodic DoS attacks is shown in Fig. 2.1, where the signal transmission between event generator (controller) and con-

D. Yue et al., *Secure Control of Networked Control Systems and Its Applications*,
https://doi.org/10.1007/978-981-33-6730-2_2

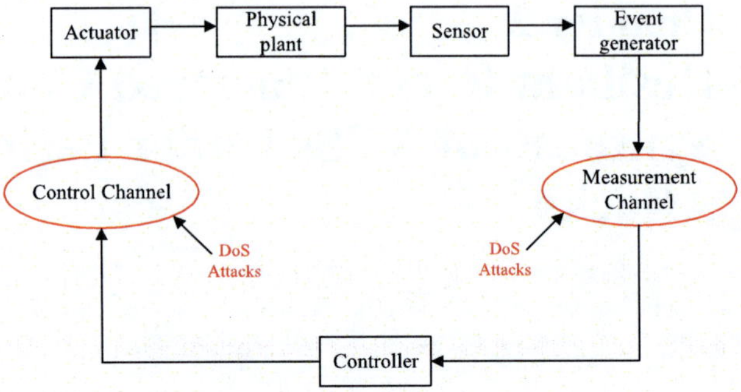

Fig. 2.1 The framework of a NCS under resilient event-triggering scheme and DoS attacks

troller (actuator) is implemented over a wireless network subjects to periodic DoS jamming attacks. In what follows, we will introduce the mathematical model of physical plant, event-triggered controller, periodic DoS jamming attacks, and resilient event-triggered communication scheme, respectively.

The state-space representation of the physical plant is given by the following linear time-invariant (LTI) system:

$$\dot{x}(t) = Ax(t) + Bu(t) \tag{2.1}$$

where $x(t) \in \mathbb{R}^n$ is the state vector of the controlled plant, $u(t) \in \mathbb{R}^m$ is the control input. A and B are known system matrices with appropriate dimensions.

As seen in Fig. 2.1, the control action is implemented over a wireless channel. In the absence of DoS attacks, let $t_k h$ $(k = 0, 1, 2, \ldots)$ be the triggering times at which it is desired to update the control action, where $t_0 \triangleq 0$, $h > 0$ is the sampling period of the sensor. In this situation, whatever the event-triggered communication scheme (it will be designed later) generates the time sequence $\{t_k h\}_{k\in\mathbb{N}}$, the triggered state data can be transmitted to the controller successfully, and the control signal can also be safely received by actuator [2], and hence the event-based control input denoted by $\tilde{u}(t)$ can be represented as

$$\tilde{u}(t) = Kx(t_k h), \forall t \in \left[t_k h, t_{k+1} h\right), k \in \mathbb{N} \tag{2.2}$$

where K is the control gain to be designed later.

To achieve the desired control objectives, some common assumptions are necessary for system (2.1).

Assumption 2.1 The pair (A, B) is controllable.

Assumption 2.2 The sensor is time-triggered, the controller and the actuator are event-triggered.

Assumption 2.3 The states of the controlled plant are fully measurable and transmitted with a single packet. The network-induced delays, data losses and disorder are not considered.

2.1.2 *Periodic DoS Attack Model and Control Objective*

In practice, the sensor periodically samples the state signal of the controlled plant and sends it to an event generator. Whether the sampled state needs to be sent to a remote controller via a wireless network channel is determined by a predefined event-triggering condition (its definition will be given later). The attacker attempts to jam the wireless communication links to destabilize the stability of the considered system (see Fig. 2.1). Following [3], we consider the following DoS jamming signal, which aims at blocking the network channels intermittently:

$$S_{DoS}(t) = \begin{cases} 1, t \in \left[nT, nT + T_{off}\right) \\ 0, t \in \left[nT + T_{off}, nT + T\right) \end{cases} \tag{2.3}$$

where the sequence $\{nT\}_{n\in\mathbb{N}}$ denotes the time instants when the DoS jamming attack is sleeping, the set $\left[nT, nT + T_{off}\right)$ represents the sleeping time interval of the DoS jamming attack, of a length $T_{off} \in \mathbb{R}_{>0}$ $\left(T_{off} < T\right)$ over which communication is possible. Further, the time interval $\left[nT + T_{off}, nT + T\right)$ denotes the set of time instants where the DoS jamming attack is active and communication is interrupted. Here, as stated in [4], there is no need to keep the parameter T_{off} time-invariant which recalls PWM jamming, and thus it is assumed that there exists an uniform real scalar $T_{off}^{\min} \in \mathbb{R}_{>0}$ such that $T_{off}^{\min} \leq T_{off} < +\infty$ holds for all $n \in \mathbb{N}$.

In general, the DoS jamming signal can affect forward and feedback channels separately. In this work, we consider the scenario that DoS jamming signals simultaneously affect both forward and feedback channels as in [2]. That is to say, under DoS jamming attacks, the released sampled state cannot reach the controller successfully, and the control signal cannot arrive at the actuator successfully too. In view of this, the output of the actuator in Fig. 2.1 $u(t)$ under DoS attacks (2.3) can be represented as

$$u(t) = \begin{cases} Kx\left(t_{k,n+1}h\right), t \in \left[t_{k,n+1}h, t_{k+1,n+1}h\right) \\ \cap\left[nT, nT + T_{off}\right) \\ 0, t \in \left[nT + T_{off}, nT + T\right) \end{cases} \tag{2.4}$$

where $\left\{t_{k,n}h\right\}$ denotes the set of successful control update instants $\left(t_{0,n+1}h \triangleq nT\right)$, which are generated by the event-triggering mechanism to be designed, $h\ (< T)$ is the sampling period, $k \in \{0, 1, \ldots, k(n)\} \triangleq \mathcal{K}(n)$ with $n \in \mathbb{N}$ and

$$k(n) = \sup\left\{k \in \mathbb{N} \mid nT + T_{off} \geq t_{k,n+1}h\right\}.$$

Remark 2.1 Different from the work [5], in which the control input is based on the latest received triggered state, in this chapter, the control input is always zero during DoS time intervals. This idea is taken from [6].

In what follows, to simplify representation, for $n \in \mathbb{N}$, define $\mathcal{I}_{1,n} \triangleq [nT, nT + T_{off})$, $\mathcal{I}_{2,n} \triangleq [nT + T_{off}, nT + T)$, $\mathcal{R}_{k,n} \triangleq [t_{k,n+1}h, t_{k+1,n+1}h)$, $k \in \mathcal{K}(n)$. Using (2.4), the state equation in (2.1) can be written as

$$\dot{x}(t) = \begin{cases} Ax(t) + BKx(t_{k,n+1}h) \\ t \in \mathcal{R}_{k,n} \cap \mathcal{I}_{1,n} \\ Ax(t), t \in \mathcal{I}_{2,n}. \end{cases} \tag{2.5}$$

Now, we formulate our main control objective: given positive scalars T and $T_{off}\left(\geq T_{off}^{\min}\right)$, jointly design a resilient event-triggered communication scheme and an event-based state-feedback controller (2.4) while guaranteeing the exponential stability of the switched system (2.5) under DoS jamming attacks (2.3).

2.1.3 Design of Resilient Event-Triggering Scheme

In this subsection, we will introduce a new RETCS which is used to remove the impact of the periodic DoS jamming attacks.

We first recall that when the effect of the jamming attacks is not considered, Yue [7] presented the following event triggering condition:

$$\begin{aligned} &[x(t_k h + lh) - x(t_k h)]^T W [x(t_k h + lh) - x(t_k h)] \\ &> \sigma x^T(t_k h) Wx(t_k h) \end{aligned} \tag{2.6}$$

where $\sigma \in (0, 1)$ is a predefined parameter, $t_k h + lh$ $(k, l \in \mathbb{N})$ represents the current sampling time, $t_k h$ denotes the last triggering instant at which the control signal is updated. Now, if the impact of DoS jamming attacks is taking into account, then the control signal is blocked over the DoS jamming time intervals, which results in the event-triggering control sequence generated by (2.6) cannot achieve the stability of the considered system. Thus, the traditional event triggering scheme is not applied directly here. Next, we will modify the above triggering scheme (2.6) to adapt to the DoS jamming attacks (2.3). Toward this end, similar to [4], we consider the jammer is constantly maintaining a "worst-case jamming scenario", i.e., $T_{off} \equiv T_{off}^{\min}$ for all $n \in \mathbb{N}$.

Definition 2.1 We define the triggering instant shown in (2.4) as:

$$t_{k,n+1}h = \left\{t_{k_l}h \text{ satisfying (2.6)} \mid t_{k_l}h \in \mathcal{I}_{1,n}\right\} \cup \{(n+1)T\} \tag{2.7}$$

where $n, t_{k_l}, k_l, l \in \mathbb{N}$, k denotes the number of event occurring in $(n+1)^{th}$ attack period.

Remark 2.2 From (2.7), it is observed that the triggering time-sequence $\{t_{k,n+1}h\}$ generated by (2.6) in $(n+1)^{th}$ jamming action period lies in either time interval $\mathcal{I}_{1,n}$

or right endpoint nT of each interval $[nT, (n+1)T)$. That is, if no event happens in $\mathcal{I}_{1,n}$, namely, $\{t_{k_l}h \text{ satisfying (2.6)} \mid t_{k_l}h \in \mathcal{I}_{1,n}\} = \emptyset$, then the triggering instant only occurs at the right endpoint $(n+1)T$. In addition, zeno phenomenon is excluded naturally in the proposed event-triggered framework.

2.1.4 Modeling of Event-Based Resilient Switched Control System

In this section, to make use of the time delay approach and switched system approach to analyze the system stability, we first convert the system (2.5) with event-triggering strategy (2.7) into an equivalent switched time delay system. For this purpose, we divide the event intervals $\mathcal{R}_{k,n}$ $(\forall k \in \mathcal{K}(n), \forall n \in \mathbb{N})$ shown in (2.5) into sampling interval-like subintervals, that is,

$$\begin{aligned}\mathcal{R}_{k,n} = & \cup_{m=1}^{\rho_{k,n}} \left[t_{k,n+1}h + (m-1)h, t_{k,n+1}h + mh\right) \\ & \cup \left[t_{k,n+1}h + \lambda_{k,n}h, t_{k+1,n+1}h\right)\end{aligned} \tag{2.8}$$

where $\rho_{k,n} \triangleq \sup\{m \in \mathbb{N} \mid t_{k,n+1}h + mh < t_{k+1,n+1}h\}$. Let

$$\begin{cases} \mathcal{D}_{k,n}^{m} = \left[t_{k,n+1}h + (m-1)h, t_{k,n+1}h + mh\right), \\ \mathcal{D}_{k,n}^{\rho_{k,n}+1} = \left[t_{k,n+1}h + \rho_{k,n}h, t_{k+1,n+1}h\right). \end{cases} \tag{2.9}$$

Note that

$$\mathcal{I}_{1,n} = \cup_{k=0}^{k(n)} \{\mathcal{R}_{k,n} \cap \mathcal{I}_{1,n}\} \subseteq \cup_{k=0}^{k(n)} \mathcal{R}_{k,n}. \tag{2.10}$$

Combining (2.8), (2.9) and (2.10), the interval $\mathcal{I}_{1,n}$ can be rewritten as

$$\mathcal{I}_{1,n} = \cup_{k=0}^{k(n)} \cup_{m=1}^{\rho_{k,n}+1} \{\mathcal{D}_{k,n}^{m} \cap \mathcal{I}_{1,n}\}$$

Set

$$\Phi_{k,n}^{m} = \mathcal{D}_{k,n}^{m} \cap \mathcal{I}_{1,n} \tag{2.11}$$

then

$$\mathcal{I}_{1,n} = \cup_{k=0}^{k(n)} \cup_{m=1}^{\rho_{k,n}+1} \Phi_{k,n}^{m}.$$

Now, two piecewise functions are defined below:

$$\eta_{k,n}(t) = \begin{cases} t - t_{k,n+1}h, \ t \in \Phi_{k,n}^{1} \\ t - t_{k,n+1}h - h, \ t \in \Phi_{k,n}^{2} \\ \vdots \\ t - t_{k,n+1}h - \rho_{k,n}h, \ t \in \Phi_{k,n}^{\rho_{k,n}+1} \end{cases} \tag{2.12}$$

and

$$\delta_{k,n}(t) = \begin{cases} 0, t \in \Phi_{k,n}^{1} \\ x\left(t_{k,n+1}h\right) - x\left(t_{k,n+1}h + h\right), t \in \Phi_{k,n}^{2} \\ \vdots \\ x\left(t_{k,n+1}h\right) - x\left(t_{k,n+1}h + \rho_{k,n}h\right), \\ t \in \Phi_{k,n}^{\rho_{k,n}+1}. \end{cases} \tag{2.13}$$

According to the definitions of $\eta_{k,n}(t)$ and $\delta_{k,n}(t)$, it can be seen that $\eta_{k,n}(t) \in [0, h), t \in \mathcal{R}_{k,n} \cap \mathcal{I}_{1,n}$. Then the successfully event-triggered sampled state $x\left(t_{k,n}h\right)$ can be described as $x\left(t_{k,n+1}h\right) = x_{k,n}(t) + \delta_{k,n}(t)$ with $x_{k,n}(t) = x\left(t - \eta_{k,n}(t)\right)$, $t \in \mathcal{R}_{k,n} \cap \mathcal{I}_{1,n}$. Based on the above analysis, the switched system (2.5) can be represented as

$$\begin{cases} \dot{x}(t) = \begin{cases} Ax(t) + BKx_{k,n}(t) + BK\delta_{k,n}(t), \\ t \in \mathcal{R}_{k,n} \cap \mathcal{I}_{1,n} \\ Ax(t), t \in \mathcal{I}_{2,n}, k \in \mathcal{K}(n) \end{cases} \\ x(t) = \varsigma(t), t \in [-h, 0] \end{cases} \tag{2.14}$$

where $\varsigma(t)$ is the supplemented initial condition of the state $x(t)$ on $[-h, 0]$ with $\varsigma(0) \triangleq \varsigma_0$, and $\delta_{k,n}(t)$ satisfies

$$\begin{aligned} \delta_{k,n}^{T}(t) W \delta_{k,n}(t) \leq \sigma \left[x_{k,n}(t) + \delta_{k,n}(t)\right]^{T} \\ W\left[x_{k,n}(t) + \delta_{k,n}(t)\right] \end{aligned} \tag{2.15}$$

In what follows, in order to construct a PLF for system analysis, inspired by [8], define $\sigma(t) = 1$ for $t \in [-h, 0] \cup \left(\bigcup_{n\in\mathbb{N}} \mathcal{I}_{1,n}\right)$, while $\sigma(t) = 2$ for $t \in \bigcup_{n\in\mathbb{N}} \mathcal{I}_{2,n}$. Then for $\sigma(t) = i \in \{1, 2\}$, let $t_{i,n} = \begin{cases} nT, i = 1 \\ nT + T_{off}, i = 2 \end{cases}$, we can obtain that $\mathcal{I}_{i,n} = \left[t_{i,n}, t_{3-i,n+i-1}\right), \sigma\left(t_{i,n}\right) = i, \sigma\left(t_{i,n}^{-}\right) = 3 - i$.

According to the definition of $\sigma(t)$, the switched system (2.14) is reformulated as

$$\begin{cases} \dot{x}(t) = A_i x(t) + \mathcal{B}_i x_{k,n}(t) + \mathcal{B}_i \delta_{k,n}(t), \\ t \in \left[t_{i,n}, t_{3-i,n+i-1}\right), n \in \mathbb{N} \\ x(t) = \varsigma(t), t \in [-h, 0] \end{cases} \tag{2.16}$$

where $A_i = A$, $\mathcal{B}_1 \triangleq BK$, $\mathcal{B}_2 \triangleq \mathbf{0}$, and $\delta_{k,n}(t)$ satisfies (2.7).

Based on the above analysis and the reports in the literature [9], we choose the following PLF for switched system (2.16):

$$\begin{aligned} V_{\sigma(t)}(t) &= x^T(t)P_{\sigma(t)}x(t) \\ &+ \int_{t-h}^{t} x^T(\theta) f(\bullet) Q_{\sigma(t)} x(\theta)\, d\theta \\ &+ \int_{-h}^{0}\int_{t+\theta}^{t} \dot{x}^T(\varpi) f(\bullet) R_{\sigma(t)} \dot{x}(\varpi)\, d\varpi\, d\theta \\ &+ \int_{-h}^{0}\int_{t+\theta}^{t} \dot{x}^T(\varpi) f(\bullet) Z_{\sigma(t)} \dot{x}(\varpi)\, d\varpi\, d\theta \end{aligned} \tag{2.17}$$

where $P_{\sigma(t)}$, $Q_{\sigma(t)}$, $R_{\sigma(t)}$, and $Z_{\sigma(t)}$ are symmetric positive definite matrices, $\alpha_{\sigma(t)}$ is a positive scalar, and $f(\bullet) \triangleq e^{2(-1)^{\sigma(t)}\alpha_{\sigma(t)}(t-s)}$.

Then, the main problems to be dealt with in this chapter are summarized below.

P1: Consider a NCS under resilient event-triggering mechanism (2.7) and DoS jamming signal $\mathcal{S}_{DoS}(t)$ (2.3) shown in Fig. 2.1. Suppose that the sequence $\{nT\}_{n\in\mathbb{N}}$, the DoS parameters $T_{off}^{\min}\ (= T_{off})$ and $T\ (T \in (T_{off}, +\infty))$ in $\mathcal{S}_{DoS}(t)$ (2.3) are known. For given system matrices A, B in (2.1), event-triggering parameter σ in (2.7), and state-feedback gain K in (2.4), how to guarantee the switched system (2.16) is exponentially stable (ES).

P2: Consider a NCS under resilient event-triggering mechanism (2.7) and DoS jamming signal $\mathcal{S}_{DoS}(t)$ (2.3) shown in Fig. 2.1. Suppose that the sequence $\{nT\}_{n\in\mathbb{N}}$, the DoS parameters $T_{off}^{\min}\ (= T_{off})$ and $T\ (T \in (T_{off}, +\infty))$ in $\mathcal{S}_{DoS}(t)$ (2.3) are known. For given system matrices A, B in (2.1), event-triggering parameter σ in (2.7), how to determine event-triggering weighting matrix W in (2.7) and state-feedback gain K in (2.4) while guaranteeing the exponential stability of the switched system (2.16).

2.2 Exponential Stability Analysis

For $\sigma(t) = i \in \{1, 2\}$, using the piecewise switching Lyapunov functional (2.17), we first give a decay estimation of $V_i(t)$ along the trajectories of the system (2.16), which is useful to derive the stability conditions in the sequel.

Lemma 2.1 *Assume the sequence $\{nT\}_{n\in\mathbb{N}}$, the parameters T and T_{off} in $\mathcal{S}_{DoS}(t)$ (2.3) are known. Consider the switched system (2.16) with a given gain matrix K and triggering parameter $\sigma \in (0, 1)$, for given scalars $h \in (0, T_{off})$ and $\alpha_i \in (0, +\infty)$, $i \in \{1, 2\}$, if there exist symmetric matrices $P_i > 0$, $Q_i > 0$, $R_i > 0$, $Z_i > 0$, $W > 0$, and matrices M_i, N_i, S_i, $i \in \{1, 2\}$, satisfying the following LMIs:*

$$\Sigma_i = \begin{bmatrix} \Sigma_{11}^i & \sqrt{h}M_i & \sqrt{h}N_i & \sqrt{h}S_i & \Sigma_{15}^i \\ \star & \Sigma_{22}^i & 0 & 0 & 0 \\ \star & \star & \Sigma_{33}^i & 0 & 0 \\ \star & \star & \star & \Sigma_{44}^i & 0 \\ \star & \star & \star & \star & \Sigma_{55}^i \end{bmatrix} < 0 \tag{2.18}$$

where $\Sigma_{11}^i = \Pi_{i1} + \Pi_{i2} + \Pi_{i2}^T$, $\Sigma_{22}^i = \Sigma_{33}^i = -e^{2(-1)^i \alpha_i \lambda_i h} R_i$, $\Sigma_{44}^i = -e^{2(-1)^i \alpha_i \lambda_i h} Z_i$, $\lambda_1 \triangleq 1$, $\lambda_2 \triangleq 0$, $\Sigma_{15}^i = \sqrt{h} \mathcal{A}_i^T (R_i + Z_i)$, $\Sigma_{55}^i = -R_i - Z_i$, $\mathcal{A}_i = \left[A_i \; \mathcal{B}_i \; 0 \; \mathcal{B}_i \right]$ *with*

$$\Pi_{11} = \begin{bmatrix} \Pi_{11}^{11} & P_1\mathcal{B}_1 & 0 & P_1\mathcal{B}_1 \\ \star & \sigma W & 0 & \sigma W \\ \star & \star & -e^{-2\alpha_1 h} Q_1 & 0 \\ \star & \star & \star & (\sigma - 1) W \end{bmatrix}$$

$$\Pi_{21} = \begin{bmatrix} He(P_2 A) + Q_2 - 2\alpha_2 P_2 & 0 & 0 \\ \star & 0 & 0 \\ \star & \star & -Q_2 \end{bmatrix}$$

$$\Pi_{11}^{11} = He(P_1 A) + Q_1 + 2\alpha_1 P_1$$
$$\Pi_{12} = \left[M_1 + S_1 \; -M_1 + N_1 \; -S_1 - N_1 \; 0 \right]$$
$$\Pi_{22} = \left[M_2 + S_2 \; -M_2 + N_2 \; -S_2 - N_2 \right]$$

then for $t \in \left[t_{i,n}, t_{3-i,n+i-1}\right)$, $i \in \{1, 2\}$, $n \in \mathbb{N}$, *the function defined in* (2.17) *satisfies*

$$V_i(t) \leq e^{2(-1)^i \alpha_i (t - t_{i,n})} V_i\left(t_{i,n}\right). \tag{2.19}$$

Proof $\forall\, k \in \mathcal{K}(n), n \in \mathbb{N}$, for $t \in \mathcal{R}_{k,n} \cap \mathcal{I}_{1,n}$, differentiating $V_1(t)$ along the trajectories of the system (2.16) with $i = 1$ gives

$$\begin{aligned} \frac{dV_1(t)}{dt} \leq & -2\alpha_1 V_1(t) + x^T(t) 2\alpha_1 P_1 x(t) \\ & + \dot{x}^T(t) 2P_1 x(t) - x_h^T(t) e^{-2\alpha_1 h} Q_1 x_h(t) \\ & + h\dot{x}^T(t)(R_1 + Z_1)\dot{x}(t) + x^T(t) Q_1 x(t) \\ & - \vartheta_1 - \vartheta_2 - \vartheta_3 \\ & + 2\xi^T(t)(S_1 \Upsilon_1 + M_1 \Upsilon_2 + N_1 \Upsilon_3) \end{aligned} \tag{2.20}$$

where $\xi(t) = \left[x^T(t) \; x_{k,n}^T(t) \; x_h^T(t) \; \delta_{k,n}^T(t) \right]^T$, $x_h(t) = x(t - h)$, $\vartheta_1 = \int_{t-h}^t \dot{x}^T(s) e^{-2\alpha_1 h} Z_1 \dot{x}(s)\, ds$, $\vartheta_2 = \int_{t-\eta_{k,n}(t)}^t \dot{x}^T(s) e^{-2\alpha_1 h} R_1 \dot{x}(s)\, ds$, $\vartheta_3 = \int_{t-h}^{t-\eta_{k,n}(t)} \dot{x}^T(s) e^{-2\alpha_1 h} R_1 \dot{x}(s)\, ds$, $0 = \Upsilon_1 \triangleq x(t) - x_h(t) - \int_{t-h}^t \dot{x}(\theta)\, d\theta$, $0 = \Upsilon_2 \triangleq x(t) - x_{k,n}(t) - \int_{t-\eta_{k,n}(t)}^t \dot{x}(\theta)\, d\theta$, $0 = \Upsilon_3 \triangleq x_{k,n}(t) - x_h(t) - \int_{t-h}^{t-\eta_{k,n}(t)} \dot{x}(\theta)\, d\theta$. Applying the similar analysis method as in [10], we have

$$\begin{aligned} \frac{dV_1(t)}{dt} \leq & -2\alpha_1 V_1(t) + \xi^T(t) \Big[\Sigma_{11}^1 + h M_1 e^{2\alpha_1 h} R_1^{-1} M_1^T \\ & + h N_1 e^{2\alpha_1 h} R_1^{-1} N_1^T \\ & + h S_1 e^{2\alpha_1 h} Z_1^{-1} S_1^T \\ & + h \mathcal{A}_1^T (R_1 + Z_1) \mathcal{A}_1 \Big] \xi(t). \end{aligned} \tag{2.21}$$

Taking the Schur complement of the matrix $\Sigma_1 < 0$ in (2.18), we get $\Sigma_{11}^1 + hM_1 e^{2\alpha_1 h} R_1^{-1} M_1^T + hN_1 e^{2\alpha_1 h} R_1^{-1} N_1^T + hS_1 e^{2\alpha_1 h} Z_1^{-1} S_1^T + h\mathcal{A}_1^T (R_1 + Z_1) \mathcal{A}_1 < 0$, which implies that $\frac{dV_1(t)}{dt} + 2\alpha_1 V_1(t) \le 0, t \in \mathcal{R}_{k,n} \cap \mathcal{I}_{1,n}$. Recalling that the arbitrary of k, it follows that $\forall t \in [t_{1,n}, t_{2,n})$,

$$V_1(t) \le e^{-2\alpha_1(t - t_{1,n})} V_1(t_{1,n}). \tag{2.22}$$

Next, differentiating $V_2(t)$ with respect to $t \in [t_{2,n}, t_{1,n+1})$ along the trajectories of the system (2.16) with $i = 2$ yields

$$\begin{aligned}\frac{dV_2(t)}{dt} \le 2\alpha_2 V_2(t) + \xi^T(t) \Big[&\Sigma_{11}^2 + hM_2 R_2^{-1} M_2^T \\ &+hN_2 R_2^{-1} N_2^T \\ &+hS_2 Z_2^{-1} S_2^T \\ &+h\mathcal{A}_2^T (R_2 + Z_2) \mathcal{A}_2 \Big] \xi(t).\end{aligned} \tag{2.23}$$

Using $\Sigma_2 < 0$ in (2.18), we deduce that $\frac{dV_2(t)}{dt} \le 2\alpha_2 V_2(t)$, which implies that $\forall t \in [t_{2,n}, t_{1,n+1})$

$$V_2(t) \le e^{2\alpha_2(t - t_{2,n})} V_2(t_{2,n}). \tag{2.24}$$

Therefore, based on the aforementioned analysis of two cases, the conditions (2.18) ensure the inequalities (2.19) are true. The proof is completed.

In what follows, on the basis of Lemma 2.1, the exponential stability criteria of the system (2.16) is proposed below.

Theorem 2.1 *Assume the sequence $\{nT\}_{n\in\mathbb{N}}$, the parameters T and T_{off} in $\mathcal{S}_{DoS}(t)$ (2.3) are known. Consider the switched system (2.16) with a given gain matrix K and triggering parameter $\sigma \in (0, 1)$, for given scalars $\mu_i \in (1, +\infty)$, $h \in (0, T_{off})$, and $\alpha_i \in (0, +\infty)$, $i \in \{1, 2\}$, satisfying (2.25)*

$$\begin{aligned} 0 < \lambda \triangleq\ & 2\alpha_1 T_{off} - 2\alpha_2 (T - T_{off}) \\ & -2(\alpha_1 + \alpha_2) h - \ln(\mu_1 \mu_2), \end{aligned} \tag{2.25}$$

if there exist symmetric matrices $P_i > 0$, $Q_i > 0$, $R_i > 0$, $Z_i > 0$, $W > 0$, and matrices M_i, N_i, S_i, $i \in \{1, 2\}$, such that the LMIs (2.18), and the following condition (2.26) is satisfied:

$$\begin{cases} P_1 \le \mu_2 P_2,\ P_2 \le \gamma \mu_1 P_1 \\ Q_i \le \mu_{3-i} Q_{3-i} \\ R_i \le \mu_{3-i} R_{3-i} \\ Z_i \le \mu_{3-i} Z_{3-i} \end{cases} \tag{2.26}$$

where $\gamma = e^{2(\alpha_1 + \alpha_2)h}$. Then the event-based switched system (2.16) under (2.3) is ES with decay rate $\rho \triangleq \frac{\lambda}{2T}$.

Proof Choose a PLF candidate as $V(t) = \begin{cases} V_1(t), t \in \mathcal{R}_{k,n} \cap \mathcal{I}_{1,n} \\ V_2(t), t \in \mathcal{I}_{2,n}. \end{cases}$ From Lemma 2.1, we obtain that for any $t \geq 0$

$$V(t) \leq \begin{cases} e^{-2\alpha_1(t-t_{1,n})} V_1(t_{1,n}), t \in [t_{1,n}, t_{2,n}) \\ e^{2\alpha_2(t-t_{2,n})} V_2(t_{2,n}), t \in [t_{2,n}, t_{1,n+1}). \end{cases} \tag{2.27}$$

According to (2.26), it is easy to show that

$$\begin{cases} \mu_2 V_2(t_{1,n}^-) - V_1(t_{1,n}) \geq 0 \\ \gamma\mu_1 V_1(t_{2,n}^-) - V_2(t_{2,n}) \geq 0. \end{cases} \tag{2.28}$$

Considering that $\forall\, t \geq 0$, we can always find a non-negative integer $n \in \mathbb{N}$ such that $t \in [t_{1,n}, t_{2,n})$ or $t \in [t_{2,n}, t_{1,n+1})$. Thus, two cases should be considered below.

Case 1: If $t \in [t_{1,n}, t_{2,n})$, it follows from (2.27) and (2.28) that

$$\begin{aligned} V(t) &\leq \mu_2 e^{-2\alpha_1(t-t_{1,n})} V_2(t_{1,n}^-) \\ &\ \vdots \\ &\leq e^{-\lambda n} V_1(t_{1,0}) = e^{-\lambda n} V_1(0). \end{aligned} \tag{2.29}$$

Notice that $t < t_{2,n} = nT + T_{off}$, i.e., $n > \frac{t-T_{off}}{T}$. Substituting this relation into (2.29) gives

$$V(t) \leq V_1(0) e^{\frac{\lambda T_{off}}{T}} e^{-\frac{\lambda}{T}t}. \tag{2.30}$$

Case 2: If $t \in [t_{2,n}, t_{1,n+1})$, by a similar procedure as above, we can obtain

$$V(t) \leq \frac{V_1(0)}{\mu_2} e^{-\frac{\lambda}{T}t}. \tag{2.31}$$

Now, define $M = \max\left\{e^{\frac{\lambda T_{off}}{T}}, \frac{1}{\mu_2}\right\}$, $\varepsilon_1 = \min\{\lambda_{\min}(P_i)\}$, $\varepsilon_2 = \max\{\lambda_{\max}(P_i)\}$, and $\varepsilon_3 = \varepsilon_2 + h\lambda_{\max}(Q_1) + \frac{h^2}{2}\lambda_{\max}(R_1 + Z_1)$, from (2.30) and (2.31), it yields that

$$V(t) \leq M e^{-\frac{\lambda}{T}t} V_1(0). \tag{2.32}$$

According to the definition of $V(t)$ in (2.17), we have

$$V(t) \geq \varepsilon_1 \|x(t)\|^2,\ V_1(0) \leq \varepsilon_3 \|\varsigma_0\|_h^2. \tag{2.33}$$

Combining (2.32) and (2.33), we have

$$\|x(t)\| \leq \sqrt{\frac{M\varepsilon_3}{\varepsilon_1}} e^{-\rho t} \|\varsigma_0\|_h, \forall t \geq 0 \tag{2.34}$$

which implies that the system (2.16) is ES with decay rate ρ.

Remark 2.3 It follows from (2.25) that the decay rate ρ can be explicitly expressed as

$$\rho = \frac{2(\alpha_1 + \alpha_2)\beta_1 - 2\alpha_2 T - \ln(\mu_1\mu_2)}{2T} \tag{2.35}$$

$$= \frac{2\alpha_1\beta_1 - 2\alpha_2\beta_2 - \ln(\mu_1\mu_2)}{2T} \tag{2.36}$$

where $\beta_1 = T_{off} - h$, $\beta_2 = T - T_{off} + h$. It is observed that the decay rate ρ in (2.35) is a monotony increase function of the parameter T_{off} (when T, α_i, h, and μ_i are fixed). The larger the T_{off} (it means that the sleeping period of jamming is becoming bigger and, correspondingly, the active period of jamming denoted by $T - T_{off}$ is becoming smaller), the bigger the decay rate ρ. Therefore, in order to seek for larger value of ρ (to guarantee the better stability performance), T_{off} should be chosen as large as possible while ensuring (2.25). This point will be validated in the simulation section. On the other hand, one can see from (2.35) that the decay rate ρ is a monotony decrease function of the sampling period h (when T, T_{off}, α_i, and μ_i are fixed) and the parameters μ_i (when T, T_{off}, α_i, and h are fixed). In addition, it can be concluded from (2.36) the decay rate ρ is a monotony increase (decrease) function of the parameter α_1 (α_2) (when T, T_{off}, α_2 (α_1), h, and μ_i are fixed). Therefore, the sampling period h, the parameters μ_i, and α_2 should be chosen small enough while the parameter α_1 should be chosen large enough to obtain some large value of ρ on the premise of guaranteeing the existence of the feasible solutions of the LMIs (2.26) with (2.25).

2.3 Resilient Controller Synthesis

Based on Theorem 2.1, the joint-design criterion of the resilient event-triggering parameter W in (2.7) and the event-based controller gain matrix K is derived below.

Theorem 2.2 *Assume the sequence $\{nT\}_{n\in\mathbb{N}}$, the parameters T and T_{off} in $S_{DoS}(t)$ (2.3) are known. Consider the switched system (2.16). For give scalars $\sigma \in (0, 1)$, $\delta_i \in (0, +\infty)$, $\varepsilon_i \in (0, +\infty)$, $\chi_i \in (0, +\infty)$, $\eta_i > 0$, $\alpha_i \in (0, +\infty)$, $\mu_i \in (1, +\infty)$, and $h \in (0, T_{off})$ satisfying (2.25), if there exist proper symmetric matrices $X_i > 0$, $\hat{Q}_i > 0$, $\hat{R}_i > 0$, $\hat{Z}_i > 0$, $\hat{W} > 0$, and matrices Y, $\hat{M}_i$, $\hat{N}_i$, $\hat{S}_i$, $i \in \{1, 2\}$, satisfying the following LMIs:*

$$\hat{\Sigma}_i < 0 \tag{2.37}$$

$$\begin{bmatrix} -\mu_2 X_2 & X_2 \\ \star & -X_1 \end{bmatrix} \le 0 \tag{2.38}$$

$$\begin{bmatrix} -\mu_1 e^{2(\alpha_1+\alpha_2)h} X_1 & X_1 \\ \star & -X_2 \end{bmatrix} \le 0 \tag{2.39}$$

$$\begin{bmatrix} -\mu_{3-i} \hat{Q}_{3-i} & X_{3-i} \\ \star & \varepsilon_i^2 \hat{Q}_i - 2\varepsilon_i X_i \end{bmatrix} \le 0 \tag{2.40}$$

$$\begin{bmatrix} -\mu_{3-i} \hat{R}_{3-i} & X_{3-i} \\ \star & \chi_i^2 \hat{R}_i - 2\chi_i X_i \end{bmatrix} \le 0 \tag{2.41}$$

$$\begin{bmatrix} -\mu_{3-i} \hat{Z}_{3-i} & X_{3-i} \\ \star & \eta_i^2 \hat{Z}_i - 2\eta_i X_i \end{bmatrix} \le 0 \tag{2.42}$$

where

$$\hat{\Sigma}_i = \begin{bmatrix} \Lambda_i & \left(\Lambda_{21}^i\right)^T & \sqrt{h}\hat{\mathcal{A}}_i^T \\ \star & \Lambda_{22}^i & 0 \\ \star & \star & \delta_i^2\left(\hat{R}_i + \hat{Z}_i\right) - 2\delta_i X_i \end{bmatrix}$$

with $\Lambda_1 = \left\{\Lambda_{1ij}\right\}$ *and* $\Lambda_2 = \left\{\Lambda_{2ij}\right\}$ *being symmetric matrices composed of the matrices*

$$\begin{aligned}
\Lambda_{111} &= 2\alpha_1 X_1 + He(AX_1) + \hat{Q}_1 + He\left(\hat{M}_{11} + \hat{S}_{11}\right) \\
\Lambda_{121} &= Y^T B^T + \hat{M}_{12} + \hat{S}_{12} - \hat{M}_{11}^T + \hat{N}_{11}^T \\
\Lambda_{122} &= \sigma \hat{W} + He\left(\hat{N}_{12} - \hat{M}_{12}\right) \\
\Lambda_{131} &= \hat{M}_{13} + \hat{S}_{13} - \hat{S}_{11}^T - \hat{N}_{11}^T \\
\Lambda_{132} &= -\hat{M}_{13} + \hat{N}_{13} - \hat{S}_{12}^T - \hat{N}_{12}^T \\
\Lambda_{133} &= -e^{-2\alpha_1 h}\hat{Q}_1 - He\left(\hat{S}_{13} - \hat{N}_{13}\right) \\
\Lambda_{141} &= Y^T B^T + \hat{S}_{14} + \hat{M}_{14} \\
\Lambda_{142} &= -\hat{M}_{14} + \hat{N}_{14} + \sigma \hat{W} \\
\Lambda_{143} &= -\hat{S}_{14} - \hat{N}_{14},\ \Lambda_{144} = \sigma \hat{W} - \hat{W} \\
\Lambda_{21}^1 &= \sqrt{h}\left[\hat{M}_1\ \hat{N}_1\ \hat{S}_1\right]^T \\
\hat{\mathcal{A}}_1 &= \left[AX_1\ BY\ 0\ BY\right] \\
\Lambda_{22}^1 &= e^{-2\alpha_1\lambda_1 h} diag\left\{-\hat{R}_1, -\hat{R}_1, -\hat{Z}_1\right\} \\
\Lambda_{211} &= -2\alpha_2 X_2 + He(AX_2) + \hat{Q}_2 + He\left(\hat{M}_{21} + \hat{S}_{21}\right) \\
\Lambda_{221} &= \hat{M}_{22} + \hat{S}_{22} - \hat{M}_{21}^T + \hat{N}_{21}^T
\end{aligned}$$

$$\Lambda_{222} = He\left(\hat{N}_{22} - \hat{M}_{22}\right)$$
$$\Lambda_{231} = \hat{M}_{23} + \hat{S}_{23} - \hat{N}_{21}^{T} - \hat{S}_{21}^{T}$$
$$\Lambda_{232} = -\hat{M}_{23} + \hat{N}_{23} - \hat{N}_{22}^{T} - \hat{S}_{22}^{T}$$
$$\Lambda_{233} = -\hat{Q}_2 - He\left(\hat{N}_{23} + \hat{S}_{23}\right)$$
$$\Lambda_{21}^{2} = \sqrt{h}\left[\,\hat{M}_2\ \hat{N}_2\ \hat{S}_2\,\right]^{T}$$
$$\hat{A}_2 = \left[\,AX_2\ 0\ 0\ 0\,\right]$$
$$\Lambda_{22}^{2} = diag\left\{-\hat{R}_2, -\hat{R}_2, -\hat{Z}_2\right\}.$$

Then, the proposed resilient controller (2.4) with $K = YX_1^{-1}$ exponentially stabilizes the system (2.1) under event-triggering scheme (2.7) with $W = X_1^{-1}\hat{W}X_1^{-1}$ and periodic DoS jamming signals (2.3).

Proof Define $X_i = P_i^{-1}$, $J_1 = diag\{J_3, X_1, X_1, X_1, I\}$,$J_2 = diag\{J_4, J_4, I\}$, $J_3 = diag\{X_1, X_1, X_1, X_1\}$, $J_4 = diag\{X_2, X_2, X_2\}$, then pre- and post-multiplying J_1 and its transpose on both sides of Σ_1 in (2.18), pre- and post-multiplying J_2 and its transpose on both sides of Σ_2 in (2.18), then introducing some new matrix variables as $\hat{Q}_i = X_iQ_iX_i$, $\hat{R}_i = X_iR_iX_i$, $\hat{Z}_i = X_iZ_iX_i$, $\hat{S}_1 = J_3S_1X_1$, $\hat{S}_2 = J_4S_2X_2$, $\hat{M}_1 = J_3M_1X_1$, $\hat{M}_2 = J_4M_2X_2$, $\hat{N}_1 = J_3N_1X_1$, $\hat{N}_2 = J_4N_2X_2$, $Y = KX_1$, and $\hat{W} = X_1WX_1$, and using the relations $R_i = X_i^{-1}\hat{R}_iX_i^{-1}$, $Z_i = X_i^{-1}\hat{Z}_iX_i^{-1}$, and $(R_i + Z_i)^{-1} = X_i\left(\hat{R}_i + \hat{Z}_i\right)^{-1}X_i \geq 2\delta_iX_i - \delta_i^2\left(\hat{R}_i + \hat{Z}_i\right)$, we find $\hat{\Sigma}_i < 0$ guarantees $\Sigma_i < 0$ holds correspondingly. Furthermore, pre- and post-multiplying the first inequality in (2.26) and the second inequality in (2.26) by X_2 and X_1, respectively, and applying the Schur complement formula, it follows that (2.38) and (2.39) are equivalent to the first inequality in (2.26) and the second inequality in (2.26), respectively. Following the preceding derivations, it can be concluded that the LMIs (2.40)–(2.42) ensure the third inequality in (2.26)-the fifth inequality in (2.26) hold, respectively. This completes the proof.

Remark 2.4 Theorem 2.2 depends nonlinearly on the design variables T, T_{off}, μ_i, α_i, σ, h, δ_i, ε_i, χ_i, and η_i, $i = 1, 2$. However, by fixing the parameters T, T_{off}, μ_i, α_i, σ, h, δ_i, ε_i, χ_i, and η_i $(i = 1, 2)$, matrix inequalities (2.37)–(2.42) become LMIs, in which case the parameters X_1, $\hat{W}$ and Y can be synthesized numerically via the standard LMI Toolbox in MATLAB, and in turn the co-design of event-triggering parameter $W = X_1^{-1}\hat{W}X_1^{-1}$ and the resilient control gain matrix $K = YX_1^{-1}$ is realized. Furthermore, earlier studies [7, 11] have shown that a bigger σ results in a bigger average release period, which implies the communication resource is saved, and thus we can expect that the switched system can tolerant more DoS attacks at the expense of higher sampling rate. Therefore, to some extent, the current method provides a meaningful option to the designer to consider a tradeoff between tolerance for DoS attacks and the data transmission reduction objectives of the closed loop system.

2.4 Simulation Example

A satellite control system (see Fig. 2.2) modeled as (2.1) with the following system parameters [12] is employed to validate the effectiveness of the developed event-based resilient control strategy:

$$A = \begin{bmatrix} 0 & 0 & 1 & 0 \\ 0 & 0 & 0 & 1 \\ -0.3 & 0.3 & -0.004 & 0.004 \\ 0.3 & -0.3 & 0.004 & -0.004 \end{bmatrix}, \; B = \begin{bmatrix} 0 \\ 0 \\ 1 \\ 0 \end{bmatrix}.$$

For this example, by simple calculations, it is found that the eigenvalues of the system matrix A are 0, 0, $-0.04 - 0.4224j$, $-0.04 + 0.4224j$, which means that this satellite system is unstable without the control input.

Joint-design of triggering parameter W and control gain K: In this section, we first show the joint-design method of the triggering parameter W and control parameter K. To this end, we consider the jammer, imposing signal $\mathcal{S}_{DoS}(t)$ (2.3), where $T = 2\,\text{s}$ and $T_{off} = 1.8\,\text{s}$. Choosing $\mu_1 = \mu_2 = 1.05$, $\alpha_1 = 0.16$, $\alpha_2 = 0.5$, $h = 0.02\,\text{s}$ satisfying (2.25). For given $\sigma = 0.2$, $\delta_i = 8$, $\varepsilon_i = 8$, $\chi_i = 8$, $\eta_i = 8$, $i = 1, 2$, applying Theorem 2.2, the matrices W and K are given by (we do not give all the obtained parameters here due to page limitation)

$$W = \begin{bmatrix} 44.9898 & -21.7786 & 89.6892 & 60.3423 \\ \star & 10.6420 & -43.5119 & -29.2242 \\ \star & \star & 179.5712 & 120.5600 \\ \star & \star & \star & 81.5863 \end{bmatrix} \tag{2.43}$$

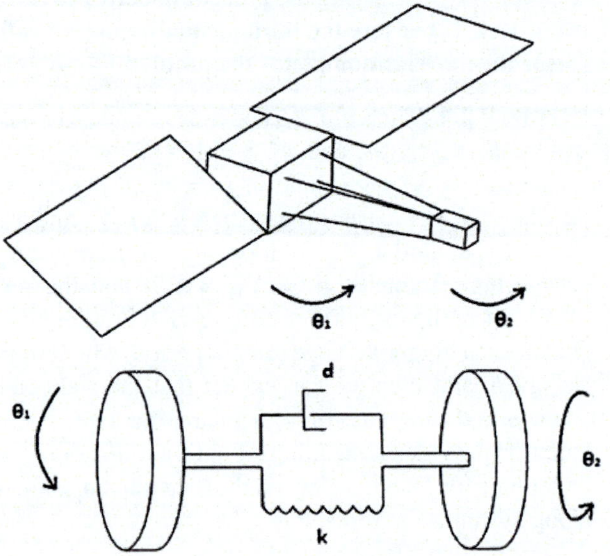

Fig. 2.2 Diagram of the satellite systems

and

$$K = \begin{bmatrix} -1.9822 \ 0.9207 \ -3.8818 \ -2.5386 \end{bmatrix}. \tag{2.44}$$

In the following, for simulation, the initial condition is set to be $x_0 = \begin{bmatrix} 1 \ 0.5 \ 0.3 \ 0.2 \end{bmatrix}^T$ and simulation time T_{end} is assumed to be 30 s. Figure 2.3 depicts the state trajectories of the satellite system controlled by the RETC (2.4) with (2.43) and (2.44). The release intervals of the event generator are depicted in Fig. 2.4. Figure 2.5 shows the control input of the controlled satellite system under DoS jamming attacks (2.3) with $T = 2$ s and $T_{off} = 1.8$ s. From Figs. 2.3, 2.4 and 2.5, it is found that (i) the satellite system is stabilized by the event-triggered controller; (ii) the proposed RETCS does eliminate the impact of the periodic DoS while reducing the amount of communication in the system.

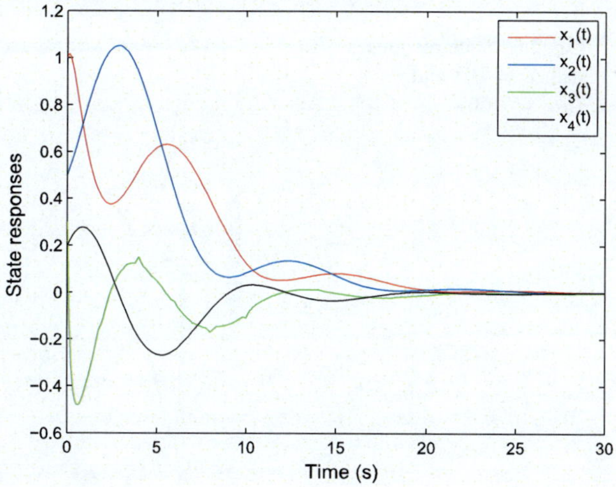

Fig. 2.3 State responses with $T = 2$ s and $T_{off} = 1.8$ s

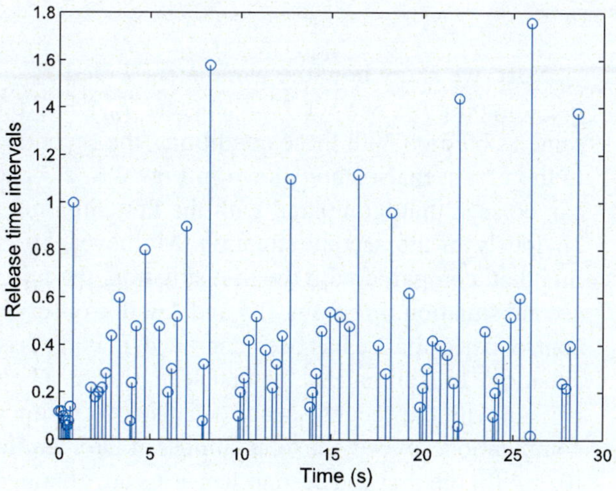

Fig. 2.4 Release time intervals with $T_{off} = 1.8$ s and $T = 2$ s

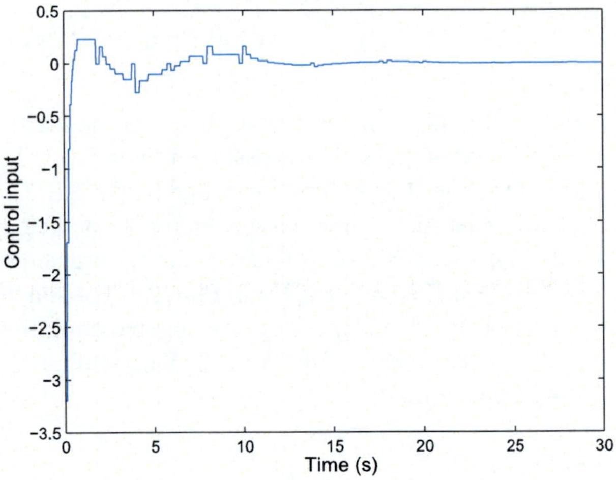

Fig. 2.5 Control input with $T = 2\,\mathrm{s}$ and $T_{off} = 1.8\,\mathrm{s}$

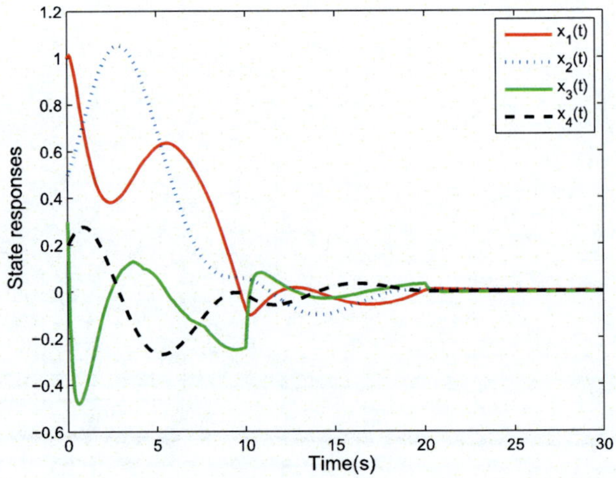

Fig. 2.6 State responses with $T = 10\,\mathrm{s}$ and $T_{off} = 7.66\,\mathrm{s}$

Next, we show how the parameters T and T_{off} affect the control performance. For this purpose, we set $T = 10\,\mathrm{s}$ and $T_{off} = 7.66\,\mathrm{s}$, other parameters remain the same as before. With these conditions, the second simulation results corresponding to the first scenarios are shown in Figs. 2.6, 2.7 and 2.8. From Figs. 2.3 and 2.6, it can be seen that, compared with the first situation, the state responses shake more obviously in the second situation. Moreover, by comparing Figs. 2.4 and 2.7, we find that, compared with the first situation, the number of events seems more in the second situation. From Fig. 2.5 and 2.8, it is observed that in the second situation the control performance has been degraded, in comparison with that achieved in the first situation. To summarize, the first scenario has a better control performance in comparison with the second scenario. This observation can also be clearly illustrated by computation. Indeed, the total number of DoS jamming attacks over the time interval $[0, 30]$ for the first and second scenarios are obtained as $n_1 = 15$ and $n_2 = 3$, respec-

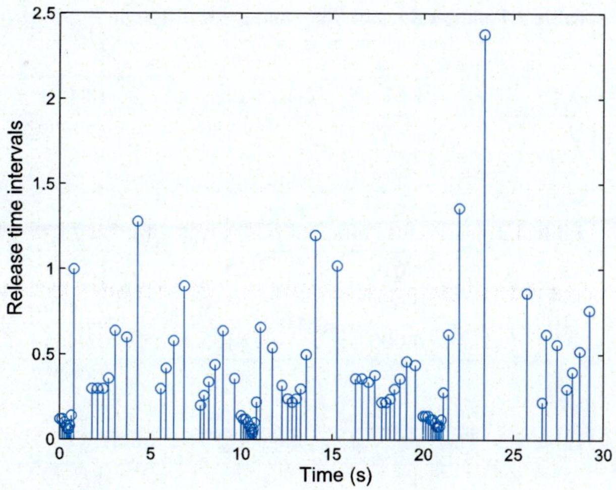

Fig. 2.7 Release time intervals with $T = 10\,\mathrm{s}$ and $T_{off} = 7.66\,\mathrm{s}$

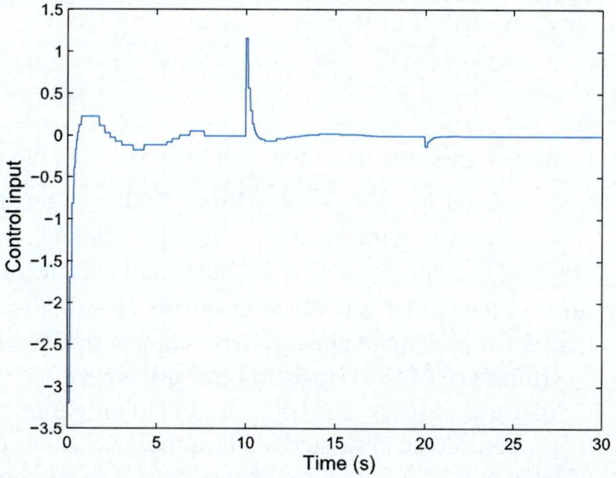

Fig. 2.8 Control input with $T = 10\,\mathrm{s}$ and $T_{off} = 7.66\,\mathrm{s}$

tively. Therefore, the DoS duration for the first and second scenarios are obtained as $\|\Xi\ (0, 30)\|_1 = n_1 \times (2 - 1.8)\, s = 3\,\mathrm{s}$ and $\|\Xi\ (0, 30)\|_2 = n_2 \times (10 - 7.66)\, s = 7.02\,\mathrm{s}$, respectively. Hence, $\|\Xi\ (0, 30)\|_2 > \|\Xi\ (0, 30)\|_1$, which implies the preceding claim is reasonable.

Influence of the parameter T:

In this section, we show the influence of the DoS parameter T on the system performance. Specifically, for given different T, we compute the corresponding minimum T_{off}. To this end, we formulate the following optimization problem (notice that the values of $\mu_i, \alpha_i, \sigma, h, \delta_i, \varepsilon_i, \chi_i$, and η_i $(i = 1, 2)$ are as before):

$$T_{off}^{\min} = \min \begin{cases} T_{off} \mid T_{off} \text{ satisfying (2.25)} \\ \text{subjects to LMIs (2.37)–(2.42).} \end{cases} \tag{2.45}$$

Table 2.1 The values of $T_{off}^{\min}$ and J for different T

T	2	4	6	8	10
$T_{off}^{\min}$	1.61	3.13	4.64	6.16	7.66
J	19.5%	21.75%	22.67%	23%	23.4%

Table 2.2 The values of λ and ρ for given different T_{off}

T_{off}	1.61	1.65	1.75	1.85	1.95
λ	0.0012	0.0540	0.1860	0.3180	0.4500
ρ	0.0003	0.0135	0.0465	0.0795	0.1125

Table 2.3 The values of $h_{\max}$ and $T_{off}^{\min}$ for different σ

σ	0.2	0.4	0.6	0.8	0.9
$h_{\max}$	0.07	0.06	0.04	0.02	0.012
$T_{off}^{\min}$	1.61	1.60	1.58	1.56	1.55

Table 2.1 gives the minimum values of T_{off} and the maximum allowable DoS attack rate denoted by $J \triangleq \frac{T-T_{off}^{\min}}{T}$. From Table 2.1, we can see that the values of $T_{off}^{\min}$ and J increase with the increasing of T. In addition, from Table 2.1 we know that $T_{off}^{\min} = 1.61$ for given $T = 2$. Table 2.2 shows the relationship between $T_{off}^{\min}$ and the decay rate ρ. The calculated results in Table 2.2 show that the bigger the $T_{off}^{\min}$, the larger the ρ, which validates the statement of Remark 2.3.

Influence of the triggering parameter σ:

We now further consider the impact of the parameter σ on the system performance of the considered system. For this purpose, let $\mu_1 = \mu_2 = 1.01, \alpha_1 = 0.16, \alpha_2 = 0.5$, $T = 2, T_{off} = 1.8, \delta_i, \varepsilon_i, \chi_i$, and η_i $(i = 1, 2)$ are chosen as the same before. Table 2.3 lists the maximum value of sampling period $h_{\max}$ and the corresponding $T_{off}^{\min}$ for various σ. According to the calculation results shown in Table 2.3, we can see that for the known DoS attacks , the values of $h_{\max}$ and $T_{off}^{\min}$ are decreasing with increasing of the triggering parameter σ.

2.5 Conclusion

In this chapter, the event-based resilient controller synthesis problem has been addressed for a linear NCS subject to periodic DoS attacks . A new periodic RETCS has been designed under the assumption that the jammer is partially known. A novel event-based switched system model has been built reflecting the effects of the proposed RECTS and periodic DoS attacks simultaneously. The exponential stability criteria of the resulting switched system under periodic DoS jamming attacks have been derived in the form of LMIs by using the PLF method. The joint-design strategy has also been given to get the resilient triggering parameter and the corresponding event-based controller gain. Finally, the efficiency of the proposed event-based resilient control scheme has been explained by a real example. There are several possible avenues for future research. It is interesting to extend the current method to deal with observer-based ETC and fault detection/filter problems subjects to nonperiodic DoS jamming attacks. Another consideration is to extend the proposed method to multi-agent systems, nonlinear systems, and smart grid by referring to [13–20] .

References

1. S. Hu, D. Yue, X. Xie, X. Chen, X. Yin, Resilient event-triggered controller synthesis of networked control systems under periodic DoS jamming attacks. IEEE Trans. Cybern. **49**(12), 4271–4281 (2019)
2. C.D. Persis, P. Tesi, Input-to-state stabilizing control under denial-of-service. IEEE Trans. Autom. Control **60**(11), 2930–2944 (2015)
3. H.S. Foroush, S. Martinez, On event-triggered control of linear systems under periodic denial-of-service jamming attacks, in *Proceedings of the 51st Annual Conference on Decision and Control* (2012), pp. 2551–2556
4. H.S. Foroush, S. Martinez, On triggering control of single-input linear systems under pulse-width modulated dos signals. SIAM J. Control Optim. **54**(6), 3084–3105 (2016)
5. C.D. Persis, P. Tesi, Resilient control under denial-of-service. IFAC Proc. Vol. **47**(3), 134–139 (2014)
6. L. Schenato, To zero or to hold control inputs with lossy links? IEEE Trans. Autom. Control **54**(5), 1093–1099 (2009)
7. D. Yue, E. Tian, Q.L. Han, A delay system method for designing event-triggered controllers of networked control systems. IEEE Trans. Autom. Control **58**(2), 475–481 (2013)
8. W.H. Chen, J. Zhong, W.X. Zheng, Delay-independent stabilization of a class of time-delay systems via periodically intermittent control EE. Automatica **71**, 89–97 (2016)
9. W.A. Zhang, L. Yu, Stabilization of sampled-data control systems with control inputs missing. IEEE Trans. Autom. Control **55**(2), 447–452 (2010)
10. X. Chen, Y. Wang, S. Hu, Event-based robust stabilization of uncertain networked control systems under quantization and denial-of-service attacks. Inf. Sci. **459**, 369–386 (2018)
11. W.P.M.H. Heemels, M.C.F. Donkers, A.R. Teel, Periodic event-triggered control for linear systems. IEEE Trans. Autom. Control **58**(4), 847–861 (2013)
12. X.-M. Zhang, Q.-L. Han, Event-triggered dynamic output feedback control for networked control systems. IET Control Theory Appl. **8**(4), 226–234 (2014)
13. D. Ding, Z. Wang, D.W.C. Ho, G. Wei, Observer-based event-triggering consensus control for multiagent systems with lossy sensors and cyber-attacks. IEEE Trans. Cybern. **47**(8), 1936–1947 (2017)
14. X. Ge, Q.L. Han, Z. Wang, A threshold-parameter-dependent approach to designing distributed event-triggered H_∞ consensus filters over sensor networks. IEEE Trans. Cybern. **49**(4), 1148–1159 (2018)
15. C.X. Dou, B. Liu, Multi-agent based hierarchical hybrid control for smart microgrid. IEEE Trans. Smart Grid **4**(2), 771–778 (2013)
16. T. Wang, Y. Zhang, J. Qiu, H. Gao, Adaptive fuzzy backstepping control for a class of nonlinear systems with sampled and delayed measurements. IEEE Trans. Fuzzy Syst. **23**(2), 302–312 (2015)

17. L. An, G.H. Yang, Decentralized adaptive fuzzy secure control for nonlinear uncertain interconnected systems against intermittent dos attacks. IEEE Trans. Cybern. **49**(3), 827–838 (2019)
18. M. Wang, J. Qiu, M. Chadli, M. Wang, A switched system approach to exponential stabilization of sampled-data T-S fuzzy systems with packet dropouts. IEEE Trans. Cybern. **46**(12), 3145–3156 (2016)
19. L. Zhang, X. Yin, Z. Ning, D. Ye, Robust filtering for a class of networked nonlinear systems with switching communication channels. IEEE Trans. Cybern. **47**(3), 671–682 (2017)
20. C. Zhang, J. Hu, J. Qiu, Q. Chen, Reliable output feedback control for T-S fuzzy systems with decentralized event triggering communication and actuator failures. IEEE Trans. Cybern. **47**(9), 2592–2602 (2017)

Chapter 3
Observer-Based Event-Triggered Control for NCSs Subject to Periodic DoS Attacks

This chapter is concerned with observer-based event-triggered control for a continuous networked linear system subjects to denial-of-service (DoS) attacks , where the attacks are launched periodically to block the data transmission in control channels. First, a new observer state-based resilient event-triggering scheme is developed in the presence of DoS attacks. Second, a novel event-based switched system model is established by considering the effect of the event-triggering scheme and DoS attacks simultaneously. By virtue of this new model combined with a piecewise Lyapunov–Krasovskii functional method, sufficient conditions are derived to guarantee exponential stability of the resultant switched system. It is shown that the proposed results can establish a quantitative relationship among the launching/sleeping periods of the attacks, the event-triggering parameters, the sampling period, and the exponential decay rate. Third, criteria for designing a desired observer-based event-triggered controller are provided and expressed in terms of a set of linear matrix inequalities. Finally, an offshore structure model is presented to illustrate the efficiency of the developed control method [1].

This chapter is organized as follows. In Sect. 3.1, a unified event-triggered NCS model under the periodic DoS attacks is proposed. In Sect. 3.2, exponential stability of the closed-loop system is analyzed. Furthermore, triggering matrix, controller gain matrix, and observer gain matrices are obtained by solving a set of LMIs. In Sect. 3.3, the effectiveness of the proposed control strategy is demonstrated by a practical example. The conclusions are drawn in Sect. 3.4.

3.1 Problem Formulation

As shown in Fig. 3.1, the NCS configuration under consideration consists of a physical plant, a smart sensor, an event generator, a controller, and an actuator. The signal transmission from sensor to controller, and controller to actuator is implemented via

D. Yue et al., *Secure Control of Networked Control Systems and Its Applications*,
https://doi.org/10.1007/978-981-33-6730-2_3

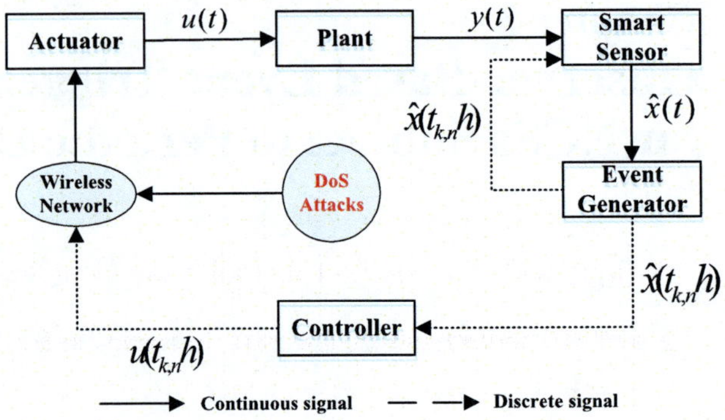

Fig. 3.1 The block diagram of an event-triggered NCS under DoS attacks

a digital channel, which can be compromised by certain DoS jamming signals. In the following, we will describe each component in detail.

3.1.1 System Description

The dynamics of the physical plant shown in Fig. 3.1 is given by

$$\begin{cases} \dot{x}(t) = Ax(t) + Bu(t) \\ y(t) = Cx(t) \\ x(0) = x_0 \end{cases} \tag{3.1}$$

where $x(t) \in \mathbb{R}^{n_x}$ is the state vector, $u(t) \in \mathbb{R}^m$ is the control input, and $y(t) \in \mathbb{R}^q$ is the output vector, n_x, m, and $q \in \mathbb{N}$, $\mathbb{N}$ represents the set of nonnegative integers. A, B and C are constant matrices of appropriate dimensions. x_0 is the initial condition. Throughout the chapter, it is assumed that (i) the state vector $x(t)$ is unmeasurable; (ii) the pairs (A, B) and (A, C) are controllable and observable, respectively, and (iii) the matrix C is of full row rank.

Here, we consider the same attack model as in Chap. 2 (see Eq. 2.3). For ease of description, we still give the DoS attack model as follows:

$$\mathcal{I}_{DoS}(t) = \begin{cases} 0, t \in \left[nT, nT + T_{off}^{min}\right) \\ 1, t \in \left[nT + T_{off}^{min}, (n+1)T\right) \end{cases} \tag{3.2}$$

where $n \in \mathbb{N}$ is the period number, $T > 0$ is action period of the jammer, $T_{off}^{\min}$ $\left(0 < T_{off}^{min} < T\right)$ denotes the sleeping period of the jammer in the nth period. Moreover, the sets $\bigcup_{n\in\mathbb{N}} \left[nT, nT + T_{off}^{min}\right)$ denote the intervals over which the jamming signal is off and communication is allowed, while the sets $\bigcup_{n\in\mathbb{N}} \left[nT + T_{off}^{min}, (n+1)T\right)$

denote the intervals over which the jamming signal is active and communication is denied, and thus no data can be transmitted in these time intervals. In the following, for notational simplicity, let $\mathcal{L}_{1,n} \triangleq \left[nT, nT + T_{off}^{min}\right), \mathcal{L}_{2,n} \triangleq \left[nT + T_{off}^{min}, (n+1)\,T\right)$.

Remark 3.1 It should be pointed out that the parameter T_{off}^{min} in (3.2) needs not to be time-invariant, and thus we assume that there exists a uniform lower bound for the sleeping period T_{off}^{min} as in [2]. Throughout the chapter, for simplicity, we use the parameter $T_{off}^{\min}$ to denote the uniform lower bound of the sleeping period in every active period T.

3.1.2 Smart Sensor Model

The smart sensor has great computational power and can pre-process measurement output $y(t)$ to get the state estimate $\hat{x}(t)$ based on the following switched full-order state observer despite the presence of the DoS jamming attacks (3.2):

$$\begin{cases} \dot{\hat{x}}(t) = \begin{cases} A\hat{x}(t) + Bu(t) + L_1\left[y(t) - \hat{y}(t)\right], \\ t \in \mathcal{L}_{1,n}, n \in \mathbb{N} \\ A\hat{x}(t) + Bu(t) + L_2\left[y(t) - \hat{y}(t)\right], \\ t \in \mathcal{L}_{2,n}, n \in \mathbb{N} \end{cases} \\ \hat{y}(t) = C\hat{x}(t), \hat{x}(0) = \hat{x}_0 \end{cases} \tag{3.3}$$

where $\hat{x}(t) \in \mathbb{R}^n$ is the observer state, $\hat{y}(t) \in \mathbb{R}^q$ is the observer output, L_1 and $L_2 \in \mathbb{R}^{n\times q}$ are observer gains to be designed later, $\hat{x}_0$ is the initial condition of the observer.

Remark 3.2 Note that in (3.3), two different observer gain matrices are introduced corresponding to the "sleeping mode" and "activated state" of the DoS jamming attacks (3.2), respectively. By doing so, the observer can adapt the duty cycle of DoS jamming attacks. Moreover, it can be inferred that as $T_{off}^{\min} \to T$, the switched state observer (3.3) reduces to the traditional Luenberger observer $\dot{\hat{x}}(t) = A\hat{x}(t) + Bu(t) + L_1\left[y(t) - \hat{y}(t)\right]$. Therefore, the traditional Luenberger observer can be regarded as a special case of the observer (3.3).

3.1.3 Design of Resilient Triggering Scheme and Observer

In this chapter, inspired by sampled-data-based event-triggered schemes [3, 4] and PETC schemes [5], we will propose a resilient triggering scheme. In doing so, we select the condition

$$e_{k_j}^T\left(t_{k_j}h\right)\Omega e_{k_j}\left(t_{k_j}h\right) > \varepsilon\hat{x}^T\left(t_kh\right)\Omega\hat{x}\left(t_kh\right) \tag{3.4}$$

where $e_{k_j}\left(t_{k_j}h\right) = \hat{x}\left(t_{k_j}h\right) - \hat{x}\left(t_kh\right)$, $\varepsilon \in (0, 1)$ is a design parameter, $t_{k_j}h \triangleq t_kh + jh$ denotes the sampling instant between two consecutive triggering instant, t_kh is the latest event-triggering instant, and Ω is a positive definite weighting matrix to be designed later.

Then in the presence of the periodic DoS jamming attacks (3.2), the triggering instant when the event generator transmits the estimate $\hat{x}(t)$ to the controller is determined by the following resilient triggering condition described by

$$t_{k,n}h = \left\{t_{k_j}h \text{ satisfying (3.4)} \mid t_{k_j}h \in \bar{\mathcal{L}}_{1,n}\right\} \cup \{nT\} \tag{3.5}$$

where $n, k_j, j \in \mathbb{N}$, k is the number of triggering times occurring in nth jammer action period. The parameter $h > 0$ denotes the sampling period. $t_{1,n}h \triangleq (n-1)T$, $0 < h < T$. $\bar{\mathcal{L}}_{1,n} \triangleq \left[(n-1)T, (n-1)T + T_{off}^{\min}\right)$, $\bar{\mathcal{L}}_{2,n} \triangleq \left[(n-1)T + T_{off}^{\min}, nT\right)$.

In what follows, inspired by the Remark 3.1 in [6], under the DoS jamming attacks, the observer-based ETC $u(t)$ can be written as

$$u(t) = \begin{cases} K\hat{x}\left(t_{k,n}h\right), t \in \left[t_{k,n}h, t_{k+1,n}h\right) \cap \bar{\mathcal{L}}_{1,n} \\ 0, t \in \bar{\mathcal{L}}_{2,n}, n \in \mathbb{N} \end{cases} \tag{3.6}$$

where K is the control gain matrix, the sequence $\left\{t_{k,n}h\right\}$ can be iteratively determined by the above event-triggering condition (3.5), $k \in \{1, \ldots, k(n)\} \triangleq K(n)$ with

$$k(n) = \sup\left\{k \in \mathbb{N} \mid t_{k,n}h \le (n-1)T + T_{off}^{\min}\right\}$$

which implies that $t_{k(n)+1,n}h > (n-1)T + T_{off}^{\min}$.

Remark 3.3 Notice that the zero-input strategy is employed when the DoS attacks are active. In doing so, the actuator does not work during the jamming period, which in turn reduces the energy consumption at the actuator side. In fact, the zero-input strategy has been widely used in the literature, see, e.g., [7–10]. However, these results do not consider the effects of the DoS attacks. In particular, when $T_{off}^{min} \to T$, it implies that the jamming signal has all but disappeared and the communication can be maintained continuously. This case has been extensively studied in the context of an event-triggered control framework, see [4] and the references therein.

3.1.4 Modeling of Observer-Based Switched Control Systems

In the sequel, for simplicity of exposition, for $n \in \mathbb{N}$, define
$\Upsilon_{k,n} \triangleq \left[t_{k,n}h, t_{k+1,n}h\right)$, $k \in K(n)$. Substituting (3.6) into (3.1) yields

$$\dot{x}(t) = \begin{cases} Ax(t) + BK\hat{x}(t_{k,n}h), t \in \Upsilon_{k,n} \cap \bar{\mathcal{L}}_{1,n} \\ Ax(t), t \in \bar{\mathcal{L}}_{2,n} \end{cases} \tag{3.7}$$

For technical analysis, we divide the event intervals $\Upsilon_{k,n}$ shown in (3.7) into sampling interval-like subintervals, that is,

$$\Upsilon_{k,n} = \cup_{l=1}^{\gamma_{k,n}} \left[t_{k,n}h + (l-1)h, t_{k,n}h + lh\right) \tag{3.8}$$

where $k \in K(n)$, $n \in \mathbb{N}$, and $\gamma_{k,n} \triangleq \inf\left\{l \in \mathbb{N} \mid t_{k,n}h + lh \geq t_{k+1,n}h\right\}$.

Let

$$\begin{cases} \mathcal{F}_{k,n}^{l} = \left[t_{k,n}h + (l-1)h, t_{k,n}h + lh\right), \\ \qquad l \in \{1, 2, \ldots, \gamma_{k,n} - 1\} \\ \mathcal{F}_{k,n}^{\gamma_{k,n}} = \left[t_{k,n}h + \left(\gamma_{k,n} - 1\right)h, t_{k+1,n}h\right). \end{cases} \tag{3.9}$$

Note that

$$\bar{\mathcal{L}}_{1,n} = \cup_{k=1}^{k(n)} \left\{\Upsilon_{k,n} \cap \bar{\mathcal{L}}_{1,n}\right\} \subseteq \cup_{k=1}^{k(n)} \Upsilon_{k,n}. \tag{3.10}$$

Combining (3.8) (3.9) and (3.10), the interval $\bar{\mathcal{L}}_{1,n}$ can be rewritten as

$$\bar{\mathcal{L}}_{1,n} = \cup_{k=1}^{k(n)} \cup_{l=1}^{\gamma_{k,n}} \left\{\mathcal{F}_{k,n}^{l} \cap \bar{\mathcal{L}}_{1,n}\right\}.$$

Set

$$\Psi_{k,n}^{l} = \mathcal{F}_{k,n}^{l} \cap \bar{\mathcal{L}}_{1,n}.$$

Then

$$\bar{\mathcal{L}}_{1,n} = \cup_{k=1}^{k(n)} \cup_{l=1}^{\gamma_{k,n}} \Psi_{k,n}^{l}.$$

Now, for $k \in K(n)$, $n \in \mathbb{N}$, we define two piecewise functions as follows:

$$\tau_{k,n}(t) = \begin{cases} t - t_{k,n}h, t \in \Psi_{k,n}^{1} \\ t - t_{k,n}h - h, t \in \Psi_{k,n}^{2} \\ \vdots \\ t - t_{k,n}h - \left(\gamma_{k,n} - 1\right)h, t \in \Psi_{k,n}^{\gamma_{k,n}} \end{cases} \tag{3.11}$$

and

$$e_{k,n}(t) = \begin{cases} 0, t \in \Psi_{k,n}^{1} \\ \hat{x}(t_{k,n}h) - \hat{x}(t_{k,n}h + h), t \in \Psi_{k,n}^{2} \\ \vdots \\ \hat{x}(t_{k,n}h) - \hat{x}\left(t_{k,n}h + \left(\gamma_{k,n} - 1\right)h\right). \\ t \in \Psi_{k,n}^{\gamma_{k,n}} \end{cases} \tag{3.12}$$

Based on the above two definitions, it can be seen that

$$\hat{x}\left(t_{k,n}h\right) = \hat{x}\left(t - \tau_{k,n}(t)\right) + e_{k,n}(t) \tag{3.13}$$

where $\tau_{k,n}(t) \in [0, h), t \in \Upsilon_{k,n} \cap \bar{\mathcal{L}}_{1,n}, k \in K(n)$.

Then, the system (3.7) can be described as

$$\begin{cases} \dot{x}(t) = \begin{cases} Ax(t) + BK\hat{x}\left(t - \tau_{k,n}(t)\right) \\ +BKe_{k,n}(t), t \in \Upsilon_{k,n} \cap \bar{\mathcal{L}}_{1,n} \\ Ax(t), t \in \bar{\mathcal{L}}_{2,n} \end{cases} \\ x(t) = \phi_1(t), t \in [-h, 0] \end{cases} \tag{3.14}$$

where $\phi_1(t)$ is the supplemented initial condition of the state $x(t)$ on $[-h, 0]$, and the error vector $e_{k,n}(t)$ satisfies

$$\begin{aligned} e_{k,n}^T(t)\,\Omega e_{k,n}(t) \le \varepsilon &\left[\hat{x}\left(t - \tau_{k,n}(t)\right) + e_{k,n}(t)\right]^T \\ &\Omega\left[\hat{x}\left(t - \tau_{k,n}(t)\right) + e_{k,n}(t)\right]. \end{aligned} \tag{3.15}$$

Substituting (3.6), (3.13) into system (3.3), the observer system under the DoS attack can be rewritten as:

$$\begin{cases} \dot{\hat{x}}(t) = \begin{cases} A\hat{x}(t) + BK\hat{x}\left(t - \tau_{k,n}(t)\right) + BKe_{k,n}(t) \\ +L_1C\left[x(t) - \hat{x}(t)\right], t \in \Upsilon_{k,n} \cap \bar{\mathcal{L}}_{1,n} \\ A\hat{x}(t) + L_2C\left[x(t) - \hat{x}(t)\right], t \in \bar{\mathcal{L}}_{2,n} \end{cases} \\ \hat{x}(t) = \phi_2(t), t \in [-h, 0] \end{cases} \tag{3.16}$$

where $\phi_2(t)$ is the supplemented initial condition of the state $\hat{x}(t)$ on $[-h, 0]$. Define $\tilde{x}(t) = x(t) - \hat{x}(t)$, then

$$\dot{\tilde{x}}(t) = \begin{cases} (A - L_1C)\tilde{x}(t), t \in \Upsilon_{k,n} \cap \bar{\mathcal{L}}_{1,n} \\ (A - L_2C)\tilde{x}(t), t \in \bar{\mathcal{L}}_{2,n} \end{cases} \tag{3.17}$$

Define $\xi^T(t) = \left[\hat{x}^T(t), \tilde{x}^T(t)\right]$. We have the following switched system

$$\begin{cases} \dot{\xi}(t) = \begin{cases} A_1\xi(t) + B_1G\xi\left(t - \tau_{k,n}(t)\right) \\ +B_1e_{k,n}(t), t \in \Upsilon_{k,n} \cap \bar{\mathcal{L}}_{1,n}, k \in K(n) \\ A_2\xi(t), \ t \in \bar{\mathcal{L}}_{2,n}, n \in \mathbb{N} \end{cases} \\ \xi(t) = \phi(t), \ t \in [-h, 0] \end{cases} \tag{3.18}$$

where $A_1 = \begin{bmatrix} A & L_1C \\ 0 & A - L_1C \end{bmatrix}$, $B_1 = \begin{bmatrix} BK \\ 0 \end{bmatrix}$, $A_2 = \begin{bmatrix} A & L_2C \\ 0 & A - L_2C \end{bmatrix}$, $\phi(t) = \phi_t = \begin{bmatrix} \phi_1(t) \\ \phi_1(t) - \phi_2(t) \end{bmatrix}$, $G = \begin{bmatrix} I & 0 \end{bmatrix}$.

3.1.5 Control Objective

Throughout the chapter, we shall use the following definition of global exponential stability for the switched system (3.18).

Definition 3.1 The switched system (3.18) is said to be globally exponentially stable (GES), if there exists a scalar $\kappa > 0$ such that the solution $\xi(t)$ of the system (3.18) satisfies $\|\xi(t)\| \leq \kappa e^{-\rho t} \|\phi_0\|_h, t \geq 0$, where $\|\phi_t\|_h \triangleq \sup_{-h \leq \theta \leq 0} \{\|x(t+\theta)\|, \|\dot{x}(t+\theta)\|\}$ and ρ is called the decay rate.

The control objective of this chapter is to jointly design the observer and event-triggering scheme such that, for all possible periodic jamming attacks $\mathcal{I}_{DoS}(t)$ (3.2), where the sequence $\{nT\}_{n\in\mathbb{N}}$ and parameters T and $T_{off}^{\min}$ are known $\left(T_{off}^{\min} \leq T < +\infty\right)$, the switched system (3.18) is GES.

3.2 Stability Analysis

The following lemma estimates the upper bounds of the chosen Lyapunov functional candidate in the absence of DoS attacks and in the presence of DoS attacks , respectively. Note that these estimations will later be used to characterize the properties of DoS attacks that guarantee the GES of the resultant switched system (3.18).

Lemma 3.1 *Given the feedback gain K and a jamming signal $\mathcal{I}_{DoS}(t)$ (3.2), where the sequence $\{nT\}_{n\in\mathbb{N}}$ and parameters T and $T_{off}^{\min}$ are known $\left(T_{off}^{\min} \leq T < +\infty\right)$. For the system (3.18), if for some prescribed constants $\alpha_i \in (0, +\infty)$, $\varepsilon \in (0, 1)$, $h \in \left(0, T_{off}^{\min}\right)$, there exist symmetric positive definite matrices P_i, Q_i, R_i, and matrices U_i $(i = 1, 2)$ of appropriate dimensions such that the following matrix inequalities hold:*

$$\Pi_1 < 0, \Re_1 > 0 \tag{3.19}$$

$$\Pi_2 < 0, \Re_2 > 0 \tag{3.20}$$

where

$$\Pi_1 = \begin{bmatrix} \Sigma_1 & * \\ hR_1GF_1 & -R_1 \end{bmatrix}, \Re_1 = \begin{bmatrix} \tilde{R}_1 & * \\ U_1 & \tilde{R}_1 \end{bmatrix}$$

$$\Pi_2 = \begin{bmatrix} \Sigma_2 & * \\ hR_2GF_2 & -R_2 \end{bmatrix}, \Re_2 = \begin{bmatrix} \tilde{R}_2 & * \\ U_2 & \tilde{R}_2 \end{bmatrix}$$

$$\tilde{R}_1 = e^{-2\alpha_1 h} \begin{bmatrix} R_1 & * \\ 0 & 3R_1 \end{bmatrix}, U_1 = \begin{bmatrix} U_{11} & U_{12} \\ U_{13} & U_{14} \end{bmatrix}$$

$$\tilde{R}_2 = \begin{bmatrix} R_2 & * \\ 0 & 3R_2 \end{bmatrix}, U_2 = \begin{bmatrix} U_{21} & U_{22} \\ U_{23} & U_{24} \end{bmatrix}$$

$$\Sigma_1 = \begin{bmatrix} \Sigma_1^{11} & * & * & * & * & * \\ \Sigma_1^{21} & \Sigma_1^{22} & * & * & * & * \\ \Sigma_1^{31} & \Sigma_1^{32} & \Sigma_1^{33} & * & * & * \\ \Sigma_1^{41} & \Sigma_1^{42} & \Sigma_1^{43} & \Sigma_1^{44} & * & * \\ \Sigma_1^{51} & \Sigma_1^{52} & \Sigma_1^{53} & \Sigma_1^{54} & \Sigma_1^{55} & * \\ \Sigma_1^{61} & \Sigma_1^{62} & 0 & 0 & 0 & \Sigma_1^{66} \end{bmatrix}$$

$$\Sigma_2 = \begin{bmatrix} \Sigma_2^{11} & * & * & * & * \\ \Sigma_2^{21} & \Sigma_2^{22} & * & * & * \\ \Sigma_2^{31} & \Sigma_2^{32} & \Sigma_2^{33} & * & * \\ \Sigma_2^{41} & \Sigma_2^{42} & \Sigma_2^{43} & \Sigma_2^{44} & * \\ \Sigma_2^{51} & \Sigma_2^{52} & \Sigma_2^{53} & \Sigma_2^{54} & \Sigma_2^{55} \end{bmatrix}$$

$$\begin{aligned}
\Sigma_1^{11} &= 2\alpha_1 P_1 + He(P_1 A_1) \\
&\quad + G^T Q_1 G - 4e^{-2\alpha_1 h} G^T R_1 G \\
\Sigma_1^{21} &= B_1^T P_1 - 2e^{-2\alpha_1 h} R_1 G - U_{11} G \\
&\quad - U_{12} G - U_{13} G - U_{14} G \\
\Sigma_1^{22} &= \varepsilon \Omega - 8e^{-2\alpha_1 h} R_1 + He\,(U_{11} - U_{12}) \\
&\quad + He\,(U_{13} - U_{14}) \\
\Sigma_1^{31} &= U_{11} G + U_{12} G - U_{13} G - U_{14} G \\
\Sigma_1^{32} &= -2e^{-2\alpha_1 h} R_1 - U_{11} + U_{12} + U_{13} - U_{14} \\
\Sigma_1^{33} &= -e^{-2\alpha_1 h} Q_1 - 4e^{-2\alpha_1 h} R_1,\ \Sigma_1^{41} = 6e^{-2\alpha_1 h} R_1 G \\
\Sigma_1^{42} &= 6e^{-2\alpha_1 h} R_1 + 2U_{12}^T + 2U_{14}^T \\
\Sigma_1^{43} &= -2U_{12}^T + 2U_{14}^T,\ \Sigma_1^{44} = -12e^{-2\alpha_1 h} R_1 \\
\Sigma_1^{51} &= 2U_{13} G + 2U_{14} G,\ \Sigma_1^{53} = 6e^{-2\alpha_1 h} R_1 \\
\Sigma_1^{52} &= 6e^{-2\alpha_1 h} R_1 - 2U_{13} + 2U_{14} \\
\Sigma_1^{54} &= -4U_{14},\ \Sigma_1^{55} = -12e^{-2\alpha_1 h} R_1 \\
\Sigma_1^{61} &= B_1^T P_1,\ \Sigma_1^{62} = \varepsilon \Omega,\ \Sigma_1^{66} = (\varepsilon - 1)\,\Omega \\
F_1 &= \begin{bmatrix} A_1 & B_1 & 0 & 0 & 0 & B_1 \end{bmatrix} \\
\Sigma_2^{11} &= -2\alpha_2 P_2 + He(P_2 A_2) + G^T Q_2 G - 4G^T R_2 G \\
\Sigma_2^{21} &= -2R_2 G - U_{21} G - U_{22} G - U_{23} G - U_{24} G \\
\Sigma_2^{22} &= -8R_2 + He\,(U_{21} - U_{22} + U_{23} - U_{24}) \\
\Sigma_2^{31} &= U_{21} G + U_{22} G - U_{23} G - U_{24} G \\
\Sigma_2^{32} &= -2R_2 - U_{21} + U_{22} + U_{23} - U_{24} \\
\Sigma_2^{33} &= -e^{2\alpha_2 h} Q_2 - 4R_2,\ \Sigma_2^{41} = 6R_2 G \\
\Sigma_2^{42} &= 6R_2 + 2U_{22}^T + 2U_{24}^T,\ \Sigma_2^{44} = -12R_2 \\
\Sigma_2^{43} &= -2U_{22}^T + 2U_{24}^T,\ \Sigma_2^{53} = 6R_2,\ \Sigma_2^{54} = -4U_{24} \\
\Sigma_2^{51} &= 2U_{23} G + 2U_{24} G,\ \Sigma_2^{52} = 6R_2 - 2\,(U_{23} - U_{24}) \\
\Sigma_2^{55} &= -12R_2,\ F_2 = \begin{bmatrix} A_2 & 0 & 0 & 0 & 0 \end{bmatrix}
\end{aligned}$$

Then along the trajectory of the system (3.18), it follows that:

$$V(t) \leq \begin{cases} \rho_1(nT)\, V(nT), t \in \bar{\mathcal{L}}_{1,n} \\ \rho_2\left(nT + T_{off}^{\min}\right) V\left(nT + T_{off}^{\min}\right), t \in \bar{\mathcal{L}}_{2,n} \end{cases} \tag{3.21}$$

where $\rho_1(s) = e^{-2\alpha_1(t-s)}$, $\rho_2(s) = e^{2\alpha_2(t-s)}$.

Proof Construct the following piecewise Lyapunov functional for system (3.18):

$$V(t) = \begin{cases} V_1(t), t \in \Upsilon_{k,n} \cap \bar{\mathcal{L}}_{1,n}, k \in K(n) \\ V_2(t), t \in \bar{\mathcal{L}}_{2,n}, n \in \mathbb{N} \end{cases} \tag{3.22}$$

where

$$\begin{aligned} V_1(t) = {} & \xi^T(t) P_1 \xi(t) \\ & + \int_{t-h}^{t} \xi^T(s) \rho_1(s) G^T Q_1 G \xi(s) ds \\ & + h \int_{-h}^{0} \int_{t+\theta}^{t} \dot{\xi}^T(s) \rho_1(s) G^T R_1 G \dot{\xi}(s) ds d\theta \end{aligned}$$

and

$$\begin{aligned} V_2(t) = {} & \xi^T(t) P_2 \xi(t) \\ & + \int_{t-h}^{t} \xi^T(s) \rho_2(s) G^T Q_2 G \xi(s) ds \\ & + h \int_{-h}^{0} \int_{t+\theta}^{t} \dot{\xi}^T(s) \rho_2(s) G^T R_2 G \dot{\xi}(s) ds d\theta. \end{aligned}$$

where $P_1 \in \mathbb{R}^{2n\times 2n}$, $Q_1 \in \mathbb{R}^{n\times n}$, $R_1 \in \mathbb{R}^{n\times n}$, $P_2 \in \mathbb{R}^{2n\times 2n}$, $Q_2 \in \mathbb{R}^{n\times n}$, $R_2 \in \mathbb{R}^{n\times n}$, $\Omega \in \mathbb{R}^{n\times n}$ are symmetric positive definite matrices. We now consider the following two cases: Case A: $\forall t \in \Upsilon_{k,n} \cap \bar{\mathcal{L}}_{1,n}$, $\forall k \in K(n)$, $n \in \mathbb{N}$, taking the derivative of $V(t)$ with respect to t along the system (3.18) yields

$$\begin{aligned} \dot{V}(t) \leq {} & -2\alpha_1 V(t) + 2\alpha_1 \xi^T(t) P_1 \xi(t) + \xi^T(t) P_1 \dot{\xi}(t) \\ & + \dot{\xi}^T(t) P_1 \xi(t) + \xi^T(t) G^T Q_1 G \xi(t) \\ & - \xi^T(t-h) e^{-2\alpha_1 h} G^T Q_1 G \xi(t-h) \\ & + h^2 \dot{\xi}^T(t) G^T R_1 G \dot{\xi}(t) \\ & - h \int_{t-h}^{t} \dot{\xi}^T(s) e^{-2\alpha_1 h} G^T R_1 G \dot{\xi}(s) ds \end{aligned} \tag{3.23}$$

Note that

$$\begin{aligned}&-h\int_{t-h}^{t}\dot{\xi}^T(s)\,G^TR_1G\dot{\xi}(s)\,ds\\&=-h\int_{t-\tau_{k,n}(t)}^{t}\dot{\xi}^T(s)\,G^TR_1G\dot{\xi}(s)\,ds\\&\quad-h\int_{t-h}^{t-\tau_{k,n}(t)}\dot{\xi}^T(s)\,G^TR_1G\dot{\xi}(s)\,ds.\end{aligned}\tag{3.24}$$

Using the inequality in [11], we obtain

$$\begin{cases}-h\int_{t-\tau_{k,n}(t)}^{t}\dot{\xi}^T(s)\,e^{-2\alpha_1 h}G^TR_1G\dot{\xi}(s)\,ds\\ \qquad\le-\frac{h}{\tau_{k,n}(t)}\Lambda_1^T\tilde{R}_1\Lambda_1\\-h\int_{t-h}^{t-\tau_{k,n}(t)}\dot{\xi}^T(s)\,e^{-2\alpha_1 h}G^TR_1G\dot{\xi}(s)\,ds\\ \qquad\le-\frac{h}{h-\tau_{k,n}(t)}\Lambda_2^T\tilde{R}_1\Lambda_2\end{cases}\tag{3.25}$$

where $\Lambda_1=diag\{G,G\}\left[\upsilon_{11}^T\ \upsilon_{12}^T\right]^T$, $\upsilon_{11}=\xi(t)-\xi\left(t-\tau_{k,n}(t)\right)$, $\upsilon_{12}=\xi(t)+\xi\left(t-\tau_{k,n}(t)\right)-2\delta_1(t)$, $\delta_1(t)=\int_{t-\tau_{k,n}(t)}^{t}\frac{\xi(s)}{\tau_{k,n}(t)}ds$, $\Lambda_2=diag\{G,G\}\left[\upsilon_{21}^T\ \upsilon_{22}^T\right]^T$, $\upsilon_{21}=\xi\left(t-\tau_{k,n}(t)\right)-\xi(t-h)$, $\upsilon_{22}=\xi\left(t-\tau_{k,n}(t)\right)+\xi(t-h)-2\delta_2(t)$, $\delta_2(t)=\int_{t-h}^{t-\tau_{k,n}(t)}\frac{\xi(s)}{h-\tau_{k,n}(t)}ds$.

Using (3.24) and (3.25) to deal with integral terms in (3.23) by Lemma 3 in [12], we have

$$-h\int_{t-h}^{t}\dot{\xi}^T(s)\,e^{-2\alpha_1 h}G^TR_1G\dot{\xi}(s)\,ds\le-\Lambda_3^T\Re_1\Lambda_3\tag{3.26}$$

where $\Lambda_3=diag\{G,G,G,G\}\left[\upsilon_{11}^T\ \upsilon_{12}^T\ \upsilon_{21}^T\ \upsilon_{22}^T\right]^T$.

Substituting (3.26) into (3.23), adding the terms $e_{k,n}^T(t)\,\Omega e_{k,n}(t)$ $-e_{k,n}^T(t)\,\Omega e_{k,n}(t)$ to the right hand side of (3.23), and using the event-triggering condition (3.15) to bound the term $e_{k,n}^T(t)\,\Omega e_{k,n}(t)$, we have

$$\begin{aligned}\dot{V}(t)&\le-2\alpha_1V(t)\\&\quad+\eta_1^T(t)\left[\Sigma_1+h^2F_1^TG^TR_1GF_1\right]\eta_1(t)\end{aligned}\tag{3.27}$$

where

$$\begin{aligned}\eta_1^T(t)=&\left[\xi^T(t)\ \xi^T\left(t-\tau_{k,n}(t)\right)G^T\ \xi^T(t-h)\,G^T\right.\\&\left.\delta_1^T(t)\,G^T\ \delta_2^T(t)\,G^T\ e_{k,n}^T(t)\right].\end{aligned}$$

Taking the Schur complement of the matrix $\Pi_1<0$ in (3.19), we obtain that $\Sigma_1+h^2F_1^TG^TR_1GF_1<0$, which implies that

$$\dot{V}(t)\le-2\alpha_1V(t).\tag{3.28}$$

Due to the arbitrary of k, it follows that $\forall t \in \bar{\mathcal{L}}_{1,n}$,

$$V(t) \leq e^{-2\alpha_1(t-nT)} V(nT). \tag{3.29}$$

Case B: Following the similar arguments as the ones in the proof of Case A, the differential of $V(t)$ with respect to $t \in \bar{\mathcal{L}}_{2,n}$ along the trajectory of system (3.18) yields

$$\begin{aligned} \dot{V}(t) \leq\ & 2\alpha_2 V(t) \\ & + \eta_2^T(t)\left[\Sigma_2 + h^2 F_2^T G^T R_2 G F_2\right]\eta_2(t) \end{aligned} \tag{3.30}$$

where

$$\begin{aligned} \eta_2^T(t) = \big[\, & \xi^T(t)\ \xi^T\left(t - \tau_{k,n}(t)\right) G^T \\ & \xi^T(t-h)\, G^T\ \delta_1^T(t)\, G^T\ \delta_2^T(t)\, G^T \,\big]. \end{aligned}$$

According to the condition $\Pi_2 < 0$ in (3.20), we obtain that $\Sigma_2 + h^2 F_2^T G^T R_2 G F_2 < 0$, which implies that $\forall t \in \bar{\mathcal{L}}_{2,n}$,

$$V(t) \leq e^{2\alpha_2\left(t-nT-T_{off}^{\min}\right)} V\left(nT + T_{off}^{\min}\right). \tag{3.31}$$

From the analysis above of Cases A and B, the conditions (3.19) and (3.20) guarantee the inequality (3.21) is satisfied. The proof is thus completed.

Based on the Lemma 3.1, we now state and establish the following stability analysis result.

Theorem 3.1 *Given the feedback gain K and a jamming signal $\mathcal{I}_{Dos}(t)$ (3.2), and the sequence $\{nT\}_{n\in\mathbb{N}}$ and parameters T and $T_{off}^{\min}$ are known $\left(T_{off}^{\min} \leq T < +\infty\right)$. For the system (3.18), if for some prescribed constants $\alpha_i \in (0, +\infty)$, $\mu_i \in (0, +\infty)$ $(i = 1, 2,\ \mu_1\mu_2 \geq 1)$, $\varepsilon \in (0, 1)$, $h \in \left(0, T_{off}^{\min}\right)$ satisfying*

$$T_{off}^{\min} > \frac{2\alpha_2 T + 2(\alpha_1+\alpha_2)h + \ln(\mu_1\mu_2)}{2(\alpha_1+\alpha_2)}, \tag{3.32}$$

there exist symmetric positive definite matrices $P_i \in \mathbb{R}^{2n\times 2n}$, $Q_i \in \mathbb{R}^{n\times n}$, $R_i \in \mathbb{R}^{n\times n}$, $\Omega \in \mathbb{R}^{n\times n}$ and matrices U_i $(i = 1, 2)$ of appropriate dimensions such that the LMIs (3.19), (3.20) and the conditions below are satisfied:

$$P_1 \leq \mu_2 P_2 \tag{3.33}$$

$$e^{-2(\alpha_1+\alpha_2)h} P_2 \leq \mu_1 P_1 \tag{3.34}$$

$$Q_1 \leq \mu_2 Q_2 \tag{3.35}$$

$$R_1 \leq \mu_2 R_2 \tag{3.36}$$

$$Q_2 \leq \mu_1 Q_1 \tag{3.37}$$

$$R_2 \leq \mu_1 R_1 \tag{3.38}$$

then the switched system (3.18) under the periodic DoS jamming attacks (3.2) is GES with the decay rate $\rho \triangleq \frac{\lambda}{2T}$, $\lambda \triangleq 2\alpha_1 T_{off}^{\min} - 2\alpha_2 \left(T - T_{off}^{\min}\right) - 2\left(\alpha_1 + \alpha_2\right) h - \ln\left(\mu_1 \mu_2\right)$.

Proof Choose a piecewise Lyapunov functional $V(t)$ as Lemma 3.1. According to (3.33)–(3.38), it follows from Lemma 3.1 that

$$\begin{aligned} V_1(nT) &= \xi^T(nT) P_1 \xi(nT) \\ &\quad + \int_{nT-h}^{nT} \xi^T(s) e^{-2\alpha_1(nT-s)} G^T Q_1 G \xi(s) ds \\ &\quad + h \int_{-h}^{0} \int_{nT+\theta}^{nT} \dot{\xi}^T(s) e^{-2\alpha_1(nT-s)} G^T \\ &\quad \times R_1 G \dot{\xi}(s) ds d\theta \\ &\leq \mu_2 V_2\left(nT^-\right). \end{aligned} \tag{3.39}$$

Similarly,

$$V_2\left(nT + T_{off}^{\min}\right) \leq \mu_1 e^{2(\alpha_1+\alpha_2)h} V_1\left[\left(nT + T_{off}^{\min}\right)^-\right]. \tag{3.40}$$

In the sequel, using (3.39), (3.40), and (3.21), by induction, when $t \in \left[nT, nT + T_{off}^{\min}\right)$, one has

$$\begin{aligned} V_1(t) &\leq \left(\mu_1\mu_2\right)^n e^{-2n\alpha_1 T_{off}^{\min}} \\ &\quad \times e^{2n\alpha_2\left(T-T_{off}^{\min}\right)} \\ &\quad \times e^{2n(\alpha_1+\alpha_2)h} V_1(0) \\ &= e^{-\lambda n} V_1(0) \\ V_1\left[\left(nT + T_{off}^{\min}\right)^-\right] &\leq \left(\mu_1\mu_2\right)^n e^{-2n\alpha_1 T_{off}^{\min}} \\ &\quad \times e^{2n\alpha_2\left(T-T_{off}^{\min}\right)} \\ &\quad \times e^{2n(\alpha_1+\alpha_2)h} \\ &\quad \times e^{-2\alpha_1 T_{off}^{\min}} V_1(0). \end{aligned} \tag{3.41}$$

Note that

$$nT \leq t \leq nT + T_{off}^{\min} \Longrightarrow \frac{t - T_{off}^{\min}}{T} < n \leq \frac{t}{T}. \tag{3.42}$$

Combining (3.41) and (3.42), we obtain

$$V(t) \leq V_1(0)\, e^{\frac{\lambda T_{off}^{\min}}{T}} e^{-\frac{\lambda}{T}t}. \tag{3.43}$$

On the other hand, when $t \in \left[nT + T_{off}^{\min}, (n+1)T\right)$, it follows from (3.32) that

$$\begin{aligned} V_2(t) &\leq \frac{(\mu_1\mu_2)^{n+1}}{\mu_2} e^{\left(-2\alpha_1 T_{off}^{\min} + 2\alpha_2\left(T - T_{off}^{\min}\right) + 2(\alpha_1+\alpha_2)h\right)n} \\ &\quad \times e^{-2\alpha_1 T_{off}^{\min} + 2\alpha_2\left(T - T_{off}^{\min}\right) + 2(\alpha_1+\alpha_2)h} V_1(0) \\ &= \frac{1}{\mu_2} e^{-\lambda(n+1)} V_1(0). \end{aligned} \tag{3.44}$$

Note that

$$nT + T_{off}^{\min} \leq t < (n+1)T \Rightarrow \frac{t}{T} < n+1 \tag{3.45}$$

Combining (3.44) and (3.45), we have

$$V(t) \leq \frac{V_1(0)}{\mu_2} e^{-\frac{\lambda}{T}t}. \tag{3.46}$$

Define $\zeta_1 = \min\{\lambda_{\min}(P_1), \lambda_{\min}(P_2)\}$, $\zeta_2 = \lambda_{\max}(P_1)$, $\zeta_3 = \zeta_2 + h\lambda_{\max}(Q_1) + \frac{h^3}{2}\lambda_{\max}(R_1)$, $\kappa = \sqrt{\frac{\chi\zeta_3}{\zeta_1}}$, and $\chi = \max\left\{e^{\frac{\lambda T_{off}^{\min}}{T}}, \frac{1}{\mu_2}\right\}$, from (3.43) and (3.46), we obtain

$$V(t) \leq \chi e^{-\frac{\lambda}{T}t} V_1(0). \tag{3.47}$$

From the definition of $V(t)$, it is not difficult to derive

$$V(t) \geq \zeta_1 \|\xi(t)\|^2, \; V_1(0) \leq \zeta_3 \|\phi_0\|_h^2. \tag{3.48}$$

Combining (3.47) and (3.48), we have

$$\|\xi(t)\| \leq \kappa e^{-\rho t} \|\phi_0\|_h, \; t \geq 0 \tag{3.49}$$

which proves that the system (3.18) is GES with decay rate ρ by Definition 3.1.

Remark 3.4 From (3.32), one can obtain the lower bound T_{off}^* of $T_{off}^{\min}$ as $T_{off}^* \triangleq \frac{2\alpha_2 T + 2(\alpha_1+\alpha_2)h + \ln(\mu_1\mu_2)}{2(\alpha_1+\alpha_2)}$, which depends on the jamming period T, the sampling period h, the convergence rate α_1 of the first subsystem of system (3.18), the divergence rate α_2 of the second subsystem of system (3.18), and two tuning parameters μ_i $(i = 1, 2)$, simultaneously. When the parameters α_i, μ_i, and h are fixed, the larger the T, the larger the lower bound T_{off}^*, which is shown in Table 3.1 of Sect. IV (its numerical solution $T_{off}^{\min}$ is given by iteration, in theory, $T_{off}^* \leq T_{off}^{\min}$). On the other hand, note that the decay rate $\rho \triangleq \frac{\lambda}{2T} = \frac{2\alpha_1 T_{off}^{\min} - 2\alpha_2\left(T - T_{off}^{\min}\right)}{2T} - \frac{2(\alpha_1+\alpha_2)h + \ln(\mu_1\mu_2)}{2T} = \frac{2(\alpha_1+\alpha_2)T_{off}^{\min}}{2T} - \frac{2\alpha_2 T + 2(\alpha_1+\alpha_2)h + \ln(\mu_1\mu_2)}{2T} = \frac{(\alpha_1+\alpha_2)\left(T_{off}^{\min} - T_{off}^{*}\right)}{T}$. One can see that when the parameters α_i, μ_i $(i = 1, 2)$, T, and h are fixed, the larger the $T_{off}^{\min}$, the larger the decay rate ρ, which is shown in Table 3.2 of Sect. IV.

Remark 3.5 Note that the inequalities (3.19)–(3.20), and (3.33)–(3.38) are linear in P_i, Q_i, R_i, U_i, and Ω for fixed $\varepsilon \in (0, 1)$, K, $T \in (0, +\infty)$, $T_{off}^{\min} \in (0, T)$, $\alpha_i \in (0, +\infty)$, $\mu_i \in (0, +\infty)$ $(i = 1, 2)$, and $h \in \left(0, T_{off}^{\min}\right)$. Therefore, when the feedback gain matrix K and triggering parameter $\varepsilon \in (0, 1)$ are given in advance, feasible solutions P_i, Q_i, R_i, U_i, and Ω for LMIs (3.19)–(3.20), (3.33)–(3.38) can be searched by iterating over a set of values for $T \in (0, +\infty)$, $T_{off}^{\min} \in (0, T)$, $\alpha_i \in (0, +\infty)$, $\mu_i \in (0, +\infty)$, and $h \in \left(0, T_{off}^{\min}\right)$ satisfying (3.32).

Remark 3.6 It should be pointed out that Theorem 3.1 provides a sufficient condition under which the system (3.18) is exponentially stable in the presence of the periodic DoS jamming attacks (3.2) if the maximum allowable jammer activity (MAJA) denoted by $J \triangleq \frac{T - T_{off}^{\min}}{T} \times 100\%$ is smaller than a certain upper bound J^*. In fact, the inequality (3.32) is equivalent to $\lambda = 2\alpha_1 T_{off}^{\min} - 2\alpha_2\left(T - T_{off}^{\min}\right) - 2\left(\alpha_1 + \alpha_2\right)h - \ln\left(\mu_1\mu_2\right) > 0$. Also note that the decay rate

$$\begin{aligned}\rho &= \frac{\lambda}{2T}\\ &= \frac{2\alpha_1 T_{off}^{\min} - 2\alpha_2\left(T - T_{off}^{\min}\right)}{2T}\\ &\quad - \frac{2\left(\alpha_1 + \alpha_2\right)h + \ln\left(\mu_1\mu_2\right)}{2T}\\ &= \alpha_1\left(1 - J\right) - \alpha_2 J\\ &\quad - \frac{2\left(\alpha_1 + \alpha_2\right)h + \ln\left(\mu_1\mu_2\right)}{2T}.\end{aligned} \tag{3.50}$$

Then due to the fact that $\lambda > 0 \Leftrightarrow \rho > 0$, one has

$$J < J^* \triangleq \frac{\alpha_1}{\alpha_1 + \alpha_2} - \frac{2h\left(\alpha_1 + \alpha_2\right) + \ln\left(\mu_1\mu_2\right)}{2T\left(\alpha_1 + \alpha_2\right)}. \tag{3.51}$$

Remark 3.7 According to expressions (3.50) and (3.51), the following three conclusions can be drawn:

(i) For given scalars $\alpha_i \in (0, +\infty)$, $\mu_i \in (0, +\infty)$ $(i = 1, 2)$, $T \in (0, +\infty)$, and $h \in (0, T)$, the exponential decay rate ρ is a linear monotonic decreasing function of the MAJA. That is, $\rho = \rho\left(J\right) = -\left(\alpha_1 + \alpha_2\right)J + \alpha_1 - \frac{2(\alpha_1+\alpha_2)h + \ln(\mu_1\mu_2)}{2T}$, which implies that the larger the J, the smaller the ρ, that is to say, the worse the stability performance.

(ii) For given scalars $\alpha_i \in (0, +\infty)$, $\mu_i \in (0, +\infty)$ $(i = 1, 2)$, $T \in (0, +\infty)$ and J, the exponential decay rate ρ is a linear monotonic decreasing function of the sampling period h. This relation can be explicitly expressed as $\rho = \rho(h) = -\frac{(\alpha_1+\alpha_2)}{T}h + \alpha_1 - \frac{\ln(\mu_1\mu_2)}{2T} - (\alpha_1 + \alpha_2)J$, from which one can see that a larger h results in a smaller ρ, and the vice versa. So in order to improve the stability performance, one can reduce the sampling period value.

(iii) For given scalars $\alpha_i \in (0, +\infty)$, $\mu_i \in (0, +\infty)$ $(i = 1, 2)$, and $T \in (0, +\infty)$, the upper bound of the MAJA denoted by J^* is a linear monotonic decreasing function of the sampling period h, which is expressed as $J^* = J^*(h) = -\frac{1}{T}h + \frac{\alpha_1}{\alpha_1+\alpha_2} - \frac{\ln(\mu_1\mu_2)}{2T(\alpha_1+\alpha_2)}$, which implies that one can improve the tolerance of the event-triggered control systems against the DoS jamming attacks by decreasing the sampling period.

3.3 Co-design of Controller and Observer

By Theorem 3.1, we now provide the following theorem for the co-design of the observer gains L_1, L_2, the controller gain K, and the weighting matrix Ω defining the resilient event-triggering scheme.

Theorem 3.2 *Consider the jamming signal $\mathcal{I}_{DoS}(t)$ (3.2), in which the sequence $\{nT\}_{n\in\mathbb{N}}$ and the parameters T and $T_{off}^{\min}$ are known and $T_{off}^{\min} \le T < +\infty$. For the system (3.18), if for given scalars $\alpha_i \in (0, +\infty)$, $\mu_i \in (0, +\infty)$, $\varepsilon \in (0, 1)$, $h \in \left(0, T_{off}^{\min}\right)$ satisfying (3.32), $\varkappa_i$ and $\varpi_i \in (0, +\infty)$, there exist symmetric positive definite matrices $X_i \in \mathbb{R}^{2n\times 2n}$, $\bar{Q}_i \in \mathbb{R}^{n\times n}$, $\bar{R}_i \in \mathbb{R}^{n\times n}$, $\bar{\Omega} \in \mathbb{R}^{n\times n}$, and matrices $\bar{U}_i$ $(i = 1, 2)$ of appropriate dimensions such that*

$$\bar{\Pi}_1 < 0, \bar{\Re}_1 > 0 \tag{3.52}$$

$$\bar{\Pi}_2 < 0, \bar{\Re}_2 > 0 \tag{3.53}$$

$$\begin{bmatrix} -\mu_2 X_2 & * \\ X_2 & -X_1 \end{bmatrix} \le 0 \tag{3.54}$$

$$\begin{bmatrix} -\mu_1 e^{2(\alpha_1+\alpha_2)h} X_1 & * \\ X_1 & -X_2 \end{bmatrix} \le 0 \tag{3.55}$$

$$\begin{bmatrix} -\mu_2 \bar{Q}_2 & * \\ X_{21} & \varpi_1^2 \bar{Q}_1 - 2\varpi_1 X_{11} \end{bmatrix} \le 0 \tag{3.56}$$

$$\begin{bmatrix} -\mu_2 \bar{R}_2 & * \\ X_{21} & \varkappa_1^2 \bar{R}_1 - 2\varkappa_1 X_{11} \end{bmatrix} \le 0 \tag{3.57}$$

$$\begin{bmatrix} -\mu_1 \bar{Q}_1 & * \\ X_{11} & \varpi_2^2 \bar{Q}_2 - 2\varpi_2 X_{21} \end{bmatrix} \le 0 \tag{3.58}$$

$$\begin{bmatrix} -\mu_1 \bar{R}_1 & * \\ X_{11} & \varkappa_2^2 \bar{R}_2 - 2\varkappa_2 X_{21} \end{bmatrix} \le 0 \tag{3.59}$$

where

$$\bar{\Pi}_1 = \begin{bmatrix} \bar{\Sigma}_1 & * \\ h\bar{F}_1 & \varkappa_1^2 \bar{R}_1 - 2\varkappa_1 X_{11} \end{bmatrix}, \bar{\mathfrak{R}}_1 = \begin{bmatrix} \bar{\bar{R}}_1 & * \\ \bar{U}_1 & \bar{\bar{R}}_1 \end{bmatrix}$$

$$\bar{\Sigma}_1 = \begin{bmatrix} \bar{\Sigma}_1^{11} & * & * & * & * & * & * \\ \bar{\Sigma}_1^{21} & \bar{\Sigma}_1^{22} & * & * & * & * & * \\ \bar{\Sigma}_1^{31} & 0 & \bar{\Sigma}_1^{33} & * & * & * & * \\ \bar{\Sigma}_1^{41} & 0 & \bar{\Sigma}_1^{43} & \bar{\Sigma}_1^{44} & * & * & * \\ \bar{\Sigma}_1^{51} & 0 & \bar{\Sigma}_1^{53} & \bar{\Sigma}_1^{54} & \bar{\Sigma}_1^{55} & * & * \\ \bar{\Sigma}_1^{61} & 0 & \bar{\Sigma}_1^{63} & \bar{\Sigma}_1^{64} & \bar{\Sigma}_1^{65} & \bar{\Sigma}_1^{66} & * \\ \bar{\Sigma}_1^{71} & 0 & \bar{\Sigma}_1^{73} & 0 & 0 & 0 & \bar{\Sigma}_1^{77} \end{bmatrix}$$

$$\bar{\Pi}_2 = \begin{bmatrix} \bar{\Sigma}_2 & * \\ h\bar{F}_2 & \varkappa_2^2 \bar{R}_2 - 2\varkappa_2 X_{21} \end{bmatrix}, \bar{\mathfrak{R}}_2 = \begin{bmatrix} \bar{\bar{R}}_2 & * \\ \bar{U}_2 & \bar{\bar{R}}_2 \end{bmatrix}$$

$$\bar{\bar{R}}_1 = e^{-2\alpha_1 h} \begin{bmatrix} \bar{R}_1 & * \\ 0 & 3\bar{R}_1 \end{bmatrix}, \bar{U}_1 = \begin{bmatrix} \bar{U}_{11} & \bar{U}_{12} \\ \bar{U}_{13} & \bar{U}_{14} \end{bmatrix}$$

$$\bar{\bar{R}}_2 = \begin{bmatrix} \bar{R}_2 & * \\ 0 & 3\bar{R}_2 \end{bmatrix}, \bar{U}_2 = \begin{bmatrix} \bar{U}_{21} & \bar{U}_{22} \\ \bar{U}_{23} & \bar{U}_{24} \end{bmatrix}$$

$$\begin{aligned}
\bar{\Sigma}_1^{11} &= 2\alpha_1 X_{11} + He\,(AX_{11}) + \bar{Q}_1 - 4e^{-2\alpha_1 h}\bar{R}_1 \\
\bar{\Sigma}_1^{21} &= C^T T_1^T,\ \bar{\Sigma}_1^{22} = 2\alpha_1 X_{12} + He\,(AX_{12} - T_1 C) \\
\bar{\Sigma}_1^{31} &= Y^T B^T - 2e^{-2\alpha_1 h}\bar{R}_1 - \bar{U}_{11} - \bar{U}_{12} - \bar{U}_{13} - \bar{U}_{14} \\
\bar{\Sigma}_1^{33} &= \varepsilon\bar{\Omega} - 8e^{-2\alpha_1 h}\bar{R}_1 + He\left(\bar{U}_{11} - \bar{U}_{12} + \bar{U}_{13} - \bar{U}_{14}\right) \\
\bar{\Sigma}_1^{41} &= \bar{U}_{11} + \bar{U}_{12} - \bar{U}_{13} - \bar{U}_{14} \\
\bar{\Sigma}_1^{43} &= -2e^{-2\alpha_1 h}\bar{R}_1 - \bar{U}_{11} + \bar{U}_{12} + \bar{U}_{13} - \bar{U}_{14} \\
\bar{\Sigma}_1^{44} &= -e^{-2\alpha_1 h}\bar{Q}_1 - 4e^{-2\alpha_1 h}\bar{R}_1,\ \bar{\Sigma}_1^{51} = 6e^{-2\alpha_1 h}\bar{R}_1 \\
\bar{\Sigma}_1^{53} &= 6e^{-2\alpha_1 h}\bar{R}_1 + 2\bar{U}_{12}^T + 2\bar{U}_{14}^T \\
\bar{\Sigma}_1^{54} &= -2\bar{U}_{12}^T + 2\bar{U}_{14}^T,\ \bar{\Sigma}_1^{55} = -12e^{-2\alpha_1 h}\bar{R}_1 \\
\bar{\Sigma}_1^{61} &= 2\bar{U}_{13} + 2\bar{U}_{14},\ \bar{\Sigma}_1^{63} = 6e^{-2\alpha_1 h}\bar{R}_1 - 2\bar{U}_{13} + 2\bar{U}_{14} \\
\bar{\Sigma}_1^{64} &= 6e^{-2\alpha_1 h}\bar{R}_1,\ \bar{\Sigma}_1^{65} = -4\bar{U}_{14},\ \bar{\Sigma}_1^{66} = -12e^{-2\alpha_1 h}\bar{R}_1 \\
\bar{\Sigma}_1^{71} &= Y^T B^T,\ \bar{\Sigma}_1^{73} = \varepsilon\bar{\Omega},\ \bar{\Sigma}_1^{77} = (\varepsilon - 1)\,\bar{\Omega} \\
\bar{F}_1 &= \begin{bmatrix} AX_{11} & T_1 C & BY & 0 & 0 & 0 & BY \end{bmatrix}
\end{aligned}$$

$$\bar{\Sigma}_2 = \begin{bmatrix} \bar{\Sigma}_2^{11} & * & * & * & * & * \\ \bar{\Sigma}_2^{21} & \bar{\Sigma}_2^{22} & * & * & * & * \\ \bar{\Sigma}_2^{31} & 0 & \bar{\Sigma}_2^{33} & * & * & * \\ \bar{\Sigma}_2^{41} & 0 & \bar{\Sigma}_2^{43} & \bar{\Sigma}_2^{44} & * & * \\ \bar{\Sigma}_2^{51} & 0 & \bar{\Sigma}_2^{53} & \bar{\Sigma}_2^{54} & \bar{\Sigma}_2^{55} & * \\ \bar{\Sigma}_2^{61} & 0 & \bar{\Sigma}_2^{63} & \bar{\Sigma}_2^{64} & \bar{\Sigma}_2^{65} & \bar{\Sigma}_2^{66} \end{bmatrix}$$

$$\begin{aligned}
\bar{\Sigma}_2^{11} &= -2\alpha_2 X_{21} + He(AX_{21}) + \bar{Q}_2 - 4\bar{R}_2 \\
\bar{\Sigma}_2^{21} &= C^T T_2^T, \bar{\Sigma}_2^{22} = -2\alpha_2 X_{22} + He(AX_{22} - T_2 C) \\
\bar{\Sigma}_2^{31} &= -2\bar{R}_2 - \bar{U}_{21} - \bar{U}_{22} - \bar{U}_{23} - \bar{U}_{24} \\
\bar{\Sigma}_2^{33} &= -8\bar{R}_2 + He\left(\bar{U}_{21} - \bar{U}_{22} + \bar{U}_{23} - \bar{U}_{24}\right) \\
\bar{\Sigma}_2^{41} &= \bar{U}_{21} + \bar{U}_{22} - \bar{U}_{23} - \bar{U}_{24} \\
\bar{\Sigma}_2^{43} &= -2\bar{R}_2 - \bar{U}_{21} + \bar{U}_{22} + \bar{U}_{23} - \bar{U}_{24} \\
\bar{\Sigma}_2^{44} &= -e^{2\alpha_2 h} \bar{Q}_2 - 4\bar{R}_2 \\
\bar{\Sigma}_2^{61} &= 2\bar{U}_{23} + 2\bar{U}_{24}, \bar{\Sigma}_2^{63} = 6\bar{R}_2 - 2\bar{U}_{23} + 2\bar{U}_{24} \\
\bar{\Sigma}_2^{64} &= 6\bar{R}_2, \bar{\Sigma}_2^{65} = -4\bar{U}_{24}, \bar{\Sigma}_2^{66} = -12\bar{R}_2 \\
\bar{F}_2 &= \left[AX_{21} \; T_2 C \; 0 \; 0 \; 0 \; 0 \right]
\end{aligned}$$

then the switched system (3.18) with

$$L_1 = T_1 \bar{X}_{12}^{-1}, L_2 = T_2 \bar{X}_{22}^{-1}, K = Y X_{11}^{-1}$$

is GES in the presence of the periodic DoS jamming attacks (3.2).

Proof Set $P_1 = \begin{bmatrix} P_{11} & * \\ 0 & P_{12} \end{bmatrix}$, $P_2 = \begin{bmatrix} P_{21} & * \\ 0 & P_{22} \end{bmatrix}$ and define $X_{11} = P_{11}^{-1}$, $X_{12} = P_{12}^{-1}$, $X_{21} = P_{21}^{-1}$, $X_{22} = P_{22}^{-1}$, $X_{11} Q_1 X_{11} = \bar{Q}_1$, $X_{11} R_1 X_{11} = \bar{R}_1$, $X_{21} Q_2 X_{21} = \bar{Q}_2$, $X_{21} R_2 X_{21} = \bar{R}_2$, $X_{11} \Omega X_{11} = \bar{\Omega}$, $X_{11} U_{11} X_{11} = \bar{U}_{11}$, $X_{11} U_{12} X_{11} = \bar{U}_{12}$, $X_{11} U_{13} X_{11} = \bar{U}_{13}$, $X_{11} U_{14} X_{11} = \bar{U}_{14}$, $X_{21} U_{21} X_{21} = \bar{U}_{21}$, $X_{21} U_{22} X_{21} = \bar{U}_{22}$, $X_{21} U_{23} X_{21} = \bar{U}_{23}$, $X_{21} U_{24} X_{21} = \bar{U}_{24}$, and $Y = K X_{11}$, $T_1 = L_1 \bar{X}_{12}$, $T_2 = L_2 \bar{X}_{22}$. For $X_{12} = W \begin{bmatrix} X_{121} & * \\ 0 & X_{122} \end{bmatrix} W^T$, $X_{22} = W \begin{bmatrix} X_{221} & * \\ 0 & X_{222} \end{bmatrix} W^T$. On the basis of Lemma 2 in [13], there exist $\bar{X}_{12} = OSX_{121} S^{-1} O^T$, $\bar{X}_{22} = OSX_{221} S^{-1} O^T$ such that $CX_{12} = \bar{X}_{12} C$ and $CX_{22} = \bar{X}_{22} C$. Define $J_1 = diag\left\{X_{11}, X_{12}, X_{11}, X_{11}, X_{11}, X_{11}, X_{11}, R_1^{-1}\right\}$, $J_2 = diag\left\{X_{21}, X_{22}, X_{21}, X_{21}, X_{21}, X_{21}, R_2^{-1}\right\}$. Then pre- and post-multiply J_1 and its transpose on both sides of (3.19), pre- and post-multiply J_2 and its transpose on both sides of (3.20). Furthermore, pre- and post-multiply (3.33) and (3.34) by $X_2 = \begin{bmatrix} X_{21} & * \\ 0 & X_{22} \end{bmatrix}$ and $X_1 = \begin{bmatrix} X_{11} & * \\ 0 & X_{12} \end{bmatrix}$, respectively, and using the Schur complement one can obtain (3.54) and (3.55). Utilizing the similar technique, pre- and post-multiplying (3.35)–(3.36) by X_{21} and (3.37)–(3.38) by X_{11}, respectively, and applying the inequalities $-X_{11} \bar{R}_1^{-1} X_{11} \le \varkappa_1^2 \bar{R}_1 - 2\varkappa_1 X_{11}$, $-X_{21} \bar{R}_2^{-1} X_{21} \le \varkappa_2^2 \bar{R}_2 - 2\varkappa_2 X_{21}$, $-X_{11} \bar{Q}_1^{-1} X_{11} \le \varpi_1^2 \bar{Q}_1 - 2\varpi_1 X_{11}$, and $-X_{21} \bar{Q}_2^{-1} X_{21} \le \varpi_2^2 \bar{Q}_2 -$

$2\varpi_2 X_{21}$ and the Schur complement one can derive (3.56)–(3.59). This completes the proof.

Remark 3.8 It should be pointed out that Theorem 3.2 is closely related to the sampling period h. From (3.13), it is seen that the upper bound of the artificial time-varying delay function $\tau_{k,n}(t)$ is dependent on h. Thus, Theorem 3.2 is a delay-dependent condition for event-triggered observed-based controllers. For a larger h, it is more likely that Theorem 3.2 fails to obtain the controller design. On the other hand, a smaller h usually results in better performance of the resulting closed loop system under consideration while increasing the network loads. Therefore, in using the proposed method, due to the fact that the event-triggering mechanism can reduce the network loads greatly, one should choose a small h to design a suitable event-triggered observed-based controllers such that the resulting closed loop system can achieve the desired system performance, which can be seen from the simulation section.

Remark 3.9 Theorem 3.2 establishes the existence condition of a resilient event-triggering scheme (3.5) and an event-triggered observer (3.3) with DoS jamming attacks (3.2). Specifically, if there exists a feasible solution of the matrix inequalities (3.52)–(3.59), then resilient event-triggering scheme (3.5) and the observer (3.3) can be designed simultaneously. Notice from (3.50) and (3.51) that ρ and J^* are monotonic increasing functions of α_1, and are monotonic decreasing functions of α_2, μ_1, and μ_2. In view of this, if the matrix inequalities (3.52)–(3.59) are feasible, α_1 should be chosen as large as possible while α_2, μ_1, and μ_2 should be chosen as small as possible to get large values of ρ and J^*. On the other hand, from (3.52)–(3.59), it is seen that a smaller α_1 and larger values of α_2, μ_1, and μ_2 are beneficial to the solvability of the inequalities (3.52)–(3.59). Therefore, one can use an iterative algorithm (see below) to obtain the maximal value of α_1, and the minimal values of α_2, μ_1, and μ_2 that guarantee the feasibility of the inequalities (3.52)–(3.59).

Step 1: Take a small initial value of $\alpha_1 > 0$ such that the inequality (3.52) is feasible. Specify an iteration step-size $\Delta\alpha_1 > 0$ for α_1.

Step 2: Set $\alpha_1 = \alpha_1 + \Delta\alpha_1$.

Step 3: Checking the feasibility of the inequality (3.52). If the inequality (3.52) is feasible, go to Step 2. Otherwise, exit and set $\alpha_1 = \alpha_1 - \Delta\alpha_1$.

Step 4: Following some similar procedures as in Steps 1–3 to obtain the minimal $\alpha_2 > 0$.

Step 5: For the obtained maximal α_1, minimal α_2, taking a large initial $\mu_1 > 0$ such that (3.55), (3.58), and (3.59) are feasible. Specify an iteration step-size $\Delta\mu_1 > 0$ for μ_1.

Step 6: Set $\mu_1 = \mu_1 - \Delta\mu_1$.

Step 7: Checking the feasibility of (3.55), (3.58), and (3.59). If these inequalities are feasible, then go to Step 6. Otherwise, exit and set $\mu_1 = \mu_1 + \Delta\mu_1$.

Step 8: Following some similar procedures as in Steps 5-7 to obtain the minimal $\mu_2 > 0$.

Step 9: Take the values of $T_{off}^{\min}$ and T satisfying $0 < T_{off}^{\min} < T$ and inequality (3.32). It ends.

3.4 Simulation Example

In this section, a practical example is utilized to demonstrate the effectiveness of the proposed method. Let the physical plant in Fig. 3.1 be an offshore structure with an active mass damper (AMD) mechanisms, which is taken from [14]. The dynamic equations of the offshore structure are given by

$$\begin{cases} m_1\ddot{z}_1(t) = -c_1\dot{z}_1(t) - k_1 z_1(t) + k_2 z_2(t) \\ -k_2 z_1(t) + c_2(\dot{z}_2(t) - \dot{z}_1(t)) - u(t) \\ m_2\ddot{z}_2(t) = -c_2(\dot{z}_2(t) - \dot{z}_1(t)) - k_2 z_2(t) \\ +k_2 z_1(t) + u(t) \end{cases} \tag{3.60}$$

where $z_1(t)$ and $z_2(t)$ are displacements of the offshore structure and the AMD, respectively; m_1 and m_2 are masses of the offshore structure and the AMD, respectively; k_1 and k_2 are stiffnesses of the offshore structure and the AMD, respectively; and c_1 and c_2 are dampings of the offshore structure and the AMD, respectively; $u(t)$ is the control force of the system. Note that here, for simplicity, the wave force acting on the offshore structure is overlooked.

Define $x_1(t) = z_1(t)$, $x_2(t) = z_2(t)$, $x_3(t) = \dot{z}_1(t)$, $x_4(t) = \dot{z}_2(t)$, $x(t) = \begin{bmatrix} x_1(t)\ x_2(t)\ x_3(t)\ x_4(t) \end{bmatrix}^T$, and assume the controlled outputs are the displacements of the offshore structure and the AMD. Thus, the state-space representation of the above Eq. (3.60) can be written as the form of the system (3.1) with

$$A = \begin{bmatrix} 0 & 0 & 1 & 0 \\ 0 & 0 & 0 & 1 \\ -\frac{k_1+k_2}{m_1} & \frac{k_2}{m_1} & -\frac{c_1+c_2}{m_1} & \frac{c_2}{m_1} \\ \frac{k_2}{m_2} & -\frac{k_2}{m_2} & \frac{c_2}{m_2} & -\frac{c_2}{m_2} \end{bmatrix}$$

$$B = \begin{bmatrix} 0\ 0 -\frac{1}{m_1}\ \frac{1}{m_2} \end{bmatrix}^T$$

$$C = \begin{bmatrix} 1\,0\,0\,0 \\ 0\,0\,1\,0 \end{bmatrix}$$

As in [14], suppose that the offshore structure is placed in the water with depths $d_0 = 218\,\text{m}$. The length L of the offshore structure is 249 m, the cylinder diameter $\tilde{D} = 1.83\,\text{m}$. The masses, natural frequencies and the damping ratios of the structure and AMD are given as $m_1 = 7825307\,\text{kg}$, $\omega_1 = 2.0466\,\text{rad/s}$, $\xi_1 = 2\%$, $m_2 = 78253\,\text{kg}$, $\omega_2 = 2.00074\,\text{rad/s}$, $\xi_2 = 20\%$.

For this simplified model, in the following, we first co-design the event-triggering parameters (ε, Ω), control gain matrix K in (3.6) and observer gain matrices L_1, L_2 in (3.3) such that the closed-loop system (3.18) is GES. To this end, we assume the jammer, imposing jamming signal $\mathcal{I}_{DoS}(t)$ (3.2) with $T = 2\,\text{s}$ and $T_{off}^{\min} = 1.6\,\text{s}$. Choosing $\mu_1 = 1.1, \mu_2 = 1.1, \alpha_1 = 0.2, \alpha_2 = 0.45, h = 0.02\,\text{s}$ satisfying (3.32). For given $\varepsilon = 0.3$, $\varkappa_1 = 1$, $\varkappa_2 = 1$, $\varpi_1 = 5$, $\varpi_2 = 5$ solving the LMIs in Theorem 3.2, it is found that the observer-based event-triggered control problem is feasible, and the obtained event-triggering parameter Ω, the controller gain matrix K, observer gain matrices L_1 and L_2 are given by

$$\Omega = 10^3 \times \begin{bmatrix} 8.1471 & -0.0298 & -0.2453 & -0.1394 \\ -0.0298 & 0.0003 & 0.0023 & 0.0006 \\ -0.2453 & 0.0023 & 0.0232 & 0.0045 \\ -0.1394 & 0.0006 & 0.0045 & 0.0025 \end{bmatrix} \quad (3.61)$$

$$K = 10^6 \times \begin{bmatrix} 4.3841 & -0.0145 & -0.1253 & -0.0709 \end{bmatrix} \quad (3.62)$$

$$L_1 = \begin{bmatrix} -0.0098 & -0.0034 \\ 0.2021 & 0.2484 \\ 0.0151 & 0.0005 \\ -1.6683 & -0.4939 \end{bmatrix}, \; L_2 = \begin{bmatrix} 0.1746 & -0.0226 \\ -0.8664 & 1.4365 \\ 0.1470 & 0.4244 \\ -4.8470 & -1.3500 \end{bmatrix} \quad (3.63)$$

Simulations are carried out by connecting the observer-based event-triggered state feedback controller (3.6) with (3.61) to the offshore structure under the event-triggering mechanism (3.5) with (3.61) in the presence of the periodic DoS jamming attack (3.2) with $T = 2\,\text{s}$ and $T_{off}^{\min} = 1.6\,\text{s}$. The initial conditions are taken as $x_0 = \begin{bmatrix} 0.2 & 0.3 & 0.6 & 0.9 \end{bmatrix}^T$, $\hat{x}_0 = \begin{bmatrix} 0.4 & 0.6 & 0.4 & 0.6 \end{bmatrix}^T$ and the simulation time is assumed to be 25 s. In the presence of DoS attacks , the state responses of the considered system, the error between system states and observed states, the release time intervals between any two consecutive release instants, and the control input are depicted in Figs. 3.2, 3.3, 3.4 and 3.5, respectively, from which we can see that (i) the observer-based error system is exponentially stable; (ii) the proposed resilient event-triggering mechanism can reduce the amount of control updates; (iii) the observer-based resilient event-triggered state-feedback controller does have counteracted the effect of the periodic jamming attacks.

In order to show the influence of the jamming period T, we solve the following optimization problem for different values of T in the time interval [0, 25 s] (the parameters μ_1, μ_2, α_1, α_2, ε, h, $\varkappa_1$, $\varkappa_2$, ϖ_1, ϖ_2 are chosen as the same before):

$$\begin{aligned} \bar{T}_{off}^{\min} = \min \{ & T_{off}^{\min} \mid T_{off}^{\min} \text{ satisfying (3.32)} \\ & \text{subjects to (3.52)–(3.59)} \} . \end{aligned} \quad (3.64)$$

Table 3.1 shows the minimum $T_{off}^{\min}$ for which the exponential stability of the closed-loop system under consideration is guaranteed, the maximum allowable jammer activity denoted by $\bar{J} \triangleq \frac{T-\bar{T}_{off}^{\min}}{T} \times 100\%$ and the corresponding upper bound of $\bar{J}$ obtained for each T chosen. From Table 3.1, it can be seen that when T increases, the values of $\bar{T}_{off}^{\min}$ and $\bar{J}$ also increase. This may be reasonable because for a larger jammer period T, the shortest time the jammer sleeps should also be larger to make the system stable correspondingly. In other words, the system can relatively tolerate more malicious attacks as well. Furthermore, it is observed that the obtained lower bound $\bar{T}_{off}^{\min}$ of $T_{off}^{\min}$ (solving (3.64) by iteration) is very close to its analytical bound T_{off}^{*},

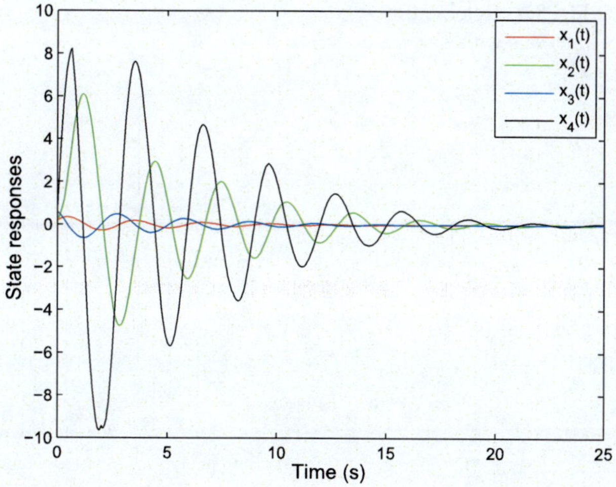

Fig. 3.2 State responses of the system with $T = 2\,\mathrm{s}$ and $T_{off}^{\min} = 1.6\,\mathrm{s}$

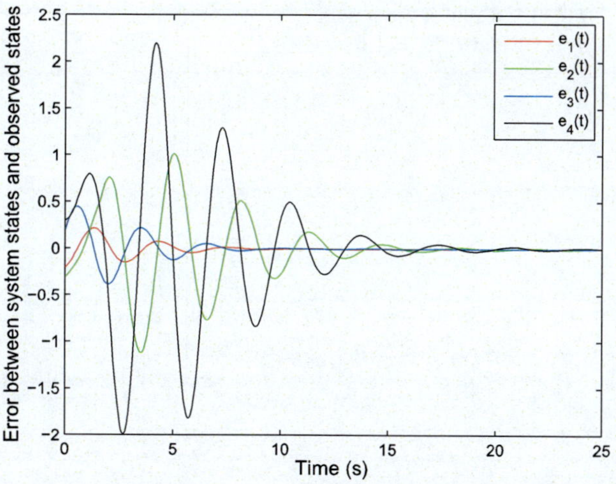

Fig. 3.3 Error responses between system states and observed states with $T = 2\,\mathrm{s}$ and $T_{off}^{\min} = 1.6\,\mathrm{s}$

which demonstrates the reasonability of the Remark 3.4. In addition, the obtained maximum allowable jammer activity $\bar{J}$ is always smaller than J^*, which illustrates the validity of the Remark 3.6. In what follows, in order to show the relationship between $\bar{T}_{off}^{\min}$ and the decay rate ρ, some calculations are listed in Table 3.2. It is observed that the bigger the $\bar{T}_{off}^{\min}$, the larger the ρ, which validates the statement of the Remark 3.4.

To further show the effect of the triggering parameter ε on the stability of the system, we assume that $\mu_1 = 1.01, \mu_2 = 1.01, \alpha_1 = 0.2, \alpha_2 = 0.45, T = 2, T_{off}^{\min} = 1.6$, and $\varkappa_1, \varkappa_2, \varpi_1, \varpi_2$ are chosen as the same above. The maximum allowable sampling period h_{max} and the corresponding $\bar{T}_{off}^{\min}$ for different ε are listed in Table 3.3. From

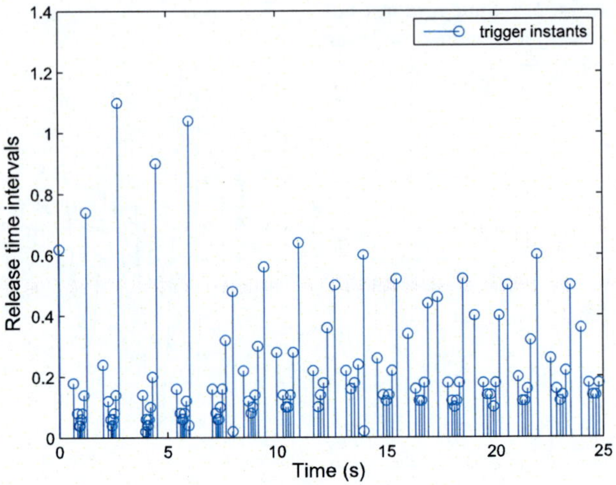

Fig. 3.4 Release time intervals with $T = 2\,\text{s}$ and $T_{off}^{\min} = 1.6\,\text{s}$

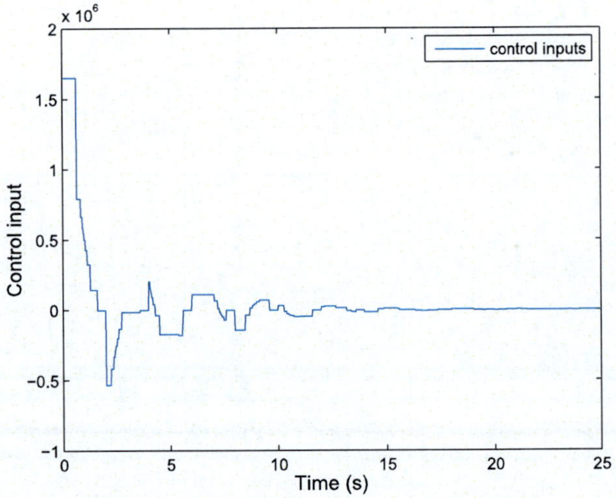

Fig. 3.5 Control input with $T = 2\,\text{s}$ and $T_{off}^{\min} = 1.6\,\text{s}$

Table 3.1 $\bar{T}_{off}^{\min}$, $\bar{J}$, and J^* for different values of T

T	2	3	5	7	9
$\bar{T}_{off}^{\min}$	1.56	2.25	3.63	5.02	6.40
T_{off}^{*}	1.55	2.24	3.63	5.01	6.40
$\bar{J}$	22%	25%	27.4%	28.3%	28.9%
J^*	22.44%	25.21%	27.44%	28.39%	28.92%

Table 3.2 λ and ρ for different values of $T_{off}^{\min}$

$T_{off}^{\min}$	1.56	1.66	1.76	1.86	1.96
λ	0.0114	0.1414	0.2714	0.4014	0.5314
ρ	0.0028	0.0353	0.0678	0.1003	0.1328

Table 3.3 h_{max} and $\bar{T}_{off}^{\min}$ for different values of σ

ε	0.3	0.2	0.1	0.05	0.01
h_{max}	0.07	0.12	0.15	0.17	0.18
$\bar{T}_{off}^{\min}$	1.47	1.52	1.55	1.57	1.58

Table 3.3, we find that for given the DoS attacks, the h_{max} and $\bar{T}_{off}^{\min}$ are becoming bigger when the triggering parameter ε is decreasing. Hence, there exists a tradeoff between system performance and resource occupancy.

3.5 Conclusion

In this chapter, we have investigated the observer-based event-triggered control for a class of networked linear systems under periodic DoS jamming attacks. A novel resilient event-triggered communication mechanism has been proposed to enhance the utilization efficiency of network resources while resisting the DoS jamming attacks. Inspired by the intermittent property of the DoS jamming attacks, a new switched observer error system model has been established, which facilitates us to consider the effects of the resilient event-triggered strategy and DoS jamming attacks in a unified framework. Tractable LMI-based stability analysis and control design criteria for the co-design of the observer and controller gains have been derived while preserving satisfactory control performance despite the presence of DoS jamming attacks. At last, a practical example has been exploited to demonstrate the effectiveness of the proposed resilient event-triggered controller design method. In our future work, we will extend the proposed method to deal with the event-based H_∞ filtering/fault detection problem and event-based adaptive control problem in the presence of DoS jamming attacks with the help of [15–17]. Besides, how to extend the proposed framework to study the event-triggered NCSs under more general DoS attacks is another interesting research venue, where the attack signals are unknown and happen aperiodically.

References

1. S. Hu, D. Yue, Q.-L. Han, X. Xie, X. Chen, C. Dou, Observer-based event-triggered control for networked linear systems subject to denial-of-service attacks. IEEE Trans. Cybern. **50**(5), 1952–1964 (2020)
2. H.S. Foroush, S. Martinez, On triggering control of single-input linear systems under pulse-width modulated dos signals. SIAM J. Control Optim. **54**(6), 3084–3105 (2016)
3. D. Yue, E. Tian, Q.L. Han, A delay system method for designing event-triggered controllers of networked control systems. IEEE Trans. Autom. Control **58**(2), 475–481 (2013)
4. X.M. Zhang, Q.L. Han, B.L. Zhang, An overview and deep investigation on sampled-data-based event-triggered control and filtering for networked systems. IEEE Trans. Industr. Inf. **13**(1), 4–16 (2017)
5. W.P.M.H. Heemels, M.C.F. Donkers, A.R. Teel, Periodic event-triggered control for linear systems. IEEE Trans. Autom. Control **58**(4), 847–861 (2013)
6. C.D. Persis, P. Tesi, Resilient control under denial-of-service. IFAC Proc. Vol. **47**(3), 134–139 (2014)
7. R. Lu, S. Peng, H. Su, Z.G. Wu, J. Lu, Synchronization of general chaotic neural networks with nonuniform sampling and packet missing: a switched system approach. IEEE Trans. Neural Netw. Learn. Syst. **29**(3), 523–533 (2018)
8. W.H. Chen, J. Zhong, W.X. Zheng, Delay-independent stabilization of a class of time-delay systems via periodically intermittent control. Automatica **71**, 89–97 (2016)
9. W.A. Zhang, L. Yu, Stabilization of sampled-data control systems with control inputs missing. IEEE Trans. Autom. Control **55**(2), 447–452 (2010)
10. L. Schenato, To zero or to hold control inputs with lossy links? IEEE Trans. Autom. Control **54**(5), 1093–1099 (2009)
11. A. Seuret, F. Gouaisbaut, Stability of linear systems with time-varying delays using Bessel-Legendre inequalities. IEEE Trans. Autom. Control **63**(1), 225–232 (2018)
12. A. Seuret, F. Gouaisbaut, Wirtinger-based integral inequality: application to time-delay systems. Automatica **49**(9), 2860–2866 (2013)
13. C. Peng, S. Ma, X. Xie, Observer-based non-PDC control for networked T-S fuzzy systems with an event-triggered communication. IEEE Trans. Cybern. **47**(8), 2279–2287 (2017)
14. B.-L. Zhang, Q.-L. Han, X.-M. Zhang, Event-triggered H_∞ reliable control for offshore structures in network environments. J. Sound Vib. **368**, 1–21 (2016)
15. Y.J. Liu, S. Tong, Adaptive fuzzy control for a class of nonlinear discrete-time systems with backlash. IEEE Trans. Fuzzy Syst. **22**(5), 1359–1365 (2014)
16. Y. Li, S. Tong, T. Li, Adaptive fuzzy output feedback dynamic surface control of interconnected nonlinear pure-feedback systems. IEEE Trans. Cybern. **45**(1), 138 (2015)
17. Y. Li, S. Tong, Adaptive fuzzy output-feedback stabilization control for a class of switched nonstrict-feedback nonlinear systems. IEEE Trans. Cybern. **47**(4), 1007–1016 (2017)

Chapter 4
Security Control of Cyber-Physical System Based on Switching Approach for Nonperiodic DoS Attacks

This chapter analyzes the effects of the intermittent denial-of-service jamming attack (I-DoS-JA) on different communication channels and communication topology transformations. Each different communication topology is taken as a subsystem of the switching system. Based on the switching system, the finite-time nonlinear system robust stability conditions are derived. Both the theoretical deduction and example simulation have corroborated that this approach has special effective in resisting the intermittent DoS attack [1].

This chapter is organized as follows. The security problem of nonlinear system is formulated in Sect. 4.1. Finite-time H_∞ analysis is given in Sect. 4.2. A simulation example is given in Sect. 4.3. Finally, Sect. 4.4 concludes the chapter.

4.1 Problem Formulation

4.1.1 System Model

In this chapter, in order to reflect the features of the CPS, the remote plant is considered as a nonlinear system

$$\begin{cases} x(k+1) = Ax(k) + Bu_a(k) + f(x(k)) + E\omega(k) \\ y(k) = Cx(k) \end{cases} \tag{4.1}$$

And the local controller is formulated via an observer framework as

$$\begin{cases} \hat{x}(k+1) = A\hat{x}(k) + Bu(k) + L\left[y_a(k) - \hat{y}(k)\right] \\ \hat{y}(k) = C\hat{x}(k) \end{cases} \tag{4.2}$$

D. Yue et al., *Secure Control of Networked Control Systems and Its Applications*,
https://doi.org/10.1007/978-981-33-6730-2_4

where $x(k) \in R^n$ and $\hat{x}(k) \in R^n$ denote the plant state and the controller state (obtained by an observer), respectively. And $y(k) \in R^m$ is the remote plant's output. $u_a(k)$ and $y_a(k)$ represent the practical control and feedback signals, while the ideal values are $u(k)$ and $y(k)$. $\omega(k) \in R^n$ stands for the outside disturbances, which satisfying $\omega(k) \in L_2 = [0, \infty)$ and we assume the disturbance is norm-bounded, which implicates $\sum_{k=0}^{N} \omega^T(k)\omega(k) \le \delta^2$. And the symbols E and L are the coefficient matrices of the disturbance and the observer.

4.1.2 Attack Model

According to the description and analysis of the attack process of I-DoS-JA, the system can be divided as follows under the attack of DoS:

$$\mathcal{T}_n = [\mathcal{T}^n_{on/off}, \mathcal{T}^{n+1}_{on/off}) = \mathcal{H}_n \cup \mathcal{D}_n, n \in \mathbb{Z}^+$$

where $\mathcal{H}_n := [\mathcal{T}^n_{on/off}, \mathcal{T}^{n+1}_{off/on})$ represent the I-DoS-JA absent intervals, while $\mathcal{D}_n := \left[\mathcal{T}^{n+1}_{off/on}, \mathcal{T}^{n+1}_{on/off}\right]$ represent the I-DoS-JA present intervals, where $\mathcal{T}^n_{on/off}$ means the ending of the n_{th} DoS attack, and $\mathcal{T}^{n+1}_{on/off}$ means the beginning of the $(n+1)_{th}$ DoS attack .

To facilitate analysis, an assumption is given for the nonlinear parts of the system.

Assumption 4.1 For the nonlinear part, we assume that, the initial case is $f(x(0)) = 0$, meanwhile, for any $x(k_i)$ and $x(k_j)$, there exists

$$\|f(x(k_i)) - f\left(x\left(k_j\right)\right)\| < \nu_{ij}\left\|x(k_i) - x\left(k_j\right)\right\|, i \ne j$$

holds. Furthermore, if there exists $x(k_i) = 0$ or $x(k_j) = 0$, then we have

$$\|f(x(k_i))\| < \nu_i \|x(k_i)\| = \nu_i\sqrt{x^T(k)x(k)}$$

Remark 4.1 From mathematical perspective, the inequalities in Assumption 4.1 are satisfied. The key point is the selection of the parameter ν_i or ν_{ij}.

To support the rationality of the defense strategy designed in this chapter, the following assumptions are made.

Assumption 4.2 In this chapter, we assume that the I-DoS-JA can be detected by software applications,such as flow monitoring. In addition, under the presented switching system framework, we assume that, when a subsystem switches from one model to another, there should be existing at least one model which is secure, and

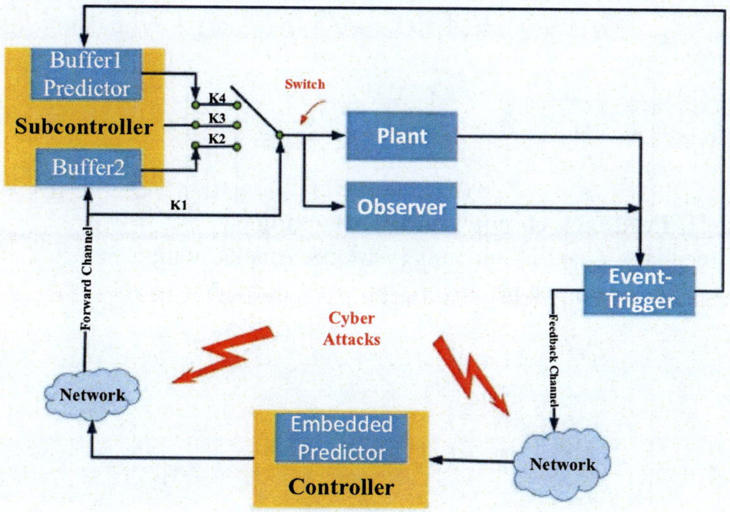

Fig. 4.1 The structure of switching system approach for I-DoS-JA schematic

this subsystem should be secure during the interval of $\tau_R + \tau_a^*$ from switching beginnings, where τ_R and τ_a^* represent the upper bound of packet loss and the average dwell-on time, respectively.

Remark 4.2 If the I-DoS-JA can not be detected, the subsystem will switch from one unsafe model to another unsafe model, the system will loss its stability or even break down. Therefore, the Assumption 4.2 is important and necessary for this chapter.

Remark 4.3 Since the index τ_R is the basic characteristic in NCS, and the dwell-on time is the important performance for for system switching from one stable and secure model to another. Thus, Assumption 4.2 is set as a constrain condition that the successful switching is guaranteed.

The I-DoS-JA usually cause data missing, in this chapter, we define $u_a(k) = u\left(k - \Delta_{ij}\right)$ and $y_a(k) = y\left(k - \Delta_{ij}\right)$ to denote the control and feedback signals, which are effected by I-DoS attacks. And $\Delta_{ij} \in \mathbb{Z}^+$ means the data missing span from node i to j. Here, we consider the special case with $\Delta_{ij} = 1$, then we can define the errors vector and state stacking vectors as $e(k) = x(k) - \hat{x}(k)$ and $\eta(k) = \left[x^T(k)\ e^T(k)\ u^T(k-1)\ \hat{y}^T(k-1)\right]^T$.

Motivated by [2], we discuss the cases that the forward and feedback channels encounter cyber-attack, respectively. And the case both channels under intermittent DoS attack is considered as well. Under these conditions, the communication topologies have been changed according to each different cyber-attack model (Fig. 4.1).

Case I: Both communication channels security

Both the forward and feedback channels are secure, which implies $u_a(k) = u(k)$ and $y_a(k) = y(k)$. Then we have

$$\eta(k+1) = \mathcal{A}_1\eta(k) + \mathcal{M}_1 f(x(k)) + \mathcal{N}_1\omega(k) \tag{4.3}$$

where $\mathcal{A}_1 = \begin{bmatrix} A+BK_1 & -BK_1 & 0 & 0 \\ 0 & A-L_1C & 0 & 0 \\ K_1 & -K_1 & 0 & 0 \\ C & 0 & 0 & 0 \end{bmatrix}$, $\mathcal{M}_1 = \begin{bmatrix} I \\ I \\ 0 \\ 0 \end{bmatrix}$, $\mathcal{N}_1 = \begin{bmatrix} E \\ E \\ 0 \\ 0 \end{bmatrix}$.

Case II: Feedback channel under attacking

The feedback channel encounters I-DoS attacks, which makes $y_a(k) = 0$ and $y_a(k) \neq y(k)$. Then, we have

$$\eta(k+1) = \mathcal{A}_2\eta(k) + \mathcal{M}_2 f(x(k)) + \mathcal{N}_2\omega(k) \tag{4.4}$$

where $\mathcal{A}_2 = \begin{bmatrix} A+BK_2 & -BK_2 & 0 & 0 \\ L_2C & A-L_2C & 0 & -L_2 \\ K_2 & -K_2 & 0 & 0 \\ C & 0 & 0 & I \end{bmatrix}$, $\mathcal{M}_2 = \mathcal{M}_1, \mathcal{N}_2 = \mathcal{N}_1$.

Case III: Forward channel under attacking

The forward channel encounters I-DoS attacks, which makes $u_a(k) = 0$, $u_a(k) \neq u(k)$ and $y_a(k) = y(k)$. Then, we have

$$\eta(k+1) = \mathcal{A}_3\eta(k) + \mathcal{M}_3 f(x(k)) + \mathcal{N}_3\omega(k) \tag{4.5}$$

where $\mathcal{A}_3 = \begin{bmatrix} A & 0 & B & 0 \\ -BK_3 & A+BK_3-L_3C & B & 0 \\ 0 & 0 & I & 0 \\ C & 0 & 0 & 0 \end{bmatrix}$, $\mathcal{M}_3 = \mathcal{M}_1, \mathcal{N}_3 = \mathcal{N}_1$.

Case IV: Double channels under attacking

Both the forward and feedback channels have encountered I-DoS attacks with leading to the data missing of both channels with leading to $y_a(k) = 0$ and $u_a(k) = 0$. They imply that $y_a(k) \neq y(k)$, $y_a(k) = \hat{y}(k)$, $u_a(k) \neq u(k)$ and $u(k) = \hat{u}(k)$. From these, we can obtain

$$\eta(k+1) = \mathcal{A}_4\eta(k) + \mathcal{M}_4 f(x(k)) + \mathcal{N}_4\omega(k) \tag{4.6}$$

where $\mathcal{A}_4 = \begin{bmatrix} A & 0 & B & 0 \\ -(BK_4-L_4C) & A+BK_4-L_4C & B & -L_4 \\ 0 & 0 & I & 0 \\ 0 & 0 & 0 & I \end{bmatrix}$, $\mathcal{M}_4 = \mathcal{M}_1, \mathcal{N}_4 = \mathcal{N}_1$.

4.1.3 A Switched System Formulation under DoS Attacks

Synthesizing (4.3)–(4.6), we can derive

$$\eta(k+1) = \mathcal{A}_{\sigma_k}\eta(k) + \mathcal{M}_{\sigma_k}\bar{f}(\eta(k)) + \mathcal{N}_{\sigma_k}\omega(k) \tag{4.7}$$

where $\sigma_k \in \mathcal{I}_{\sigma_k} := \{1, 2, 3, 4\}$.

Definition 4.1 ([3, 4]) For given system (4.7), when the system is under intermittent DoS attack and which also caused $u(k) = 0$, together with the influence of outside disturbance $\omega(k)$, if there exist parameters $0 < \alpha < \phi$, and positive number $N \in \mathbb{Z}^+$ and positive matrix $\mathcal{R}$, such that

$$x^T(k)\mathcal{R}x(k) < \varphi^2, \forall k \in \{1, 2, \ldots, N\} \tag{4.8}$$

and $x_0^T\mathcal{R}x_0 < \alpha^2$, then it is determined that the system (4.7) is finite-time bounded respect to $\{\alpha, \varphi, \delta, \mathcal{R}, N\}$.

Definition 4.2 ([4, 5]) For given system (4.7), when the system is under intermittent DoS attack and which also caused $u(k) = 0$, together with the influence of outside disturbance $\omega(k)$, if there exist parameters $0 < \alpha < \phi$, $\gamma > 0$, and positive number $N \in \mathbb{Z}^+$ and positive matrix $\mathcal{R}$, such that

(i) the system (4.7) is finite-time bounded;

(ii) under zero initial condition, the synthesis system (4.7) output satisfied

$$\sum_{k=0}^{N} z^T(k)z(k) \le \gamma^2 \sum_{k=0}^{N} \omega^T(k)\omega(k) \tag{4.9}$$

the system (4.7) is said to be H_∞ finite-time bounded.

Lemma 4.1 *For given any $x, y \in \mathbb{R}^n$ and arbitrary positive definite matrix $P \in \mathbb{R}^{N\times N}$, then we have*

$$2x^T y \le x^T P x + y^T P^{-1} y \tag{4.10}$$

Lemma 4.2 ([6]) *The redfollowing linear matrix inequality*

$$\mathcal{Y} = \begin{bmatrix} \mathcal{Y}_{11} & \mathcal{Y}_{12} \\ \mathcal{Y}_{21}^T & \mathcal{Y}_{22} \end{bmatrix}$$

where $\mathcal{Y}_{11} = \mathcal{Y}_{11}^T$ and $\mathcal{Y}_{22} = \mathcal{Y}_{22}^T$ is equivalent to

$$\mathcal{Y}_{22} < 0, \mathcal{Y}_{11} - \mathcal{Y}_{12}\mathcal{Y}_{22}^{-1}\mathcal{Y}_{12}^T \tag{4.11}$$

4.2 Stability Analysis

Theorem 4.1 *For given switched system (4.7), during the interval* $[0, N)$ *and the I-DoS-JA, which will caused* $u_a(k) \equiv 0$ *or* $y_a(k) \equiv 0$, *under* ***Assumption*** *4.2, if there exist positive matrices* $P_i, Y_{ij}, \mathcal{W}_{ij}, \mathcal{Z}_{ij}$, $i, j \in \mathcal{I}_k = \{1, 2, 3, 4\}$ *and positive symmetric matrix* $\mathcal{R}$, *and positive scalars* $\tau \geq 0, \lambda_1 > 0, \lambda_2 > 0, 0 < \varphi < \zeta, \kappa_{ij} > 0$, *satisfying the follow inequalities*

$$\lambda_1 \mathcal{R} \leq P_i \leq \lambda_2 \mathcal{R} \tag{4.12}$$

$$\begin{bmatrix} -(1+\tau) P_i + \kappa_{ij} I & 0 & \mathcal{A}_i^T P_j & \mathcal{A}_i^T P_j & 0 \\ \star & -Y_{ij} & \mathcal{N}_i^T & 0 & \mathcal{N}_i^T \\ \star & \star & -P_j & 0 & 0 \\ \star & \star & \star & -\left(\mathcal{W}_{ij} + \mathcal{Z}_{ij}\right) & 0 \\ \star & \star & \star & \star & -\mathcal{Z}_{ij} \end{bmatrix} < 0 \tag{4.13}$$

$$(1+\tau)^N \left\{ \lambda_2 \varphi^2 + \max_{i,j \in \mathcal{I}_k} \left\{ \lambda_{\max}\left(Y_{ij}\right)\right\} \delta^2 \right\} - \lambda_1 \xi^2 \leq 0 \tag{4.14}$$

such that the system (4.7) is finite-time bounded stable respect to $\{\varphi, \zeta, \delta, \mathcal{R}, N\}$.

Proof The Lyapunov function is chosen as

$$V_{\sigma_k}(\eta(k)) = \eta^T(k) P_{\sigma_k} \eta(k), \forall \sigma_k \in \mathcal{I}_k = \{1, 2, 3, 4\}$$

Along with the trajectory of system (4.7), the difference equation can be presented as

$$\begin{aligned} \Delta V(\eta(k)) &= V(\eta(k+1)) - V(\eta(k)) \\ &= \eta^T(k+1) P_j \eta(k+1) - \eta^T(k) P_i \eta(k) \\ &= \left[\mathcal{A}_i \eta(k) + \mathcal{M}_i \bar{f} \eta(x(k)) + \mathcal{N}_i \omega(k)\right]^T P_j \\ &\quad \left[\mathcal{A}_i \eta(k) + \mathcal{M}_i \bar{f}(\eta(k)) + \mathcal{N}_i \omega(k)\right] - \eta^T(k) P_i \eta(k) \\ &= \eta^T(k)\left[\mathcal{A}_i^T P_j \mathcal{A}_i - P_i\right]\eta(k) + \bar{f}^T(\eta(k)) \mathcal{M}_i^T P_j \mathcal{M}_i \bar{f}(\eta(k)) \\ &\quad + \omega^T(k) \mathcal{N}_i^T P_j \mathcal{N}_i \omega(k) + 2\eta^T(k) \mathcal{A}_i^T P_j \mathcal{M}_i f(x(k)) \\ &\quad + 2\eta^T(k) \mathcal{A}_i^T P_j \mathcal{N}_i \omega(k) + 2\bar{f}^T(\eta(k)) \mathcal{M}_i^T P_j \mathcal{N}_i \omega(k), \quad i, j \in \mathcal{I}_k \end{aligned} \tag{4.15}$$

According to Lemma 4.1, we can obtain that

$$\begin{aligned} & 2\eta^T(k) \mathcal{A}_i^T P_j \mathcal{M}_i \bar{f}(\eta(k)) - \bar{f}^T(\eta(k)) \mathcal{M}_i^T \mathcal{W}_{ij} \mathcal{M}_i \bar{f}(\eta(k)) \\ & \leq \eta^T(k) \mathcal{A}_i^T P_j \mathcal{W}_{ij}^{-1} \mathcal{A}_i \eta(k) \end{aligned} \tag{4.16}$$

$$
\begin{aligned}
&2\bar{f}^T(\eta(k))\mathcal{M}_i^T P_j \mathcal{N}_i \omega(k) \\
&\le \bar{f}^T(\eta(k))\mathcal{M}_i^T \mathcal{Z}_{ij}\mathcal{M}_i \bar{f}(\eta(k)) + \omega^T(k)\mathcal{N}_i^T \mathcal{Z}_{ij}^{-1} P_j \mathcal{N}_i \omega(k)
\end{aligned}
\tag{4.17}
$$

$$
\begin{aligned}
\Delta V(\eta(k)) &= V(\eta(k+1)) - V(\eta(k)) \\
&= \eta^T(k+1) P_j \eta(k+1) - \eta^T(k) P_i \eta(k) \\
&= \eta^T(k)\left[\mathcal{A}_i^T P_j \mathcal{A}_i - P_i\right]\eta(k) + \bar{f}^T(\eta(k))\mathcal{M}_i^T P_j \mathcal{M}_i \bar{f}(\eta(k)) \\
&\quad + \omega^T(k)\mathcal{N}_i^T P_j \mathcal{N}_i \omega(k) + 2\eta^T(k)\mathcal{A}_i^T P_j \mathcal{M}_i f(x(k)) \\
&\quad + 2\eta^T(k)\mathcal{A}_i^T P_j \mathcal{N}_i \omega(k) + 2\bar{f}^T(\eta(k))\mathcal{M}_i^T P_j \mathcal{N}_i \omega(k), \quad i, j \in \mathcal{I}_k
\end{aligned}
$$

Based on previous definitions and assumptions, the nonlinear parts are limited with

$$
\begin{aligned}
&\bar{f}^T(\eta(k))\mathcal{M}_i^T P_j \mathcal{M}_i \bar{f}(\eta(k)) = f^T(x(k))\mathcal{M}_i^T P_j \mathcal{M}_i f(x(k)) \\
&\le \lambda_{\max}\left\{\mathcal{M}_i^T P_j \mathcal{M}_i\right\}\|f(\eta(k))\| \le \nu_{ij}\|\eta(k)\| = \nu_{ij}\eta^T(k)\eta(k)
\end{aligned}
\tag{4.18}
$$

Meanwhile, we have

$$
\bar{f}^T(\eta(k))\mathcal{M}_i^T \mathcal{W}_{ij}\mathcal{M}_i \bar{f}(\eta(k)) \le \mu_{ij}\eta^T(k)\eta(k)
$$

$$
\bar{f}^T(\eta(k))\mathcal{M}_i^T \mathcal{Z}_{ij}\mathcal{M}_i \bar{f}(\eta(k)) \le \eta_{ij}\eta^T(k)\eta(k)
$$

where $\lambda_{\max}\left\{\mathcal{M}_i^T \mathcal{W}_{ij}\mathcal{M}_i\right\} = \mu_{ij}$, $\lambda_{\max}\left\{\mathcal{M}_i^T \mathcal{Z}_{ij}\mathcal{M}_i\right\} = \eta_{ij}$.

Synthesis (4.16)–(4.18), we can obtain

$$
\begin{aligned}
&\Delta V(\eta(k)) \\
&= \varpi^T(k)\begin{bmatrix} \mathcal{A}_i^T P_j \mathcal{A}_i - P_i & \mathcal{A}_i^T P_j \mathcal{N} \\ \star & \mathcal{N}_i^T P_j \mathcal{N} \end{bmatrix}\varpi(k) + \bar{f}^T(\eta(k))\mathcal{M}_i^T P_j \mathcal{M}_i \bar{f}(\eta(k)) \\
&+ 2\eta^T(k)\mathcal{A}_i^T P_j \mathcal{M}_i f(x(k)) + 2\bar{f}^T(\eta(k))\mathcal{M}_i^T P_j \mathcal{N}_i \omega(k) \\
&\le \varpi^T(k)\begin{bmatrix} \mathcal{A}_i^T P_j \mathcal{A}_i - P_i + \mathcal{A}_i^T P_j\left(\mathcal{W}_{ij}^{-1} + \mathcal{Z}_{ij}^{-1}\right)P_j \mathcal{A}_i & \mathcal{A}_i^T P_j \mathcal{N} \\ \star & \mathcal{N}_i^T\left(P_j + \mathcal{Z}_{ij}^{-1}\right)\mathcal{N} \end{bmatrix} \\
&\varpi(k) + \bar{f}^T(\eta(k))\mathcal{M}_i^T\left(P_j + \mathcal{W}_{ij} + \mathcal{Z}_{ij}\right)\mathcal{M}_i \bar{f}(\eta(k)) \\
&\le \varpi^T(k)\Omega\varpi(k)
\end{aligned}
\tag{4.19}
$$

where $\varpi(k) = \begin{bmatrix} \eta(k) \\ \omega(k) \end{bmatrix}$,

$$
\Omega = \begin{bmatrix} \mathcal{A}_i^T P_j \mathcal{A}_i - P_i + \mathcal{A}_i^T P_j\left(\mathcal{W}_{ij}^{-1} + \mathcal{Z}_{ij}^{-1}\right)P_j \mathcal{A}_i + \kappa_{ij} I & \mathcal{A}_i^T P_j \mathcal{N} \\ \star & \mathcal{N}_i^T\left(P_j + \mathcal{Z}_{ij}^{-1}\right)\mathcal{N} \end{bmatrix}.
$$

Define $\nu_{ij} = \lambda_{\max}\left\{\mathcal{M}_i^T P_j \mathcal{M}_i\right\}$, $\mu_{ij} = \lambda_{\max}\left\{\mathcal{M}_i^T \mathcal{W}_{ij} \mathcal{M}_i\right\}$ and $\eta_{ij} = \lambda_{\max}\left\{\mathcal{M}_i^T \mathcal{Z}_{ij} \mathcal{M}_i\right\}$. Then, we have $\kappa_{ij} = \lambda_{\max}\left\{\mathcal{M}_i^T\left(P_j + \mathcal{W}_{ij} + \mathcal{Z}_{ij}\right)\mathcal{M}_i\right\} = \nu_{ij} + \mu_{ij} + \eta_{ij}$.

For given matrix $\Gamma = \begin{bmatrix} -\tau P_i & 0 \\ 0 & -Y_{ij} \end{bmatrix}$, where P_i, Y_{ij} are positive matrix, and $\tau \geq 0$. According to (4.19) and Lemma 4.1, we can obtain

$$\tilde{\Omega} = \begin{bmatrix} \tilde{\Omega}_{11} & \mathcal{A}_i^T P_j \mathcal{N} \\ * & \mathcal{N}_i^T\left(P_j + \mathcal{Z}_{ij}^{-1}\right)\mathcal{N} - Y_{ij} \end{bmatrix} < 0 \tag{4.20}$$

where $\tilde{\Omega}_{11} = \mathcal{A}_i^T P_j \mathcal{A}_i - (1+\tau) P_i + \mathcal{A}_i^T P_j \left(\mathcal{W}_{ij}^{-1} + \mathcal{Z}_{ij}^{-1}\right) P_j \mathcal{A}_i + \kappa_{ij} I$.

For further, we have

$$\Delta V\left(\eta(k)\right) \leq \tau V\left(\eta(k)\right) + \tau \eta^T(k) P_j \eta(k) + \omega^T(k) Y_{ij} \omega(k) \tag{4.21}$$

According to (4.21) and the condition $\sum_{k=1}^{N} \omega^T(k)\omega(k) \leq \delta^2$, we can get

$$\begin{aligned} V(k) &\leq (1+\tau)^k V(0) + \max_{i,j\in\mathcal{I}_k}\left\{\lambda_{\max}\left(Y_{ij}\right)\right\} \sum_{s=0}^{k-1} (1+\tau)^{k-s-1} \omega^T(s)\omega(s) \\ &\leq (1+\tau)^k V(0) + \max_{i,j\in\mathcal{I}_k}\left\{\lambda_{\max}\left(Y_{ij}\right)\right\} (1+\tau)^k \sum_{s=0}^{k-1} \omega^T(s)\omega(s) \\ &\leq (1+\tau)^k \left\{V(0) + \max_{i,j\in\mathcal{I}_k}\left\{\lambda_{\max}\left(Y_{ij}\right)\right\}\delta^2\right\} \end{aligned} \tag{4.22}$$

In addition, according to the definition of Lyapunov function, there exists symmetric matrices $\mathcal{R}$ and $\mathcal{S}_i$, $i \in \mathcal{I}_k$, $k \in \{1, \ldots, N\}$, satisfying

$$\begin{aligned} V(k) &= \eta^T(k) P_i \eta(k) = \eta^T(k) \mathcal{R}^{1/2} \mathcal{S}_i \mathcal{R}^{1/2} \eta(k) \\ &\geq \min_{i,j\in\mathcal{I}_k}\left\{\lambda_{\min}\left(\mathcal{S}_i\right)\right\} \eta^T(k) \mathcal{R} \eta(k) \end{aligned} \tag{4.23}$$

In another way, we can also get

$$V(0) \leq \max_{i,j\in\mathcal{I}_k}\left\{\lambda_{\max}\left(\mathcal{S}_i\right)\right\} \eta^T(0) \mathcal{R} \eta(0) \tag{4.24}$$

Due to $(1+\tau)^k \geq 1$, there exists an inequality

$$(1+\tau)^k V(0) \leq (1+\tau)^k \max_{i,j\in\mathcal{I}_k}\left\{\lambda_{\max}\left(\mathcal{S}_i\right)\right\} \eta^T(0) \mathcal{R} \eta(0)$$

holds. Based on $\tau \geq 0$ and $\eta^T(0) \mathcal{R} \eta(0) \leq \varphi^2$, $k \in \{1, 2, \ldots, N\}$, then we have

$$(1+\tau)^k V(0) \leq (1+\tau)^N \max_{i,j\in\mathcal{I}_k}\left\{\lambda_{\max}\left(\mathcal{S}_i\right)\right\} \varphi^2 \tag{4.25}$$

Synthesis (4.22)–(4.25), we can obtain

$$\begin{aligned} V(\eta(k)) &\leq (1+\tau)^k \left\{ V(0) + \max_{i,j\in\mathcal{I}_k} \left\{\lambda_{\max}\left(Y_{ij}\right)\right\}\delta^2 \right\} \\ &\leq (1+\tau)^N \left\{ \max_{i,j\in\mathcal{I}_k} \{\lambda_{\max}(\mathcal{S}_i)\}\bar{\delta}^2 + \max_{i,j\in\mathcal{I}_k} \left\{\lambda_{\max}\left(Y_{ij}\right)\right\}\delta^2 \right\} \end{aligned} \tag{4.26}$$

From (4.26), we have

$$\eta^T(k)\mathcal{R}\eta(k) \leq \frac{(1+\tau)^N \left\{ \max\limits_{i,j\in\mathcal{I}_k} \{\lambda_{\max}(\mathcal{S}_i)\}\bar{\delta}^2 + \max\limits_{i,j\in\mathcal{I}_k} \left\{\lambda_{\max}\left(Y_{ij}\right)\right\}\delta^2 \right\}}{\min\limits_{i,j\in\mathcal{I}_k} \{\lambda_{\min}(\mathcal{S}_i)\}} \tag{4.27}$$

Define $\lambda_1 = \min\limits_{i,j\in\mathcal{I}_k} \{\lambda_{\min}(\mathcal{S}_i)\}$ and $\lambda_2 = \max\limits_{i,j\in\mathcal{I}_k} \{\lambda_{\max}(\mathcal{S}_i)\}$, such that

$$\begin{aligned} &\lambda_1 I \leq \mathcal{S}_i \leq \lambda_2 I \\ &\Rightarrow \lambda_1 I \leq \left(\mathcal{R}^{-1/2}\right)^T P_i \left(\mathcal{R}^{-1/2}\right) \leq \lambda_2 I \\ &\Rightarrow \lambda_1 \mathcal{R} \leq P_i \leq \lambda_2 \mathcal{R} \end{aligned} \tag{4.28}$$

Meanwhile, we can consequently get

$$\eta^T(k)\mathcal{R}\eta(k) \leq \frac{(1+\tau)^N \left\{ \lambda_2\varphi^2 + \max\limits_{i,j\in\mathcal{I}_k} \left\{\lambda_{\max}\left(Y_{ij}\right)\right\}\delta^2 \right\}}{\lambda_1} \tag{4.29}$$

In above equations, scalars N, λ_1, λ_2, ϕ, δ and $\max\limits_{i,j\in\mathcal{I}_k} \left\{\lambda_{\max}\left(Y_{ij}\right)\right\}$ are all bounded, such that

$$\eta^T(k)\mathcal{R}\eta(k) \leq \frac{(1+\tau)^N \left\{ \lambda_2\varphi^2 + \max\limits_{i,j\in\mathcal{I}_k} \left\{\lambda_{\max}\left(Y_{ij}\right)\right\}\delta^2 \right\}}{\lambda_1} \leq \xi^2 \tag{4.30}$$

According to the **Definition** 4.1, the switching system (4.7) is said to be finite-time bounded stable subjected to $\{\varphi, \zeta, \delta, \mathcal{R}, N\}$.

This completes the proof of Theorem 4.1. □

Theorem 4.2 *The system (4.7) is under the I-DoS-JA, which has caused $u(k) \equiv 0$ during the interval $[0, N]$. If there exist positive matrices P_i, Y_{ij}, $\mathcal{W}_{ij}$, $\mathcal{Z}_{ij}$, $i, j \in \mathcal{I}_k = \{1, 2, 3, 4\}$, and positive symmetric matrix $\mathcal{R}$ and positive scalar $\tau \geq 0$, $\lambda_1 > 0$, $\lambda_2 > 0$, $0 < \varphi < \zeta$, $\gamma > 0$, $\kappa_{ij} > 0$, meanwhile the **Assumption** 4.2 is true, which make the following inequalities*

$$\lambda_1 \mathcal{R} \leq P_i \leq \lambda_2 \mathcal{R} \tag{4.31}$$

$$\begin{bmatrix} -P_i & 0 & \mathcal{A}_i^T P_j & \mathcal{A}_i^T P_j & \mathcal{H}_i^T & \mathcal{A}_i^T P_j & I \\ \star & -\frac{\gamma^2}{(1+\tau)^N} I & 0 & \mathcal{N}_i^T & F_i^T & 0 & 0 \\ \star & \star & -P_j & 0 & 0 & 0 & 0 \\ \star & \star & \star & -\mathcal{Z}_{ij} & 0 & 0 & 0 \\ \star & \star & \star & \star & -I & 0 & 0 \\ \star & \star & \star & \star & \star & -\mathcal{W}_{ij} & 0 \\ \star & \star & \star & \star & \star & \star & -\frac{1}{\kappa_{ij}} I \end{bmatrix} < 0 \tag{4.32}$$

$$(1+\tau)^N \lambda_2 \varphi^2 + \gamma^2 \delta^2 - \lambda_1 \xi^2 \leq 0 \tag{4.33}$$

hold, such that the switching system (4.7) is H_∞ finite-time stable subject to $\{\varphi, \zeta, \delta, \mathcal{R}, N\}$.

Proof The Lyapunov candidate function is chosen as $V_i(k) = \eta^T(k) P_i \eta(k)$, where $i, j \in \mathcal{I}_k = \{1, 2, 3, 4\}$. We consider the performance index function

$$J(k) = z^T(k) z(k) - \gamma^2 \omega^T(k) \omega(k) + \Delta V(k) \tag{4.34}$$

In order to further reduce conservatism, based on the proof of Theorem 4.1 and the condition $(1+\tau)^N \geq 1$, we choose $J(k) \leq z^T(k) z(k) - \frac{\gamma^2}{(1+\tau)^N} \omega^T(k) \omega(k) + \Delta V(k)$, which equals to $Y_{ij} = \frac{\gamma^2}{(1+\tau)^N} I$ in **Theorem** 4.1. Then, we can obtain

$$z^T(k) z(k) - \frac{\gamma^2}{(1+\tau)^N} \omega^T(k) \omega(k) + \Delta V(k) \leq \tau V(k) \tag{4.35}$$

And further transformation can be obtained as

$$\Delta V(k) \leq \tau V(k) - z^T(k) z(k) + \frac{\gamma^2}{(1+\tau)^N} \omega^T(k) \omega(k) \tag{4.36}$$

According to the definition, (4.36) equals to

$$V(k+1) - V(k) \leq \tau V(k) - z^T(k) z(k) + \frac{\gamma^2}{(1+\tau)^N} \omega^T(k) \omega(k) \tag{4.37}$$

Then, we can obtain

$$\begin{aligned} &V(k) \leq (1+\tau)^k V(0) + \\ &\sum_{s=0}^{k-1} (1+\tau)^{k-s-1} \left\{ \tfrac{\gamma^2}{(1+\tau)^N} \omega^T(s) \omega(s) - z^T(s) z(s) \right\} \end{aligned} \tag{4.38}$$

Because we are considering zero initial conditions, we have

$$V(k) \le \sum_{s=0}^{k-1} (1+\tau)^{k-s-1} \left\{ \frac{\gamma^2}{(1+\tau)^N} \omega^T(s)\,\omega(s) - z^T(s)\,z(s) \right\} \tag{4.39}$$

According to **Definitions** 4.1 and 4.2 and the condition $V(k) \ge 0$, thus, we have

$$\sum_{s=0}^{k-1} (1+\tau)^{k-s-1} \left\{ \frac{\gamma^2}{(1+\tau)^N} \omega^T(s)\,\omega(s) - z^T(s)\,z(s) \right\} \ge 0 \tag{4.40}$$

Equals to

$$\begin{aligned} &\sum_{s=0}^{k-1} (1+\tau)^{k-s-1} \left\{ z^T(s)\,z(s) \right\} \\ &\le \sum_{s=0}^{k-1} (1+\tau)^{k-s-1} \left\{ \tfrac{\gamma^2}{(1+\tau)^N} \omega^T(s)\,\omega(s) \right\} \end{aligned} \tag{4.41}$$

Since there exist

$$\sum_{s=0}^{k-1} (1+\tau)^{k-s-1} \left\{ z^T(s)\,z(s) \right\} \ge \sum_{s=0}^{k-1} \left\{ z^T(s)\,z(s) \right\}$$

and

$$\begin{aligned} &\sum_{s=0}^{k-1} (1+\tau)^{k-s-1} \left\{ \tfrac{\gamma^2}{(1+\tau)^N} \omega^T(s)\,\omega(s) \right\} \\ &\le (1+\tau)^N \tfrac{\gamma^2}{(1+\tau)^N} \sum_{s=0}^{k-1} \left\{ \omega^T(s)\,\omega(s) \right\} = \gamma^2 \sum_{s=0}^{k-1} \left\{ \omega^T(s)\,\omega(s) \right\} \end{aligned}$$

Then, we can derive

$$\sum_{s=0}^{k-1} \left\{ z^T(s)\,z(s) \right\} \le \gamma^2 \sum_{s=0}^{k-1} \left\{ \omega^T(s)\,\omega(s) \right\} \tag{4.42}$$

In light of **Definition** 4.2, the switching system (4.7) is said to be H_∞ exponential stable (ES) under I-DoS-JA. In addition, from the inequality $J(k) = z(k)^T z(k) - \frac{\gamma^2}{(1+\tau)^N} \omega^T(k)\,\omega(k) + \Delta$
$V(k) < 0$, we can derive the linear matrix inequality (4.32). According to **Theorem** 4.1, we know that

$$(1+\tau)^N \left\{ \lambda_2 \varphi^2 + \max_{i,j \in \mathcal{I}_k} \left\{ \lambda_{\max}\left(Y_{ij}\right) \right\} \delta^2 \right\} - \lambda_1 \xi^2 \le 0,$$

substitute $\max_{i,j \in \mathcal{I}_k} \left\{ \lambda_{\max}\left(Y_{ij}\right) \right\} = \frac{\gamma^2}{(1+\tau)^N}$ into the above inequality, then can obtain

$$\begin{aligned}&(1+\tau)^N\left\{\lambda_2\varphi^2+\tfrac{\gamma^2}{(1+\tau)^N}\delta^2\right\}-\lambda_1\xi^2\le 0\\&\Rightarrow (1+\tau)^N\lambda_2\varphi^2+\gamma^2\delta^2-\lambda_1\xi^2\le 0\end{aligned}\tag{4.43}$$

This is the end of the proof! □

4.3 Controller Design

Theorem 4.3 *Given a switched system (4.7), in the interval* $[0, N]$, *the DoS attack will cause* $u(k)\equiv 0$, *under the* ***Assumption*** *4.2, if there exist positive matrices* X_i, Y_{ij}, $\bar{\mathcal{W}}_{ij}$ *and symmetric matrix* $\mathcal{R}$, *and positive scalars* $\tau\ge 0$, $\lambda_1>0$, $\lambda_2>0$, $0<\varphi<\zeta$, $\gamma>0$, $\kappa_{ij}>0$, $i,j\in\mathcal{I}_k=\{1,2,3,4\}$ *satisfying the following inequalities*

$$\begin{bmatrix}-\lambda_1\mathcal{R} & X_i\\ \star & -X_i\end{bmatrix}<0,\ \begin{bmatrix}-\lambda_2\mathcal{R} & I\\ \star & -X_i\end{bmatrix}<0\tag{4.44}$$

$$\begin{bmatrix}-X_i & 0 & X_i\mathcal{A}_i^T & X_i\mathcal{A}_i^T & \mathcal{H}_i^T & X_i\mathcal{A}_i^T & I\\ \star & -\frac{\gamma^2}{(1+\tau)^N}I & 0 & \mathcal{N}_i^T & F_i^T & 0 & 0\\ \star & \star & -X_j & 0 & 0 & 0 & 0\\ \star & \star & \star & -\bar{\mathcal{Z}}_{ij} & 0 & 0 & 0\\ \star & \star & \star & \star & -I & 0 & 0\\ \star & \star & \star & \star & \star & -\bar{\mathcal{W}}_{ij} & 0\\ \star & \star & \star & \star & \star & \star & -\frac{1}{\kappa_{ij}}I\end{bmatrix}<0\tag{4.45}$$

$$(1+\tau)^N\lambda_2\varphi^2+\gamma^2\delta^2-\lambda_1\xi^2\le 0\tag{4.46}$$

where $\bar{\mathcal{Z}}_{ij}=X_j\mathcal{Z}_{ij}X_j$, $\bar{\mathcal{W}}_{ij}=X_j\mathcal{W}_{ij}X_j$, $X_i=diag\{X_{i1}\ X_{i2}\ X_{i3}\ X_{i4}\}$ *and* $X_j=diag\{X_{j1}\ X_{j2}\ X_{j3}\ X_{j4}\}$, $i,j\in\mathcal{I}_k=\{1,2,3,4\}$

And there exists state feedback $K_i=Y_{ij}X_{ij}^{-1}$, *which makes the closed loop switching system (5-5) is* H_∞ *bounded stable under finite time subject to* $\{\varphi,\zeta,\delta,\mathcal{R},N\}$.

Proof Based on the proof of **Theorem** 4.2, we can get the inequality (4.32). Since the matrix blocks $\mathcal{A}_i^T P_j$ are nonlinear, the toolbox of linear matrix inequality (LMI) can not be adopted directly. Thus, it is necessary to transform inequalities. Define $P_i=diag\{P_{i1}\ P_{i2}\ P_{i3}\ P_{i4}\}$ and $P_j=diag\{P_{j1}\ P_{j2}\ P_{j3}\ P_{j4}\}$, then pre- and post-multiplying (4.32) with $diag\{P_i^{-1}\ I\ P_j^{-1}\ P_j^{-1}\ I\ P_j^{-1}\ I\}$, we can obtain

$$\begin{bmatrix} -X_i & 0 & X_i\mathcal{A}_i^T & X_i\mathcal{A}_i^T & \mathcal{H}_i^T & X_i\mathcal{A}_i^T & I \\ * & -\frac{\gamma^2}{(1+\tau)^N}I & 0 & \mathcal{N}_i^T & F_i^T & 0 & 0 \\ * & * & -X_j & 0 & 0 & 0 & 0 \\ * & * & * & -\bar{\mathcal{Z}}_{ij} & 0 & 0 & 0 \\ * & * & * & * & -I & 0 & 0 \\ * & * & * & * & * & -\bar{\mathcal{W}}_{ij} & 0 \\ * & * & * & * & * & * & -\frac{1}{\kappa_{ij}}I \end{bmatrix} < 0 \tag{4.47}$$

where $X_i = P_i^{-1} = diag\{X_{i1}\ X_{i2}\ X_{i3}\ X_{i4}\}$, $\bar{\mathcal{Z}}_{ij}=P_j^{-1}\mathcal{Z}_{ij}P_j^{-1}$, $\bar{\mathcal{W}}_{ij}=P_j^{-1}\mathcal{W}_{ij}P_j^{-1}$,$X_j = P_j^{-1} = diag\{X_{j1}\ X_{j2}\ X_{j3}\ X_{j4}\}$. From above modeling process, we have

$$\mathcal{A}_1X_1 = \begin{bmatrix} AX_{11}+BY_{11} & -BY_{12} & 0\ 0 \\ 0 & AX_{12}-L_1CX_{12} & 0\ 0 \\ Y_{11} & -Y_{12} & 0\ 0 \\ CX_{11} & 0 & 0\ 0 \end{bmatrix},$$

$$\mathcal{A}_2X_2 = \begin{bmatrix} AX_{21}+BY_{21} & -BY_{22} & 0 & 0 \\ L_2CX_{21} & AX_{22}-L_2CX_{22} & 0 & -L_2X_{24} \\ Y_{21} & -Y_{22} & 0 & 0 \\ CX_{21} & 0 & 0 & I \end{bmatrix},$$

$$\mathcal{A}_3X_3 = \begin{bmatrix} AX_{31} & 0 & BX_{33} & 0 \\ -BY_{31} & AX_{32}+BY_{32}-L_3CX_{32} & BX_{33} & 0 \\ 0 & 0 & X_{33} & 0 \\ CX_{31} & 0 & 0 & 0 \end{bmatrix} \text{ and}$$

$$\mathcal{A}_4X_4 = \begin{bmatrix} AX_{41} & 0 & BX_{43} & 0 \\ -(BY_{41}-L_4CX_{41}) & AX_{42}+BY_{42}-L_4CX_{42} & BX_{43} & -L_4X_{44} \\ 0 & 0 & X_{43} & 0 \\ 0 & 0 & 0 & X_{44} \end{bmatrix}$$

where $Y_{ij} = K_iX_{ij}, \forall i \in \{1, 2, 3, 4\}, j \in \{1, 2\}$.

Thus, the linear matrix inequality is obtained as presented in (4.45).

From the condition (4.44), we can derive that $\begin{bmatrix} -X_i^{-1} & X_i^{-1} \\ * & -\lambda_1\mathcal{R}^{-1} \end{bmatrix} < 0 \Rightarrow -X_i^{-1}+\lambda_1\mathcal{R} < 0 \Rightarrow \lambda_1\mathcal{R} < P_i$, $\begin{bmatrix} -\lambda_2\mathcal{R} & I \\ * & -X_i \end{bmatrix} < 0 \Rightarrow -\lambda_2\mathcal{R}+X_i^{-1} = -\lambda_2\mathcal{R}+P_i < 0 \Rightarrow$, $P_i < \lambda_2\mathcal{R}$, where $\lambda_1=\lambda_{\min}\{\mathcal{S}_i\}$, $\lambda_2=\lambda_{\max}\{\mathcal{S}_i\}$, $\mathcal{S}_i = \mathcal{R}^{-1/2}P_i\mathcal{R}^{-1/2}$.

This is the end of the proof! □

Theorem 4.4 *For given system (4.7), the system is under I-DoS-JA with zero disturbance, if there exists positive matrices P_i, $\mathcal{W}_{ij}$, $(i, j \in I_k = \{1, 2, 3, 4\})$ and scalars $\lambda > 0$, $\bar{\phi} \geq 1$, which satisfy*

$$P_i < \bar{\phi}P_j \quad \forall i, j \in I_k \tag{4.48}$$

$$\mathcal{A}_i^TP_j\mathcal{A}_i - P_i + \mathcal{A}_i^TP_j\mathcal{W}_{ij}^{-1}P_i\mathcal{A}_i + \eta_{ij}I < 0 \tag{4.49}$$

then, the switching system (4.7) is global asymptotic stability (GAS), and the decay rate is $\chi = \bar{\phi}^{\frac{1}{2\tau_a}} \lambda$. *In addition, the average dwell-on time of the switching system satisfy*

$$\tau_a > \tau_a^* \triangleq \frac{\ln \bar{\phi}}{2 \ln \lambda^{-1}} \tag{4.50}$$

where $\bar{\phi} = \max \left\{ \phi_{\sigma(k_j)}, \phi_{\sigma(k_{j-1})}, \ldots, \phi_{\sigma(k_0)} \right\}$.

Proof The Lyapunov function is chosen as

$$V(k) = \eta^T(k) P_{\sigma_k} \eta(k) \tag{4.51}$$

According to zero input and zero disturbance under Assumption 4.1, we can obtain

$$\begin{aligned}
\Delta V(k) = &\eta^T(k) \left(\mathcal{A}_i^T P_j \mathcal{A}_i - P_i \right) \eta(k) + \eta^T(k) \mathcal{A}_i^T P_j \mathcal{W}_{ij}^{-1} P_i \mathcal{A}_i \eta(k) \\
&+ 2\eta^T(k) \mathcal{A}_i^T P_j \mathcal{M}_i \bar{f}(\eta(k)) \\
\leq &\eta^T(k) \left(\mathcal{A}_i^T P_j \mathcal{A}_i - P_i + \mathcal{A}_i^T P_j \mathcal{W}_{ij}^{-1} P_i \mathcal{A}_i \right) \eta(k) \\
&+ \bar{f}^T(\eta(k)) \mathcal{M}_i^T \left(P_j + \mathcal{W}_{ij} \right) \mathcal{M}_i \bar{f}(\eta(k)) \\
\leq &\eta^T(k) \left(\mathcal{A}_i^T P_j \mathcal{A}_i - P_i + \mathcal{A}_i^T P_j \mathcal{W}_{ij}^{-1} P_i \mathcal{A}_i + \eta_{ij} I \right) \eta(k)
\end{aligned} \tag{4.52}$$

where $\eta_{ij} = \max\limits_{i,j \in I_k} \left\{ \lambda_{\max} \left(\mathcal{M}_i^T \left(P_j + \mathcal{W}_{ij} \right) \mathcal{M}_i \right) \right\}$.

If the system is stable, it requires $\Delta V(k) < 0$, which equals to $\mathcal{A}_i^T P_j \mathcal{A}_i - P_i + \mathcal{A}_i^T P_j \mathcal{W}_{ij}^{-1} P_i \mathcal{A}_i + \eta_{ij} I < 0$, and the condition $\lambda > 0$ satisfying

$$\begin{aligned}
&\lambda^2 \left(\mathcal{A}_i^T P_j \mathcal{A}_i + \mathcal{A}_i^T P_j \mathcal{W}_{ij}^{-1} P_i \mathcal{A}_i + \eta_{ij} I \right) - \lambda^2 P_i < 0 \\
&\Rightarrow \mathcal{A}_i^T P_j \mathcal{A}_i - \lambda^2 P_i < 0
\end{aligned} \tag{4.53}$$

According to the proof process of **Theorems** 4.1 and 4.2, the stability is restricted to a certain limit and the stability condition is relaxed as

$$\mathcal{A}_i^T P_j \mathcal{A}_i - (1 + \tau) P_i + \mathcal{A}_i^T P_j \mathcal{W}_{ij}^{-1} P_i \mathcal{A}_i + \eta_{ij} I < 0, \quad \tau > 0 \tag{4.54}$$

Thus, $\tau \to 0$, which can ensure $\mathcal{A}_i^T P_j \mathcal{A}_i - \lambda^2 P_i < 0$.

Because the state and output of the system do not change at the switching point, there are two adjacent switched systems

$$V_{\sigma(k_j)}(k_j) \leq \phi_{\sigma(k_j)} V_{\sigma(k_{j-1})}(k_j) \tag{4.55}$$

Based on this, we can obtain

$$\begin{aligned}
V_{\sigma(k_i)}(k_i) &\le \lambda^{2(k_i-k_j)} V_{\sigma(k_j)}(k_j) \\
&\le \phi_{\sigma(k_j)} \lambda^{2(k_i-k_j)} V_{\sigma(k_j-1)}(k_j) \\
&\le \phi_{\sigma(k_j)} \phi_{\sigma(k_{j-1})} \lambda^{2(k_i-k_j)} \lambda^{2(k_i-k_{j-1})} V_{\sigma(k_{j-1})}(k_{j-1}) \\
&\vdots \\
&\le \phi_{\sigma(k_j)} \phi_{\sigma(k_{j-1})} \cdots \phi_{\sigma(k_0)} \lambda^{2(k_i-k_j)} \lambda^{2(k_i-k_{j-1})} \cdots \lambda^{2(k_1-k_0)} V_{\sigma(k_0)}(k_0) \\
&= \bar{\phi}^{N(0,k)} \lambda^{2(k-k_0)} V_{\sigma(k_0)}(k_0) \\
&= \bar{\phi}^{\frac{k-k_0}{\tau_a}} \lambda^{2(k-k_0)} V_{\sigma(k_0)}(k_0) \\
&= \chi^{2(k-k_0)} V_{\sigma(k_0)}(k_0)
\end{aligned} \tag{4.56}$$

where $\bar{\phi} = \max\left\{\phi_{\sigma(k_j)}, \phi_{\sigma(k_{j-1})}, \ldots, \phi_{\sigma(k_0)}\right\}$. According to the definition, we can get $V(k) \ge \lambda_{\min\limits_{i,j\in\mathcal{I}_k}}(P_i)\,\eta^T(k)\,\eta(k) = \bar{\lambda}_1 \|\eta(k)\|_2^2$. On the other hand, we have $V(k) \le \lambda_{\max\limits_{i,j\in\mathcal{I}_k}}(P_i)\,\eta^T(k)\,\eta(k) = \bar{\lambda}_2 \|\eta(k)\|_2^2$, combine with (4.50), we can get

$$\|\eta(k)\| \le \sqrt{\frac{\bar{\lambda}_2}{\bar{\lambda}_1}} \left(\bar{\phi}^{\frac{1}{2\tau_a}} \lambda\right)^{k-k_0} \|\eta(k_0)\| \tag{4.57}$$

Then, we can derive

$$\|\eta(k)\| \le \sqrt{\frac{\bar{\lambda}_2}{\bar{\lambda}_1}} \left(\bar{\phi}^{\frac{1}{2\tau_a}} \lambda\right)^{k-k_0} \|\eta(k_0)\| \tag{4.58}$$

$$\tau_a > \tau_a^* \triangleq \frac{\ln \bar{\phi}}{2 \ln \lambda^{-1}} \tag{4.59}$$

This is the end of the proof! □

4.4 Simulation Example

An example of switching system is borrowed from[4],
$A = \begin{bmatrix} 0.8133 & 0.4413 \\ 0.6953 & 0.4707 \end{bmatrix}$, $B = \begin{bmatrix} 0.126 \\ 0.5361 \end{bmatrix}$, $C = \begin{bmatrix} 0.0223 & 0.1438 \end{bmatrix}$, $E = \begin{bmatrix} 0.7573 & 0.8858 \end{bmatrix}$, $F = 0.2857$. By solving the linear matrix inequality, we can obtain the intermediate variable $Y_{11} = \begin{bmatrix} -0.2205 \\ -0.0621 \end{bmatrix}$, $Y_{12} = \begin{bmatrix} -0.4030 \\ -2.4552 \end{bmatrix}$, $Y_{21} = \begin{bmatrix} -0.7508 \\ 1.7152 \end{bmatrix}$, $Y_{22} = \begin{bmatrix} -1.7201 \\ 3.2943 \end{bmatrix}$, $Y_{31} = \begin{bmatrix} 5.7354 \\ -9.7622 \end{bmatrix}$, $Y_{32} = \begin{bmatrix} -3.6990 \\ -1.2059 \end{bmatrix}$, $Y_{41} = \begin{bmatrix} -0.7004 \\ 1.1922 \end{bmatrix}$, $Y_{42} = \begin{bmatrix} -1.0459 \\ 0.9415 \end{bmatrix}$ and free-weighting matrix

$$X_1 = \begin{bmatrix} 0.4328 & -0.7366 & 0 & 0 & 0 & 0 \\ -0.7366 & 1.2537 & 0 & 0 & 0 & 0 \\ 0 & 0 & 6.6193 & 4.3074 & 0 & 0 \\ 0 & 0 & 4.3074 & 6.2805 & 0 & 0 \\ 0 & 0 & 0 & 0 & 9.6304 & 0 \\ 0 & 0 & 0 & 0 & 0 & 9.6547 \end{bmatrix},$$

$$X_2 = \begin{bmatrix} 0.2495 & -0.4247 & 0 & 0 & 0 & 0 \\ -0.4247 & 0.7228 & 0 & 0 & 0 & 0 \\ 0 & 0 & 6.7440 & 5.3359 & 0 & 0 \\ 0 & 0 & 5.3359 & 6.6309 & 0 & 0 \\ 0 & 0 & 0 & 0 & 9.6304 & 0 \\ 0 & 0 & 0 & 0 & 0 & 0.0453 \end{bmatrix},$$

$$X_3 = \begin{bmatrix} 0.0392 & -0.0667 & 0 & 0 & 0 & 0 \\ -0.0667 & 0.1136 & 0 & 0 & 0 & 0 \\ 0 & 0 & 1.2748 & 0.9098 & 0 & 0 \\ 0 & 0 & 0.9098 & 1.0270 & 0 & 0 \\ 0 & 0 & 0 & 0 & 0.0000 & 0 \\ 0 & 0 & 0 & 0 & 0 & 0.9657 \end{bmatrix},$$

$$X_4 = \begin{bmatrix} 0.036 & -0.0613 & 0 & 0 & 0 & 0 \\ -0.0613 & 0.1043 & 0 & 0 & 0 & 0 \\ 0 & 0 & 1.5439 & 0.9793 & 0 & 0 \\ 0 & 0 & 0.9793 & 0.8465 & 0 & 0 \\ 0 & 0 & 0 & 0 & 0.0000 & 0 \\ 0 & 0 & 0 & 0 & 0 & 0.0043 \end{bmatrix}$$

. Based on these, we can obtain the control gain of subsystem as $K_1 = [-1.5822 \ -1.0247]$, $K_2 = \begin{bmatrix} -3.1513 & -1.8515 \end{bmatrix}$ and $K_3 = \begin{bmatrix} -3.4376 & -2.0204 \end{bmatrix}$.

Substituting the control gain of the subsystem, we can obtain the system state output with switching security control scenario under the I-DoS-JA (Fig. 4.2).

The simulations indicate that the different control strategies for each subsystem modeled from different communication topology transformation under I-DoS-JA can be achieved stable control. In this chapter, the switching scenarios is limited by the upper bound of the average dwell-on time. As to the further investigation of switching trigger conditions, we will do our best to solve this problem in the future work.

4.5 Conclusion

This chapter takes the I-DoS-JA process as an example to analyze its attack process, aiming at the different effects of control loop and feedback loop on the communication topology of the system when the control loop and feedback loop are separately and simultaneously encountered with I-DoS-JA. It is modeled as a control system with switching structure, and this method of switching control is adopted to realize the CPS security and stability control under attack by means of control theory. In

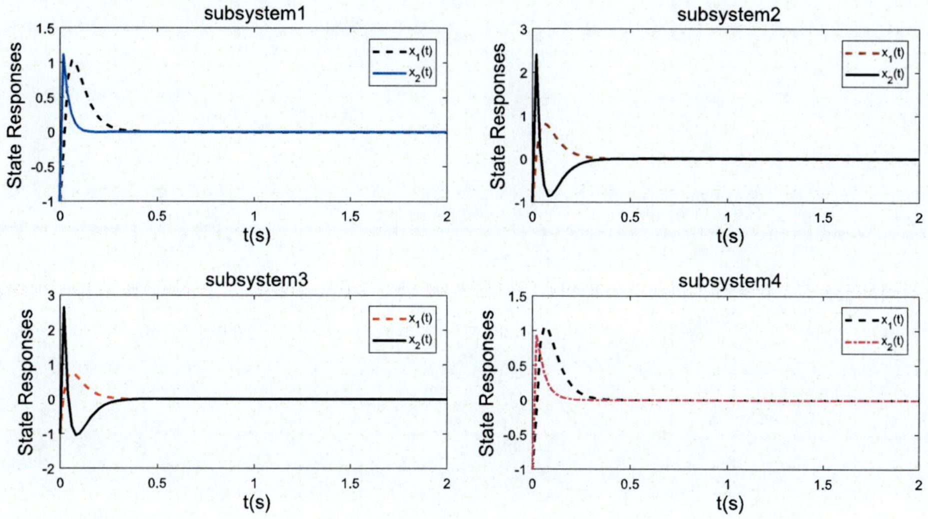

Fig. 4.2 Security control for I-DoS-JA schematic with switching system approach

order to reflect the complexity of the system, in this chapter, taking nonlinear as an example, under the strategy of switching control, the condition that the system can achieve bounded stability after intermittent DoS attack under constraints is derived. At the same time, the average residence time condition is derived for secure handover of handoff systems under network attack.

References

1. H. Ge, D. Yue, X. Xie, C. Dou, S. Wang, Security control of cyber-physical system based on switching approach for intermittent denial-of-service jamming attack. ISA Trans. **20**, 700–714 (2019)
2. W.A. Zhang, L. Yu, Output feedback stabilization of networked control systems with packet dropouts. IEEE Trans. Autom. Control **52**(9), 1705–1710 (2007)
3. D. Yue, Q.L. Han, J. Lam, Network-based robust control of systems with uncertainty. Automatica **41**(6), 999–1007 (2005)
4. W. Xiang, J. Xiao, H∞ finite-time control for switched nonlinear discrete-time systems with norm-bounded disturbance. J. Franklin Inst. **348**(2), 331–352 (2011)
5. D. Yue, E. Tian, Q.L. Han, A delay system method for designing event-triggered controllers of networked control systems. IEEE Trans. Autom. Control **58**(2), 475–481 (2013)
6. Y.S. Moon, P. Park, W.H. Kwon, Delay-dependent robust stabilization of uncertain state-delayed systems. Int. J. Control **74**(14), 1447–1455 (2001)

Part II
Resilient Consensus Control of Multi-agent Systems

Chapter 5
Observer-Based Distributed Secure Consensus Control of Linear Multi-agent Systems Subject to Random Attacks

This chapter is concerned with an observer-based distributed secure consensus control strategy for a class of linear multi-agent systems (MASs) with random attacks. Due to the fact that not all state information are available, observers are employed to estimate the internal states. The communication topology is randomly switching under the attacks, and a distributed secure consensus strategy is proposed using the output information. The feedback gains are calculated via solving the Riccati equation and Riccati inequality, and stability analysis is proven that the MAS achieves secure consensus tracking in the mean square sense. Finally, two examples are provided to verify the effectiveness of the observer-based control strategy [1].

This chapter is structured as below. The system model of this chapter is described in Sect. 5.1 and the distributed secure consensus protocol is designed in Sect. 5.2. Section 5.3 proved the stability analysis of the proposed scheme and explained the switching topology. In Sect. 5.4, the simulation results of numerical and example are presented to verify the feasibility of the secure consensus protocol. Finally, Sect. 5.5 concludes the chapter.

5.1 Problem Formulation

5.1.1 Communication Topology

The MAS consists of one leader and N followers, and the communication topology among these agents is represented by a directed graph $\mathcal{G}\{\mathcal{V}, \varepsilon_{r(t)}, \mathcal{A}_{r(t)}\}$, where $\mathcal{V} = \{\nu_1, \nu_2, \nu_3, \ldots, \nu_N\}$ is the set of nodes. The neighbor node of ν_i is defined as $\mathcal{N}_{r(t)} = \{\nu_i \in \mathcal{V}, (\nu_j, \nu_i) \in \varepsilon_r(t), j \neq i\}$, where $\varepsilon_{r(t)}$ is the edge set, $(j, i) \in \varepsilon_{r(t)}$ means that the information of Agent i can be sent to Agent j. A graph is said to be balanced if $\sum_{j=1}^{N} a_{ij} = \sum_{j=1}^{N} a_{ji}$ for all i. The adjacency matrix associated with

D. Yue et al., *Secure Control of Networked Control Systems and Its Applications*,
https://doi.org/10.1007/978-981-33-6730-2_5

$\mathcal{G}$ is recorded as $\mathcal{A}_{r(t)} = [a_{ij}^{r(t)}] \in R^{N\times N}$, where $a_{ij}^{r(t)} > 0$ for $(n_j, n_i) \in \varepsilon_{r(t)}$, and $a_{ij}^{r(t)} = 0$, otherwise. $\mathcal{L}_{r(t)} = \mathcal{D}_{r(t)} - \mathcal{A}_{r(t)}$ is denoted as the Laplacian matrix for $\mathcal{G}_{r(t)}$, where $\mathcal{D}_{r(t)} = \text{diag}\{d_1^{r(t)}, d_2^{r(t)}, \ldots, d_N^{r(t)}\}$ is the row sum of the adjacency matrix, and $d_i^{r(t)} = \sum_{j=1}^{N} a_{ij}^{r(t)}$. The information-exchange matrix for consensus control is expressed as $\mathcal{Q}_{r(t)} = \mathcal{L}_{r(t)} + \mathcal{B}_{r(t)}$, where $\mathcal{B}_{r(t)}$ represents the communication between leader and followers under the attack, and $\mathcal{B}_{r(t)} = \text{diag}\{b_1^{r(t)}, b_2^{r(t)}, \ldots, b_N^{r(t)}\}$. If the ith follower can access the information of the leader, $b_i^{r(t)} = 1$, and $b_i^{r(t)} = 0$ otherwise.

5.1.2 Random DoS Attack Model

Inspired by the Markov chain with its applications in [2] and compared with other stochastic processes, the Markov process is considered to represent random switching graphs caused by cyber-attacks. This process holds the memoryless property that its future situation is only of dependence on current information with emphasis on data evolutions, and it is suitable to describe the random switching process of the topology. In this chapter, $\theta(t)$ is used to represent the condition of the system, and it is determined by the attack behavior att and the defense behavior def. The two behaviors are associated with time [3]. For a group (att, def), let $\theta(t)$ be a right continuous, homogeneous, random Markov process [4, 5], and it takes values in the set $\mathcal{S} = \{1, 2, \ldots, s\}$, where s is a given positive integer representing one of topologies, $att \in \mathcal{A}$ and $def \in \mathcal{D}$, and $\mathcal{A} = \{att_1, att_2, \ldots, att_m\}$ and $\mathcal{D} = \{def_1, def_2, \ldots, def_n\}$ are the attack space composed of m attack actions and the defense space composed of n defense behaviors, respectively. Assuming that the set of nodes is invariant, and the connections among the nodes may be disconnected or connected as the topology changes. $\Xi = [\xi_{pq}]$ is treated as the transition rate matrix, where the sum of each line of Ξ is zero and all elements are positive except for diagonal elements. For $p, q \in \mathcal{S}$, when the state is switched from p to state q, we have

$$P_{pq}(t) = \text{Prob}\{r(t+\Delta) = q | r(t) = p\}$$
$$= \begin{cases} \xi_{pq}\Delta + o(\Delta), & \text{if } p \neq q, \\ 1 + \xi_{pq}\Delta + o(\Delta), & \text{if } p = q, \end{cases} \tag{5.1}$$

where $r(t)$ is the switching signal of the topology when the system is attacked, and $o(\Delta)/\Delta = 0$ as $\Delta \to 0$. $f_p(k)$ and $g_q(k)$ are the probabilities of choosing $def_p(k) \in \mathcal{D}$ and $att_q(k) \in \mathcal{A}$, respectively, with $f(k) = [f_p(k)]_{p=1}^{n}$ and $g(k) = [g_q(k)]_{q=1}^{m}$. For $\forall p, q \in \mathcal{S}$, when $p \neq q$, $\xi_{pq} > 0$ denotes the transition rate, and this means the state p at time t switches to state q at time $t + \Delta$, when $p = q$, $\xi_{pp} = -\sum_{q=1, p\neq q} \xi_{pq}$. ξ_{pq} is from the average transition rates $\tilde{\xi}_{pq}(k) = \tilde{\xi}_{pq}(att_q(k), def_p(k))$

$$\xi_{pq}(f(k), g(k)) = \sum_{p=1}^{n}\sum_{q=1}^{m} f_p(k)g_q(k)\tilde{\xi}_{pq}(k). \tag{5.2}$$

5.1.3 System Description

In the MAS, the dynamics of the followers are, $1 \leq i \leq N$,

$$\begin{cases} \dot{x}_i(t) = Ax_i(t) + Bu_i(t), \\ y_i(t) = Cx_i(t), \end{cases} \tag{5.3}$$

and that of the leader are

$$\begin{cases} \dot{x}_0(t) = Ax_0(t), \\ y_0(t) = Cx_0(t), \end{cases} \tag{5.4}$$

where $x_i(t) \in R^n$ and $x_0(t) \in R^n$ are the state vectors of the followers and the leader, respectively, $u_i(t) \in R^l$ is the control input, $A \in R^{n\times n}$, $B \in R^{n\times l}$ and $C \in R^{p\times n}$ are constant matrices with the appropriate dimensions, and $y_i(t) \in R^p$ and $y_0(t) \in R^p$ are the output signals of the followers and the leader, respectively. The dynamics of (5.3) can represent a large class of practical applications in circuits and systems, such as RLC circuits, armature-controlled DC motors, direct current-direct current (DC-DC) buck converters [6]. In this chapter, the state vector $x_i(t)$ is not available while the output ones are measurable.

Assumption 5.1 The pairs (A, B) and (A, C) in the system (5.3) are stabilizable and detectable, respectively.

For the MAS consisting of the followers (5.3) and the leader (5.4) in the presence of random attacks, the objective in this chapter is to design an observer-based distributed control strategy with the aim that followers are able to track the leader and achieve consensus in mean square sense. The system achieves the secure consensus tracking performance, if there exist positive scalars ϵ and ι satisfying for all $t > t_0$, $i \in \mathcal{V}$,

$$E\{\|x_i(t)-x_0(t)\|^2\}\leq\epsilon\exp(-\iota(t-t_0))E\{\|x_i(t_0)-x_0(t_0)\|^2\}. \tag{5.5}$$

In fact, the control strategy in this chapter is not suitable for all attacks, and suitable parameters and conditions should be introduced. Inspired by the switching mechanism [7], we consider the following two concepts, attack frequency and attack length rate, in this chapter, and it is convenient to analyze how the the attacks affect on the performance of the MAS.

Definition 5.1 For switching signal $\sigma(t)$ and $\forall t_2 > t_1 \geq t_0$, the number of $\sigma(t)$ changes of MASs over $[t_1, t_2)$ is defined as $N_f(t_1, t_2)$, then define $F_f(t_1, t_2) = N_f(t_1, t_2)/(t_2 - t_1)$ as the attack frequency on $[t_1, t_2)$.

Definition 5.2 For $\forall t > 0$, the whole time for the attack on a MAS at $[t_0, t)$ is defined as $T_a(t_0, t)$, then define $R_a = T_a(t_0, t)/(t - t_0)$ as the attack length rate on $[t_0, t)$.

The conditions about attacks will be presented in the following context.

5.2 Observer-Based Distributed Secure Control Strategy

In this section, an observer-based distributed control strategy is proposed for the problem of secure consensus tracking control of the MAS under random attacks. The control architecture with observers, control strategy and random attacks is shown in Fig. 5.1. The attacks are driven by the random Markov process, and the connection among nodes will be interrupted and the system will be paralyzed.

Due to the facts that the not all internal states of the system are available and Assumption 5.1, the observer is designed as

$$\begin{cases} \dot{\hat{x}}_i(t) = A\hat{x}_i(t) + Bu_i + K_{ob}(y_i(t) - \hat{y}_i(t)), \\ \hat{y}_i(t) = C\hat{x}_i(t), \end{cases} \tag{5.6}$$

where $\hat{x}_i(t) \in R^n$ is the observer state of $x_i(t)$, $\hat{y}_i(t) \in R^p$ is the measured output of the observer, $K_{ob} \in R^{n\times p}$ is a constant matrix, and the selection of K_{ob} satisfies $K_{ob}C - A > 0$.

For $t \in [t_{2k}, t_{2(k+1)})$ and an infinite sequence $k = 0, 1, \ldots$, it is assumed that t_{2k+1} is the time instant when MAS is attacked. That is to say, the network is in a normal state

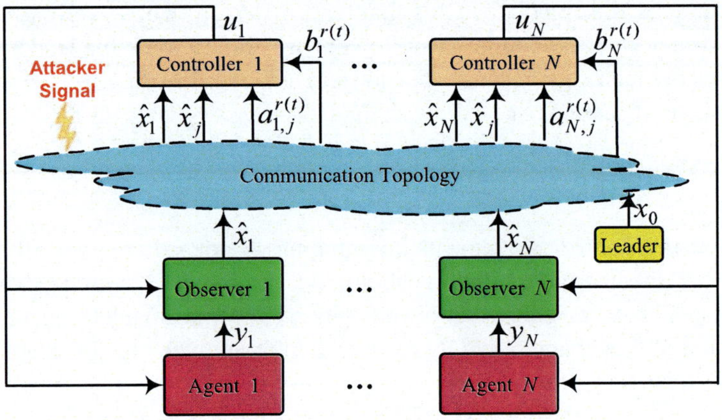

Fig. 5.1 The architecture of the control strategy

during the period $[t_{2k}, t_{2k+1})$ without any attacks connected though an initial graph $\mathcal{G}_0$, and the communication network $\mathcal{G}_{r(t)}$ is paralyzed over $t \in [t_{2k+1}, t_{2(k+1)})$. On the basis of the system restoration mechanism, it is assumed that multiagent network can return to a connectivity-maintained topology at the instant $t = t_{2k}^{+}$. Define the ith follower's measurement error $\tilde{x}_i(t) = x_i(t) - \hat{x}_i(t)$ and $\tilde{x}(t) = \left[\tilde{x}_1^T, \tilde{x}_2^T, \ldots, \tilde{x}_N^T\right]^T$. Define the observer-based tracking error for the ith agent

$$e_i(t) = \hat{x}_i(t) - x_0(t), \tag{5.7}$$

and taking its time derivative yields

$$\dot{e}_i(t) = \begin{cases} [I_N \otimes A - \varrho(Q_0 \otimes BL)]e(t) + (I_N \otimes K_{ob}C)\tilde{x}(t), \\ \quad \text{if } t \in [t_{2k}, t_{2k+1}), \\ [I_N \otimes A - \vartheta(Q_{r(t)} \otimes BJ)]e(t) + (I_N \otimes K_{ob}C)\tilde{x}(t), \\ \quad \text{if } t \in [t_{2k+1}, t_{2(k+1)}), \end{cases} \tag{5.8}$$

where $e(t) = \left[e_1^T, e_2^T, \ldots, e_N^T\right]^T$, Q_0 is a nonsingular matrix determined by the initial graph $\mathcal{G}_0$ (the graph without any attacks), and $Q_{r(t)}$ is denoted as the information exchange matrix of the graph $\mathcal{G}_{r(t)}$ (the graph under malicious attacks).

The control strategy for the ith follower is designed as

$$u_i(t) = \begin{cases} \varrho L\left[\sum_{j=1}^{N} a_{ij}^0(\hat{x}_j(t) - \hat{x}_i(t)) + b_i^0(x_0(t) - \hat{x}_i(t))\right], \\ \quad \text{if } t \in [t_{2k}, t_{2k+1}), \\ \vartheta J\left[\sum_{j=1}^{N} a_{ij}^{r(t)}(\hat{x}_j(t) - \hat{x}_i(t)) + b_i^{r(t)}(x_0(t) - \hat{x}_i(t))\right], \\ \quad \text{if } t \in [t_{2k+1}, t_{2(k+1)}), \end{cases} \tag{5.9}$$

where ϱ and ϑ are positive constants representing the coupling strength, and $L \in R^{l \times n}$ and $J \in R^{l \times n}$ are the control gain parameters to be determined later.

5.3 Consensus Analysis

The system is switched between non-attacked and attacked scenarios, and, in this part, we introduce the stability lemmas of the MAS in the presence of the normal condition and attack one. Then, we further proof the the consensus of the system in mean square sense.

To move on, the tracking error for the ith follower is $z_i(t) = x_i(t) - x_0(t)$, and denote $z(t) = \left[z_1^T, z_2^T, \ldots, z_N^T\right]^T$. Then, it follows that

$$z_i(t) = \tilde{x}_i(t) + e_i(t), \tag{5.10}$$

and

$$\dot{\tilde{x}}(t) = (A - K_{ob}C)\tilde{x}(t). \tag{5.11}$$

For the analysis of the whole system, we introduce the following two lemmas in normal and attacked cases.

5.3.1 Attack-Free Case

In this subsection, we mainly deal with the MAS that are not exposed to attacks, and the corresponding topology is not paralyzed.

Lemma 5.1 *For the MAS (5.3) with (5.4) suffered any attacks, $R > 0$ and $Y > I$, there exists a unique symmetric positive definite matrix $P > 0$ such that*

$$PA + A^T P - PBR^{-1}B^T P + Y = 0. \tag{5.12}$$

Then, for the Lyapunov function $V_c(t) = V_a(e(t)) + V_{ob}(\tilde{x}(t))$ and a positive constant $\alpha_1 = \min\left(|2G_1|, |\frac{G_2}{\lambda_{\min}(\theta_i^{-1}P)}|\right)$, $\Theta = \mathrm{diag}\{\theta_1^{-1}, \theta_2^{-1}, \ldots, \theta_N^{-1}\}$ with $\theta = [\theta_1^{-1}, \theta_2^{-1}, \ldots, \theta_N^{-1}]^T = (Q_0^T)^{-1}\mathbf{1}$, $\Phi = \Theta Q_0 + Q_0^T \Theta > 0$, $G_1 = \lambda_{\min}(K_{ob}C - A) - \|K_{ob}C\|^2$ and $G_2 = \alpha\lambda_{\min}(\theta_i^{-1}P) - \lambda_{\max}(\theta_i^{-1}P)$, the following property

$$V_c(t) \leq \exp(-\alpha_1(t - t_0))V_c(t_0) \tag{5.13}$$

holds under the control strategy (5.9) with $L = R^{-1}B^T P$.

Proof The Lyapunov function is chosen as

$$V_a(e(t)) = \sum_{i=1}^{N} e_i^T(t)\theta_i^{-1}Pe_i(t). \tag{5.14}$$

Taking its time derivative, from (5.14),

$$\begin{aligned}\dot{V}_a(e(t)) = & [e^T(t)(I_N \otimes A^T - \sigma(Q_0^T \otimes (BL)^T)) \\ & + ((I_N \otimes K_{ob}C)\tilde{x}(t))^T](\Theta \otimes P)e(t) \\ & + e^T(t)(\Theta \otimes P)[I_N \otimes A \\ & - \sigma(Q_0 \otimes BL)e(t) + (I_N \otimes K_{ob}C)\tilde{x}(t)].\end{aligned} \tag{5.15}$$

Denoting $M = \tilde{x}^T(t)(\Theta \otimes (K_{ob}C)^T P)e(t) + e^T(t)(\Theta \otimes PK_{ob}C)\tilde{x}(t)$ and taking $L = R^{-1}B^T P$ into (5.15), we obtain

$$\begin{aligned}\dot{V}_a(e(t)) \leq & e^T(t)(\Theta \otimes (A^T P + PA))e(t) - \sigma\lambda_{\min}\theta_{\min} e^T(t) \\ & (\Phi \otimes PBR^{-1}B^T P)e(t) + M,\end{aligned} \tag{5.16}$$

where $\lambda_{\min} = \lambda_{\min}(\Phi)$ and $\theta_{\min} = \min_{i=1,2,\ldots,N} \theta_i$.

Assuming $\sigma > 1/(\lambda_{\min}\theta_{\min})$ and according to (5.12), (5.16) can be written as

$$\begin{aligned}\dot{V}_a(e(t)) &= e^T(t)(\Theta \otimes (A^T P + PA - PBR^{-1}B^T P))e(t) + M \\ &= e^T(t)(\Theta \otimes (-Y))e(t) + M.\end{aligned} \tag{5.17}$$

From $-Y \leq -\lambda_{\min}(Y)I \leq -\alpha\lambda_{\max}(P)$ and $\alpha = \lambda_{\min}(Y)/\lambda_{\max}(P)$, (5.17) is

$$\begin{aligned}\dot{V}_a(e(t)) &\leq -\alpha e^T(t)(\Theta \otimes P)e(t) + M \\ &\leq -\alpha V_a(e(t)) + M.\end{aligned} \tag{5.18}$$

Integrating (5.18) over $[t_0, t]$, one obtains

$$V_a(e(t)) \leq \exp(-\alpha(t - t_0))V_a(e(t_0)) + \frac{M}{\alpha}\exp(\alpha t_0). \tag{5.19}$$

We choose

$$V_{ob}(\tilde{x}(t)) = \frac{1}{2}\sum_{i=1}^{N} \tilde{x}_i^T(t)\tilde{x}_i(t), \tag{5.20}$$

and its time derivative yields

$$\begin{aligned}\dot{V}_{ob}(\tilde{x}(t)) &= \sum_{i=1}^{N} \tilde{x}_i^T(t)\dot{\tilde{x}}_i(t) \\ &= -\tilde{x}^T(t)(K_{ob}C - A)\tilde{x}(t).\end{aligned} \tag{5.21}$$

The above equation can be further written as

$$\dot{V}_{ob}(\tilde{x}(t)) \leq -2\lambda_{\min}(K_{ob}C - A)V_{ob}(\tilde{x}(t)), \tag{5.22}$$

and then, taking integration on both sides, we have

$$V_{ob}(\tilde{x}(t)) \leq \exp(-2\lambda_{\min}(K_{ob}C - A)(t - t_0))V_{ob}(\tilde{x}(t_0)). \tag{5.23}$$

It can be derived, from the definition of V_c, (5.14) and (5.20), that

$$\begin{aligned}\dot{V}_c(t) \leq & -\alpha e^T(t)(\Theta \otimes P)e(t) + \tilde{x}^T(t)(\Theta \otimes (K_{ob}C)^T P)e(t) \\ & + e^T(t)(\Theta \otimes P K_{ob}C)\tilde{x}(t) - \lambda_{\min}(K_{ob}C - A)\tilde{x}^T(t)\tilde{x}(t) \\ \leq & [\|K_{ob}C\|^2 - \lambda_{\min}(K_{ob}C - A)]\|\tilde{x}(t)\|^2 \\ & - [\alpha\lambda_{\min}(\theta_i^{-1}P) - \lambda_{\max}(\theta_i^{-1}P)]\|e(t)\|^2. \end{aligned} \tag{5.24}$$

According to $G_1 = \lambda_{\min}(K_{ob}C - A) - \|K_{ob}C\|^2$ and $G_2 = \alpha\lambda_{\min}(\theta_i^{-1}P) - \lambda_{\max}(\theta_i^{-1}P)$, and then, it follows, from (5.24), that

$$\begin{aligned}\dot{V}_c(t) \leq & -\min\left(|2G_1|, |\frac{G_2}{\lambda_{\min}(\theta_i^{-1}P)}|\right)(V_{ob}(\tilde{x}(t)) + V_a(e(t))) \\ \leq & -\alpha_1 V_c(t), \end{aligned} \tag{5.25}$$

where $\alpha_1 = \min\left(|2G_1|, |\frac{G_2}{\lambda_{\min}(\theta_i^{-1}P)}|\right)$. Accordingly, (5.13) holds, and this completes the proof.

5.3.2 *Attack Case*

In this subsection, we will take the case that the network system suffers from attacks into account. This kind of attacks is driven by the Markov random process.

Assumption 5.2 The switching signal $r(t)$ depends on the initial distribution $\pi = [\pi_1, \pi_2, \ldots, \pi_s]$.

Assumption 5.3 The topology $\mathcal{G}_{r(t)}$ subject to attacks is balanced, and the union of digraphs consisting of the leader and followers contains a spanning tree.

Lemma 5.2 *For the MAS (5.3) with (5.4), Assumptions 5.2–5.3 and a symmetric positive matrix T, there exists a symmetric positive define matrix $S > 0$, satisfying that*

$$SA + A^T S - SBT^{-1}B^T S - \rho S < 0. \tag{5.26}$$

Then, for a Lyapunov function $V_d(t) = V_b(e(t)) + V_{ob}(\tilde{x}(t))$, the following property

$$E\{V_d(t)\} \leq \exp(\rho_1(t - t_0))E\{V_d(t_0)\} \tag{5.27}$$

holds under the control strategy (5.9) with $J = T^{-1}B^T S$, where $\rho_1 = G_3/\lambda_{\min}(S)$, and $G_3 = \rho\lambda_{\max}(S) + \lambda_{\max}(S)$.

Proof Consider the Lyapunov function candidate

$$V_b(e(t)) = \sum_{i=1}^{N} e_i^T(t) S e_i(t). \tag{5.28}$$

Combining the random Markov jump process, the Lyapunov equation (5.28) can be written as

$$V_b^p(e(t)) = E[e^T(t)(I_N \otimes S)e(t)\mathcal{W}_{r(t)=p}], \forall p \in \mathcal{S}, \tag{5.29}$$

where $r(t)$ represents the switching signal driven by the Markov process under attack, and $\mathcal{W}_{r(t)=p} = 1$, if $r(t) = p$, and $\mathcal{W}_{r(t)=p} = 0$, otherwise.

The derivative of $V_b^p(e(t))$ yields

$$\begin{aligned}\frac{\mathrm{d}V_b^p(e(t))}{\mathrm{d}t} &= E\{e^T(t)[I_N \otimes (SA + A^T S) \\ &- \vartheta(Q_p \otimes SBJ + (Q_p \otimes SBJ)^T)]e(t) \\ &+ ((I_N \otimes (K_{ob}C))\tilde{x}(t))^T (I_N \otimes S)e(t) \\ &+ e^T(t)(I_N \otimes S)(I_N \otimes K_{ob}C)\tilde{x}(t) \\ &+ \sum_{q=1}^{s} \xi_{qp} V_b^q(e(t))\mathrm{d}t + o(\mathrm{d}t)\},\end{aligned} \tag{5.30}$$

where $V_b^q(e(t))$ is the expression of the Lyapunov function when $r(t) = q$. On the basis of Assumption 5.2, we choose $\vartheta \geq (2\pi_{\min}\lambda_{\min}(\hat{Q}_{un}))^{-1}$, where $\pi_{\min} = \min_{p\in\mathcal{S}}\{\pi_p\}$ with $\mathcal{S} = \{1, 2, \ldots, s\}$. Under the premise of $\pi_p \geq \pi_{\min}$ and Assumption 5.3, substituting $J = T^{-1}B^T S$ into (5.30) to get

$$\begin{aligned}\frac{\mathrm{d}V_b(e(t))}{\mathrm{d}t} \leq & E\{e^T(t)[I_N \otimes (SA + A^T S) \\ & - (\tilde{Q} \otimes SBT^{-1}B^T S)]e(t) + \breve{N}\},\end{aligned} \tag{5.31}$$

where $\breve{N} = \tilde{x}(t)^T(I_N \otimes (K_{ob}C)^T S)e(t) + e^T(t)(I_N \otimes SK_{ob}C)\tilde{x}(t)$, $\tilde{Q} = (Q_{un} + Q_{un}^T)/(2\lambda_{\min}(\hat{Q}_{un}))$, the information-exchange matrix Q_{un} of the union of digraphs $\mathcal{G}_{un} = \bigcup_{p\in\mathcal{S}} \mathcal{G}_p$ is denoted as $Q_{un} = \sum_{p=1}^{s} Q_p$, $p \in \mathcal{S}$, and $\hat{Q}_{un}$ is the corresponding information-exchange of the mirror of $\mathcal{G}_{un}$.

An identify matrix is constructed as $\Psi = [\Psi_1, \Psi_2, \ldots, \Psi_N]$, where Ψ_i is one of orthonormal eigenvectors of $\hat{Q}_{un}$. The eigenvalue of $\hat{Q}_{un}$ is denoted as $\lambda_i(\hat{Q}_{un})$ and holds the property that $\Psi_i^T \hat{Q}_{un} = \lambda_i(\hat{Q}_{un})\Psi_i^T$, $i = 1, 2, \ldots, N$. By defining $\tilde{e}(t) = (\Psi^T \otimes I_N)e(t)$, we can be further obtain that

$$\sum_{j=1}^{N}\tilde{e}_j^T(t)\left(SA + A^T S - \frac{\lambda_j(\hat{Q}_{un})}{\lambda_{\min}(\hat{Q}_{un})} SBT^{-1}B^T S\right)\tilde{e}_j(t) + \check{N}$$
$$\leq \tilde{e}_j^T(t)(SA + A^T S - SBT^{-1}B^T S)\tilde{e}_j(t) + \check{N}. \tag{5.32}$$

According to $SA + A^T S - SBT^{-1}B^T S - \rho S < 0$ and combining (5.31) with (5.32), it follows that

$$E\{\dot{V}_b(e(t))\} \leq \rho E\{e^T(t)(I_N \otimes S)e(t) + \check{N}\}$$
$$\leq \rho E\{V_b(e(t)) + \check{N}\}. \tag{5.33}$$

Thus, taking the time derivative of V_d is

$$\dot{V}_d = \dot{V}_b(e(t)) + \dot{V}_{ob}(\tilde{x}(t))$$
$$\leq E\{(\|K_{ob}C\|^2 - \lambda_{\min}(K_{ob}C - A))\|\tilde{x}(t)\|^2$$
$$+ (\rho\lambda_{\max}(S) + \lambda_{\max}(S))\|e(t)\|^2\}. \tag{5.34}$$

Taking $G_1 = \lambda_{\min}(K_{ob}C - A) - \|K_{ob}C\|^2$ and $G_3 = \rho\lambda_{\max}(S) + \lambda_{\max}(S)$, (5.34) indicates that

$$E\{\dot{V}_d\} = E\{-G_1\|\tilde{x}(t)\|^2 + G_3\|e(t)\|^2\}$$
$$\leq E\left\{-2G_1 V_{ob}(\tilde{x}(t)) + \frac{G_3}{\lambda_{\min}(S)} V_a(e(t))\right\}$$
$$\leq E\left\{\rho_1(V_{ob}(\tilde{x}(t)) + V_a(e(t)))\right\}, \tag{5.35}$$

where $\rho_1 = \frac{G_3}{\lambda_{\min}(S)}$. Then, (5.27) holds. This completes the proof.

5.3.3 *Secure Consensus in Mean Square Sense*

Theorem 5.1 *Under Assumptions 5.1–5.3, the MAS with (5.3) and (5.4) reaches the secure consensus tracking in the mean square sense with the observer (5.6) and distributed control strategy (5.9) if there exist constants $\tau^* \in (0, \alpha_1)$ and $\tau \in (0, \tau^*)$ such that*

$$\frac{T_a(t_0, t)}{t - t_0} \leq \frac{\alpha_1 - \tau^*}{\alpha_1 + \rho_1} \tag{5.36}$$

and

$$F_{f(t_0,t)} = \frac{N_f(t_0, t)}{t - t_0} \leq F_f^* = \frac{\tau^* - \tau}{2\ln\mu}, \tag{5.37}$$

where $\mu = \max\{\max\{\theta_{\min}\lambda_{\min}(P), \lambda_{\max}(0.5I)\}/\min\{\lambda_{\min}(S), \lambda_{\min}(0.5I)\}, \max\{\max(S), \max(0.5I)\}/\min\{\theta_{\min}\lambda_{\min}(P), \lambda_{\min}(0.5I)\}\} \geq 1$. *Further, the tracking error of the MAS satisfies*

$$E\{\|z_i(t)\|^2\} \leq \varsigma \exp(-\tau(t-t_0))E\{\|z_i(t_0)\|^2\}, \tag{5.38}$$

where $c = \max\{\max\{\lambda_{\max}(\theta_i^{-1}P), \lambda_{\max}(0.5I)\}, \max\{\lambda_{\max}(S), \lambda_{\max}(0.5I)\}\}$, $d = \min\{\min\{\lambda_{\min}(\theta_i^{-1}P), \lambda_{\min}(0.5I)\}, \min\{\lambda_{\min}(S), \lambda_{\min}(0.5I)\}\}$ *and* $\varsigma = c/d$,.

Proof On the basis of Lemmas 5.1 and 5.2, we choose the Lyapunov functional candidate as $V_{\sigma(t)}(e(t))$. The switching signal $\sigma(t)$ is a piecewise function that changes over time, and it is set as c or d which will be given later. Then, the entire running time of the controller is expressed as

$$V_{\sigma(t)}(e(t)) = \begin{cases} V_c(e(t)), \text{ if } t \in [t_{2k}, t_{2k+1}), \\ V_d(e(t)), \text{ if } t \in [t_{2k+1}, t_{2(k+1)}). \end{cases} \tag{5.39}$$

According to Lemmas 5.1 and 5.2 and assuming that $V_c(e(t))$ is valid in $[t_{2k}, t_{2k+1})$ and $V_d(e(t))$ is valid in $[t_{2k+1}, t_{2(k+1)})$, it comes the result that

$$E\{V_{\sigma(t)}(e(t))\} \leq \begin{cases} \exp(-\alpha_1(t-t_{2k}))E\{V_c(t_{2k})\}, \\ \quad \text{if } t \in [t_{2k}, t_{2k+1}), \\ \exp(\rho_1(t-t_{2k+1}))E\{V_d(t_{2k+1})\}, \\ \quad \text{if } t \in [t_{2k+1}, t_{2(k+1)}). \end{cases} \tag{5.40}$$

It means that, when $t = t_{2k}^+$ and $t = t_{2k+1}^+$, the closed-loop error switches. We are going to talk about the two conditions that are going on $t \in [t_{2k}, t_{2k+1})$ and $t \in [t_{2k+1}, t_{2(k+1)})$, respectively.

Case 1: when $t \in [t_{2k}, t_{2k+1})$, according to (5.40), one has

$$\begin{aligned} E\{V(t)\} \leq& \exp(-\alpha_1(t-t_{2k}))E\{V_c(t_{2k})\} \\ \leq& \mu \exp(-\alpha_1(t-t_{2k}))E\{V_d(t_{2k}^-)\} \\ \leq& \mu \exp(-\alpha_1(t-t_{2k-1})) \\ &\cdot [\exp(\rho_1(t_{2k}-t_{2k+1}))E\{V_d(t_{2k-1})\}] \\ \leq& \cdots \\ \leq& \mu^{2k} \exp(-\alpha_1(t-t_0-T_a(t_0, t))) \\ &\cdot \exp(\rho_1 T_a(t_0, t))E\{V_c(t_0)\}. \end{aligned} \tag{5.41}$$

Case 2: when $t \in [t_{2k+1}, t_{2(k+1)})$, according to (5.40), one has

$$\begin{aligned}E\{V(t)\} \leq &\exp(\rho_1(t-t_{2k+1}))E\{V_d(t_{2k})\}\\ \leq &\mu\exp(\rho_1(t-t_{2k}))E\{V_c(t_{2k+1}^-)\}\\ \leq &\mu\exp(\rho_1(t-t_{2k+1}))\\ &\cdot[\exp(-\alpha_1(t_{2k+1}-t_{2k}))E\{V_c(t_{2k})\}]\\ \leq &\cdots\\ \leq &\mu^{2k+1}\exp(-\alpha_1(t-t_0-T_a(t_0,t)))\\ &\cdot\exp(\rho_1 T_a(t_0,t))E\{V_c(t_0)\}.\end{aligned} \tag{5.42}$$

For $t \in [t_{2k}, t_{2k+1})$ and $t \in [t_{2k+1}, t_{2(k+1)})$, we can define $N_f(t_0, t) = k$ and $N_f(t_0, t) = k+1$, respectively.

For $\forall t \geq t_0$, from (5.41) and (5.42), we can get

$$\begin{aligned}E\{V(t)\} \leq &\mu^{2N_f(t_0,t)}\exp(-\alpha_1(t-t_0-T_a(t_0,t)))\\ &\cdot\exp(\rho_1 T_a(t_0,t))E\{V(t_0)\}.\end{aligned} \tag{5.43}$$

From (5.36) and (5.37), the following inequalities

$$\begin{aligned}&\exp(-\alpha_1(t-t_0-T_a(t_0,t)))\exp(\rho_1 T_a(t_0,t))\\ &\leq\exp(-\tau^*(t-t_0))\end{aligned} \tag{5.44}$$

and

$$\exp\left(2N_f(t_0,t)\ln\mu\right) \leq \exp((\tau^*-\tau)(t-t_0)) \tag{5.45}$$

are obtained, respectively. Utilizing the results (5.44) and (5.45), (5.43) further gives

$$E\{V(t))\} \leq \exp(-\tau(t-t_0))E\{V(t_0)\}. \tag{5.46}$$

From (5.39), it follows that

$$dE\{\|e_i(t)\|^2\} \leq E\{V(t)\},\ E\{V(t_0)\} \leq cE\{\|e_i(t_0)\|^2\}. \tag{5.47}$$

Combining (5.46) with (5.47), one has

$$E\{\|z_i(t)\|^2\} \leq \varsigma\exp(-\tau(t-t_0))E\{\|z_i(t_0)\|^2\}, \tag{5.48}$$

and this indicates that $x_i(t) \to x_0(t)$ as $t \to +\infty$.

Remark 5.1 The calculation process of the control strategy is briefly described as below. P and S can be obtained by calculating (5.12) and (5.26), and the feedback gains L and J are from $L = R^{-1}B^T P$ and $J = T^{-1}B^T S$, respectively. It is worth mentioning that $\vartheta \geq (2\pi_{\min}\lambda_{\min}(\hat{Q}_{un}))^{-1}$ and $\sigma > 1/(\lambda_{\min}\theta_{\min})$ should be satisfied.

Remark 5.2 It can be seen, from the stability analysis, that $E\{\|x_i(t) - x_0(t)\|^2\} \leq 2(E\{\|x_i(t) - \hat{x}_i(t)\|^2\} + E\{\|\hat{x}_i(t) - x_0(t)\|^2\})$. Compared with the SF result in [8], the upper bound of the tracking error is higher in this chapter when $t \geq T_0$. As $t \to +\infty$, it yields that $E\{\|x_i(t) - x_0(t)\|^2\} \to 0$ since $E\{\|x_i(t) - \hat{x}_i(t)\|^2\} \to 0$ and $E\{\|\hat{x}_i(t) - x_0(t)\|^2\} \to 0$. Therefore, though it is cost-saving for the output control strategy in this chapter, we should make promise between the tracking precision and redundant equipment.

5.4 Simulation Example

Two simulation examples are conducted in this section to demonstrate our secure consensus control. The first one is on a numerical example in tracking consensus, and the other one is on a practical application, a group of unmanned aerial vehicles (UAVs), in the stabilizing case. The system matrix in the first example is not Hurwitz while the one in Example B is Hurwitz. To illustrate the behavior of the system, we investigate comparison results on SF/OF and with/without secure consensus strategy, in Example A and Example B, respectively.

The connections among agents are represented in Figs. 5.2 and 5.3, and they are divided into the following cases. (i) Fig. 5.2 represents the communication topology of the network system without attack, and this topology of the network system is expressed by $\mathcal{G}$. It can be seen that it contains a directed spanning tree, and the leader is the root. (ii) Fig. 5.3, including (a), (b) and (c), represents interruptions of the link between nodes suffer from attacks, and these three different topologies in this figure are denoted by $\mathcal{G}_{r(t)}$, where $r(t) = 1, 2, 3$.

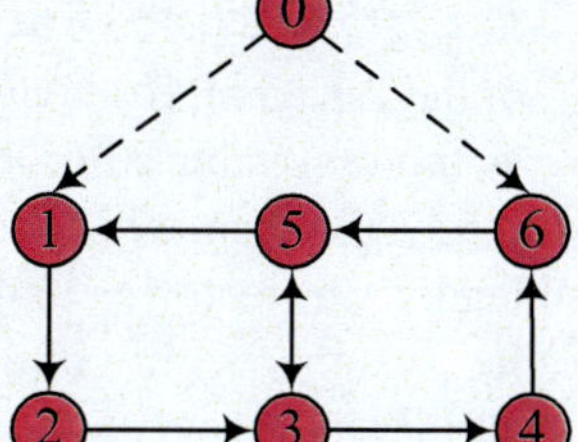

Fig. 5.2 Initial topology

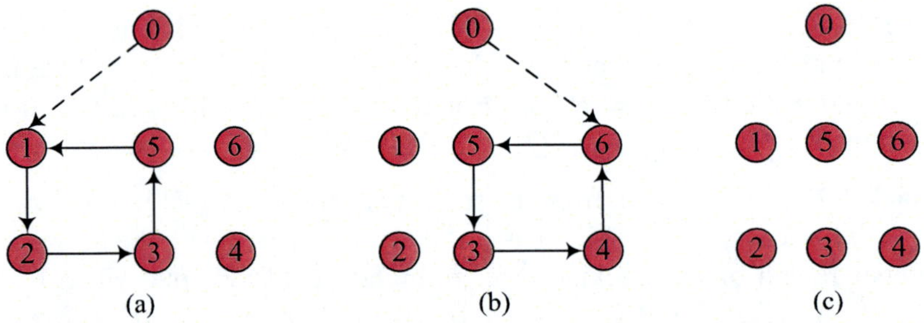

Fig. 5.3 Three possible topologies under attacks

5.4.1 A Numerical Example

Consider a MAS consisting of one leader and six follower agents with $A = \begin{bmatrix} 0 & 1 & 0 & 0 \\ -2.27 & 1 & 0.9 & 0 \\ 0 & 0 & 0 & 1 \\ 1.1 & 0 & -0.9 & 0 \end{bmatrix}$, $B = \begin{bmatrix} 0 \\ 0.9 \\ 0 \\ 0 \end{bmatrix}$, and $C = \begin{bmatrix} 0, 1, 0, 1 \end{bmatrix}$. The generation matrix $\Xi = \begin{bmatrix} -0.1 & 0.02 & 0.08 \\ 0.3 & -0.5 & 0.2 \\ 0.1 & 0.1 & -0.2 \end{bmatrix}$ is used to generate the attack model, and the invariant distribution $\pi = [0.59, 0.15, 0.31]$ gives the initial distribution of Markov jump progress. $f_p(k) = g_q(k) = 1/3$, $p, q = 1, 2, 3$, are the probabilities of the system being attacked and defended, respectively. According to Markov process, $\sigma(t)$ in Fig. 5.4 represents whether the MASs are attacked during the entire running process. In other words, the switching of $\sigma(t)$ indicates whether the system is under attack or not. $r(t)$ in Fig. 5.4 describes the switching the communication topologies among $\mathcal{G}_1$(a), $\mathcal{G}_2$(b) and $\mathcal{G}_3$(c) under the attack. The $\gamma(t)$ in Fig. 5.4 represents the switching of MASs topology during the entire running. The parameters are set as follows $K_{ob} = [0.1, 1.8, 0.5, 2.1]^T$, $R = 2$, $T = 0.051$, $Y = 10I$, and we obtain that $L = [3.24, 3.48, -1.25, 2.42]$ and $J = [0.036, 0.090, -0.016, 0.034]$ by denoting $\alpha_1 = 4.75$ and $\rho_1 = 0.25$, $\varrho = 15$, $\vartheta = 25$, $\lambda_{\min} = 0.08$, $\theta_{\min} = 2.54$, $\lambda_{\min}(\hat{Q}_{un}) = 0.15$, $\mu = 11.2$. Figure 5.5 shows the result of the observer on the state of the 1st agent, from which we can see that the observation value tends to be the same as the real value after a period time. It is shown, in Figs. 5.6, 5.7, 5.8 and 5.9, that the system can achieve secure consensus tracking ultimately and followers track the leader's trajectories under the control law. The comparison of tracking errors $E_c(t)$ for systems with/without observers is presented in Fig. 5.10. This indicates that, although the two methods are able achieve consensus, the OF one, adding observers, does affect the consensus performance.

According to (5.36) and (5.37), we can get

$$\frac{T_a(t_0, t)}{t - t_0} \leq \frac{\alpha_1 - \tau^*}{\alpha_1 + \rho_1} = \frac{4.75 - 3.75}{0.25 + 4.75} = 0.2, \tag{5.49}$$

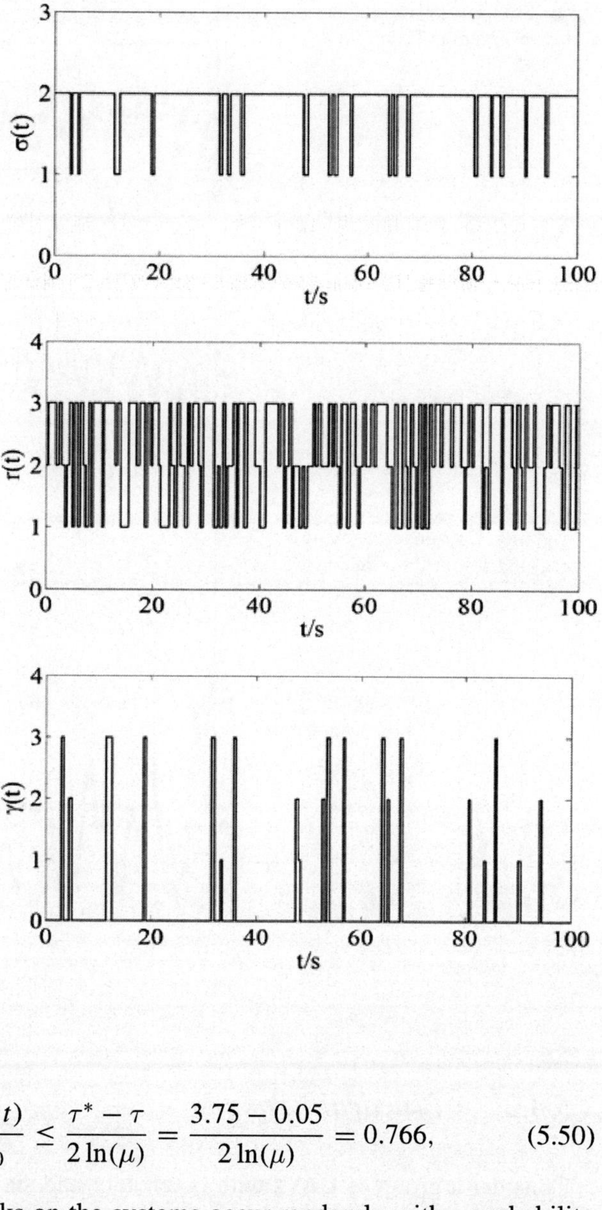

Fig. 5.4 Switching signals $\sigma(t)$, $r(t)$ and $\gamma(t)$ in Example A

and

$$F_{f(t_0,t)} = \frac{N_f(t_0, t)}{t - t_0} \leq \frac{\tau^* - \tau}{2\ln(\mu)} = \frac{3.75 - 0.05}{2\ln(\mu)} = 0.766, \tag{5.50}$$

which states clearly that attacks on the systems occur randomly with a probability of less than 0.766 during a unit of time and the average recovery time based on the frequency and length rate of the attack is calculated as

$$\frac{\alpha_1 - \tau^*}{(\alpha_1 + \rho_1)F_{f(t_0,t)}} = 0.261, \tag{5.51}$$

and this indicates that the average recovery time is no more than 0.261 in a unit of time.

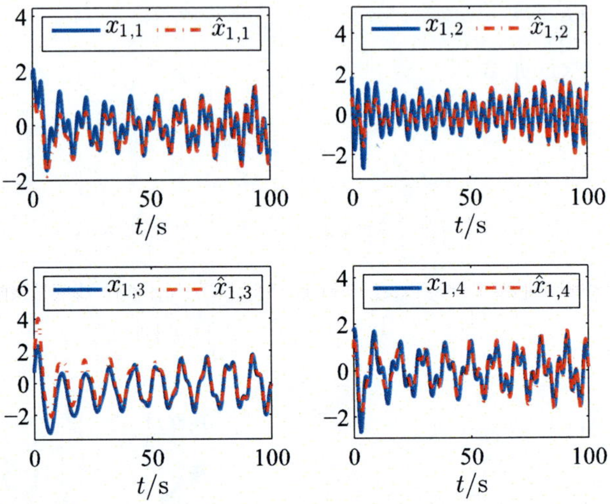

Fig. 5.5 The performance of the observers in Example A

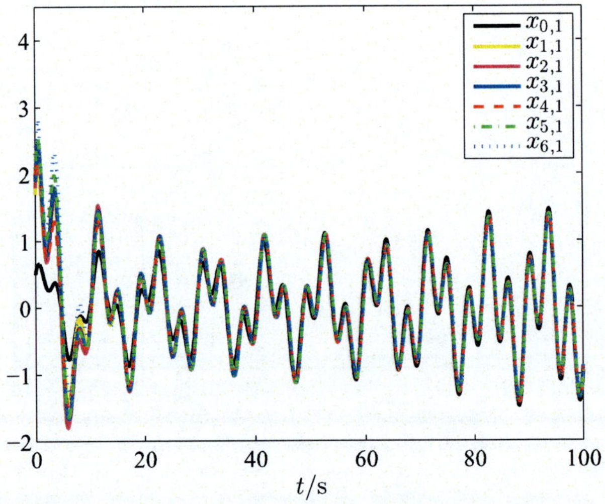

Fig. 5.6 Consensus tracking $x_{0,1}$ and $x_{i,1}$ in Example A, $i = 1, \ldots, 6$

5.4.2 A Practical Example

Consider a group of UAVs with one leader and six followers, and the dynamics of the ith follower are [9]

$$\begin{cases} \dot{p}_{f_i} = -p_{f_i} + 0.2132 v_{f_i} + s_{pi} u_{pf_i}, \\ \dot{v}_{f_i} = -0.5 v_{f_i} + s_{vi} u_{vf_i}, \\ y_i(t) = C[p_{f_i}, v_{f_i}]^T, \end{cases} \tag{5.52}$$

where p_{f_i} and v_{f_i} are the position and velocity information, s_{pi} and s_{vi} are set as 0.3153 and 0.4759, respectively, and y_i is the measured output. Choosing $x_i =$

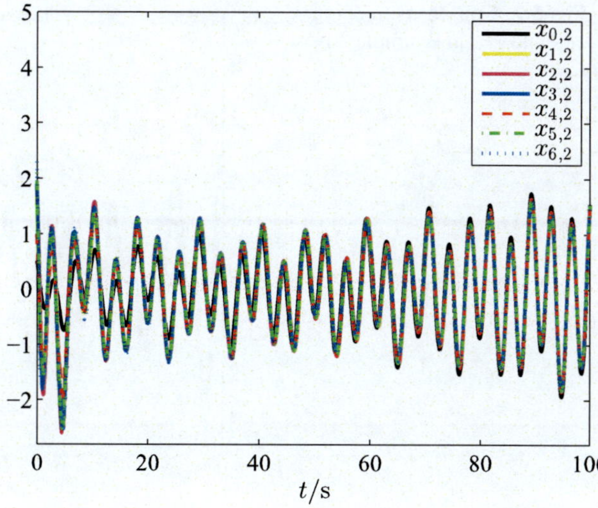

Fig. 5.7 Consensus tracking $x_{0,2}$ and $x_{i,2}$ in Example A, $i = 1, \ldots, 6$

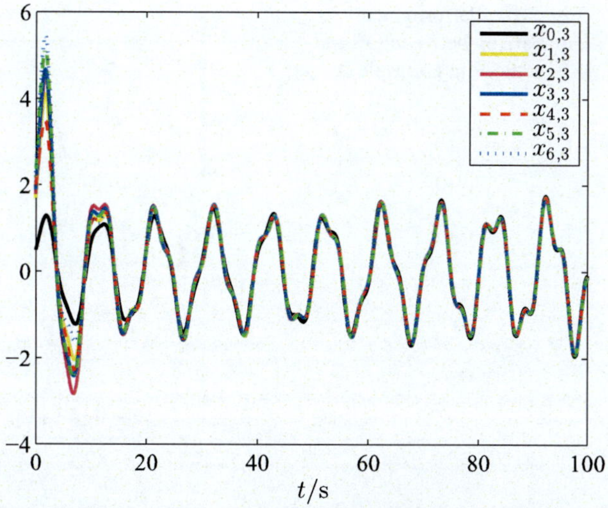

Fig. 5.8 Consensus tracking $x_{0,3}$ and $x_{i,3}$ in Example A, $i = 1, \ldots, 6$

$[p_{f_i}, v_{f_i}]^T, u_i = [u_{pf_i}, u_{vf_i}]^T, i = 0, 1, \ldots, 6$, the followers' dynamics (5.52) can be transformed into the form of (5.3), where $A = \begin{bmatrix} -1 & 0.2132 \\ 0 & -0.5 \end{bmatrix}, B = \begin{bmatrix} 0.3153 & 0 \\ 0 & 0.4759 \end{bmatrix}$, $C = \begin{bmatrix} 1, 0 \end{bmatrix}$.

The parameters of the control strategy are $K_{ob} = \begin{bmatrix} 2.8, & 3.9 \end{bmatrix}^T$, $R = 5$, $T = 0.1$, $Y = 4I$, and we obtain that $L = [0.1351, 0.3390]$ and $J = [0.4186, 0.5251]$ by denoting $\alpha_1 = 4.25$ and $\rho_1 = 0.55$, $\varrho = 15$, $\vartheta = 25$, $\pi = [0.21, 0.79]$, $\lambda_{\min} = 0.13$, $\theta_{\min} = 1.55$, $\lambda_{\min}(\hat{Q}_{un}) = 0.17$, $\mu = 12.3$. Figure 5.11 shows the switching of $r(t)$, $\sigma(t)$ and $\gamma(t)$, respectively. It is obvious, from Fig. 5.12, that all the agents are able to achieve consensus in spite of the random attacks. Figures 5.13, 5.14 and 5.15 despite consensus tracking and the performance of the observer, respectively. To

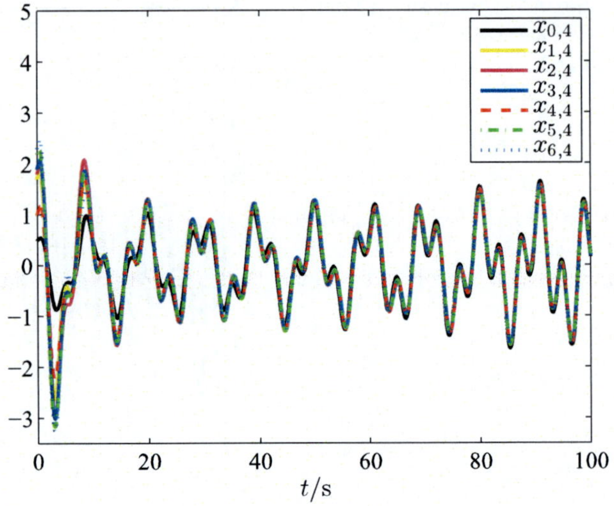

Fig. 5.9 Consensus tracking $x_{0,4}$ and $x_{i,4}$ in Example A, $i = 1, \ldots, 6$

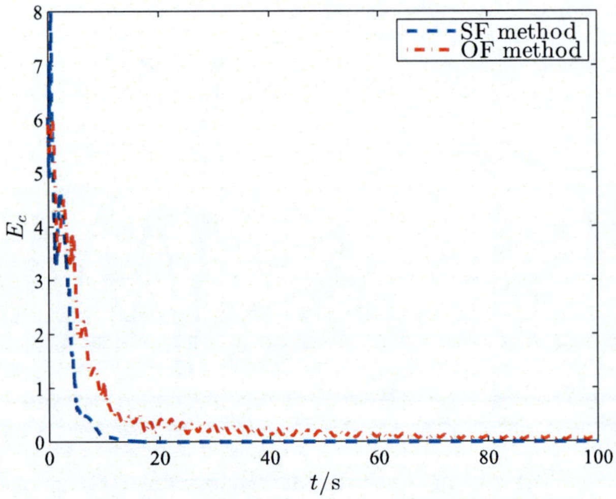

Fig. 5.10 Performance comparison between OF and SF methods in Example A

further illustrate the validity of the control strategy, we present the case that the controller without anti-attacks in Fig. 5.16 in the same initial and network conditions. The results reveal the decrease of the UAVs performance, even divergence under the attacks.

5.5 Conclusion

In this chapter, the observer-based distributed secure control strategy is designed where the randomness of the attack is described by the Markov process. Given this reality that some information of the system under random attack cannot be obtained,

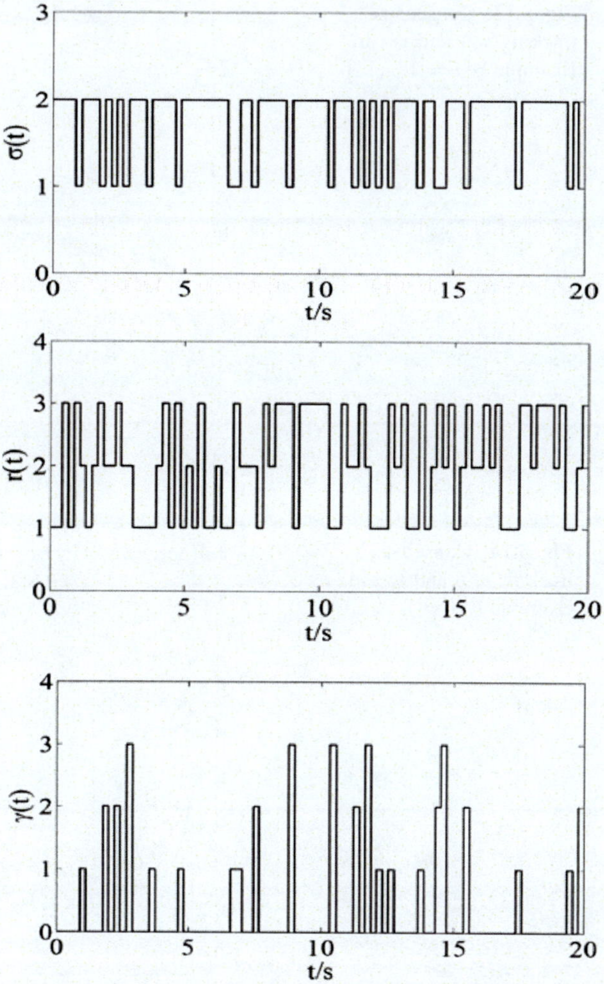

Fig. 5.11 Switching signal $\sigma(t)$, $r(t)$ and $\gamma(t)$ in Example B

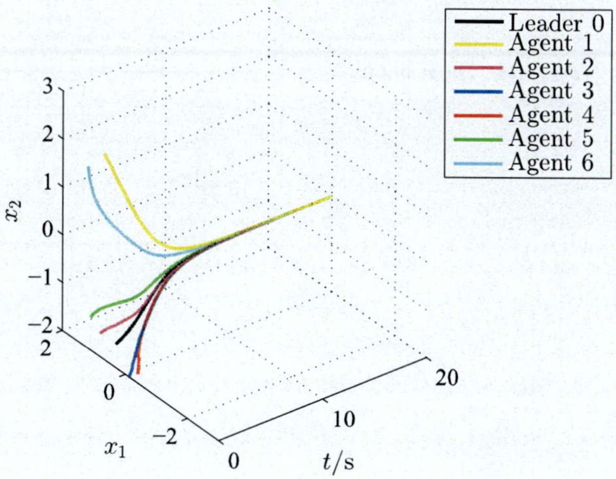

Fig. 5.12 State trajectories of the UAVs

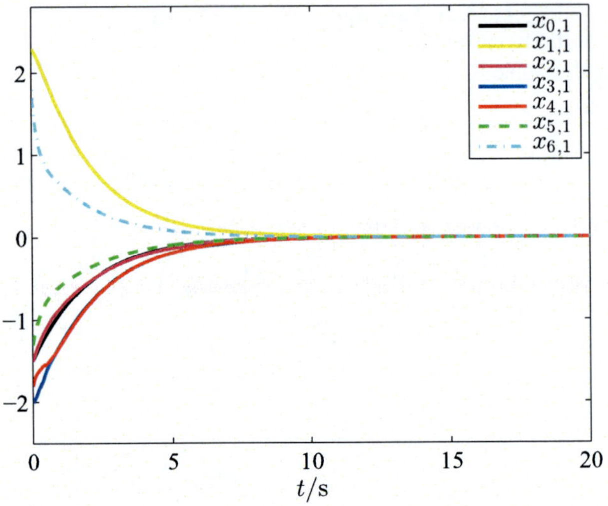

Fig. 5.13 Consensus tracking $x_{0,1}$ and $x_{i,1}$ in Example B, $i = 1, \ldots, 6$

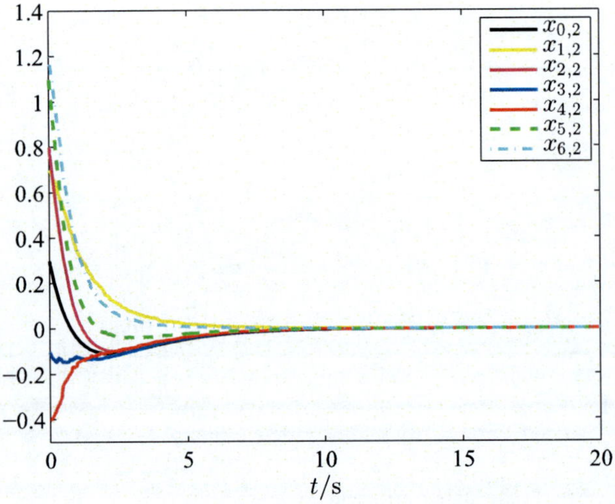

Fig. 5.14 Consensus tracking $x_{0,2}$ and $x_{i,2}$ in Example B, $i = 1, \ldots, 6$

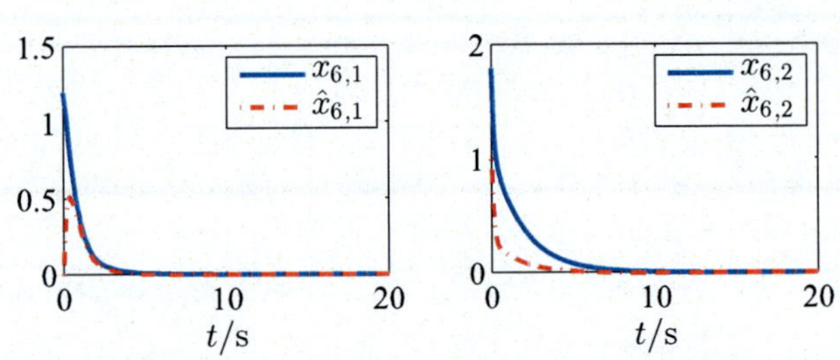

Fig. 5.15 The result of observer

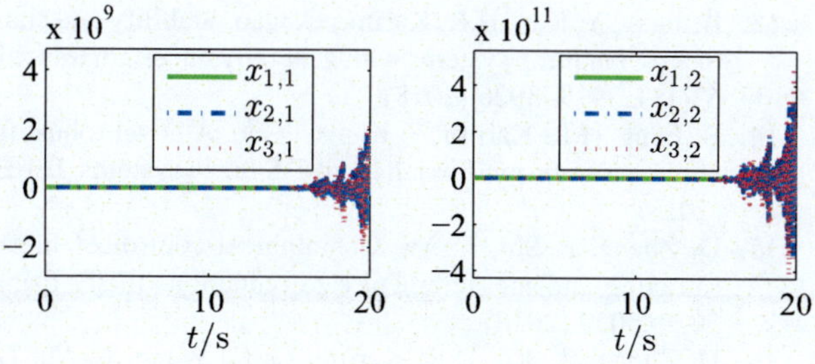

Fig. 5.16 Trajectories comparisons

distributed observers are employed to reconstruct the system states. Under two sufficient conditions on attack frequency and attack length rate, the secure consensus tracking in mean square sense can be obtained. The stability analysis is performed by the Lyapunov theory, and the simulation results demonstrate the validity of the proposed strategy. Our future research would be directed to heterogeneous cases, reduced-order/intercommunication-free protocols [10–12], semi-Markovian jump processes [13, 14] and the static OF approaches [15, 16].

References

1. Y. Yang, H. Xu, D. Yue, Observer-based distributed secure consensus control of a class of linear multi-agent systems subject to random attacks. IEEE Trans. Circuits Syst. I: Regul. Pap. **66**(8), 3089–3099 (2019)
2. N. Ye, Y. Zhang, C.M. Borror, Robustness of the markov-chain model for cyber-attack detection. IEEE Trans. Reliab. **53**(1), 116–123 (2004)
3. Q. Zhu, T. Basar, Game-theoretic methods for robustness, security, and resilience of cyber-physical control systems: games-in-games principle for optimal cross-layer resilient control systems. IEEE Control Syst. **35**(1), 46–65 (2015)
4. Y. Wei, J.H. Park, J. Qiu, L. Wu, H.Y. Jung, Sliding mode control for semi-markovian jump systems via output feedback. Automatica **81**, 133–141 (2017)
5. Y. Wei, J. Qiu, H.R. Karimi, W. Ji, A novel memory filtering design for semi-markovian jump time-delay systems. IEEE Trans. Syst., Man, Cybern.: Syst. **48**(12), 2229–2241 (2018)
6. H. Yan, H. Zhang, X. Zhan, Z. Li, C. Yang, Event-based H_∞ fault detection for buck converter with multiplicative noises over network. IEEE Trans. Circuits Syst. I: Regul. Pap. **66**(6), 2361–2370 (2019)
7. K. Zhang, B. Jiang, V. Cocquempot, Distributed fault estimation observer design for multi-agent systems with switching topologies. IET Control Theory Appl. **11**(16), 2801–2807 (2017)
8. Z. Feng, G. Wen, G. Hu, Distributed secure coordinated control for multiagent systems under strategic attacks. IEEE Trans. Cybern. **47**(5), 1273–1284 (2017)
9. W. Xu, D.W. Ho, Clustered event-triggered consensus analysis: an impulsive framework. IEEE Trans. Indust. Electron. **63**(11), 7133–7143 (2016)
10. X. Li, Y.C. Soh, L. Xie, A novel reduced-order protocol for consensus control of linear multi-agent systems. IEEE Trans. Autom. Control **64**(7), 3005–3012 (2019)
11. X. Li, L. Xie, Output-feedback protocols without controller interaction for consensus of homogeneous multi-agent systems: a unified robust control view. Automatica **81**, 37–45 (2017)
12. X. Li, Y.C. Soh, L. Xie, F.L. Lewis, Cooperative output regulation of heterogeneous linear multi-agent networks via H_∞ performance allocation. IEEE Trans. Autom. Control **64**(2), 683–696 (2019)

13. B. Jiang, Y. Kao, H.R. Karimi, C. Gao, Stability and stabilization for singular switching semi-markovian jump systems with generally uncertain transition rates. IEEE Trans. Autom. Control **63**(11), 3919–3926 (2018)
14. B. Jiang, H.R. Karimi, Y. Kao, C. Gao, A novel robust fuzzy integral sliding mode control for nonlinear semi-markovian jump T-S fuzzy systems. IEEE Trans. Fuzzy Syst. **26**(6), 3594–3604 (2018)
15. D. Zhang, P. Shi, L. Yu, Containment control of linear multiagent systems with aperiodic sampling and measurement size reduction. IEEE Trans. Neural Netw. Learn. Syst. **29**(10), 5020–5029 (2018)
16. D. Zhang, Z. Xu, H.R. Karimi, Q.-G. Wang, L. Yu, Distributed H_∞ output-feedback control for consensus of heterogeneous linear multiagent systems with aperiodic sampled-data communications. IEEE Trans. Indust. Electron. **65**(5), 4145–4155 (2018)

Chapter 6
Event-Based Secure Leader-Following Consensus Control for Multiagent Systems With Multiple Cyber Attacks

This chapter concentrates on event-based secure leader-following consensus control for multi-agent systems (MASs) with multiple cyber-attacks, which contain replay attacks and denial-of-service (DoS) attacks. A multiple cyber-attacks model is built by considering replay attacks and DoS attacks simultaneously. The changes of communication topologies caused by DoS attacks are considered for MASs. Besides, an event-triggered mechanism is adopted for mitigating the load of network bandwidth by scheduling transmission of sampled data. Then an event-based consensus control protocol is first developed for MASs subjected to multiple cyber-attacks. In view of this, by using the Lyapunov stability theory, sufficient conditions are obtained to ensure the mean-square exponential consensus of MASs. Furthermore, the event-based controller gain is derived by solving a set of linear matrix inequalities. Finally, an example is simulated for confirming the effectiveness of the theoretical results [1].

The rest of this chapter is organized as follows. The graph theory and problem formulation are provided in Sect. 6.1. The mains results of the event-based leader-following consensus control issue for the discussed MASs are given in Sect. 6.2. In Sect. 6.3, the usefulness of designed method is illustrated through a simulated example. At last, Sect. 6.4 presents the conclusions.

6.1 Problem Formulation

6.1.1 Graph Theory

In a directed graph $\mathbb{G} = (\mathbb{V}, \mathbb{E}, \mathbb{A})$, $\mathbb{V} = \{1, 2, \ldots, N\}$, $\mathbb{E} \subseteq \{(i, j), i, j \in \mathbb{V}\}$ and $\mathbb{A} = [a_{ij}]_{N\times N}$ denote a set of nodes, a set of edges and a weighted adjacency matrix with non-negative elements a_{ij}, $i, j = 1, 2, \ldots, N$, respectively. An edge $(j, i) \in \mathbb{E}$

D. Yue et al., *Secure Control of Networked Control Systems and Its Applications*,
https://doi.org/10.1007/978-981-33-6730-2_6

in graph $\mathbb{G}$ denotes that agent i can acquire the information of agent j. $a_{ij} > 0$ only when $(i, j) \in \mathbb{E}$; $a_{ij} = 0$ when $(i, j) \notin \mathbb{E}$. $D = diag\{indeg_1, indeg_2, \ldots, indeg_N\}$ is represented as the in-degree matrix, where $indeg_i = \sum_{j \in \mathbb{M}_i} a_{ij}$ for agent i. $\mathbb{M}_i = \{j|(i, j) \in \mathbb{E}\}$ is the set of neighbors of agent i. Define Laplacian matrix $L = [l_{ij}]_{N \times N}$ of $\mathbb{G}$ as $L = D - \mathbb{A}$.

6.1.2 System Description

Consider the following MAS composed of N followers and one leader labeled as node 0. The dynamics of the i_{th} agent and the leader are described as

$$\begin{cases} \dot{x}_i(t) = Ax_i(t) + Fh(x_i(t), t) + Bu_i(t) \\ \dot{x}_0(t) = Ax_0(t) + Fh(x_0(t), t) \end{cases} \tag{6.1}$$

where $x_i(t) \in \mathrm{R}^a$ and $h(x_i(t), t) \in \mathrm{R}^a$ are the state vector and its nonlinear dynamics of the i_{th} agent, respectively; $u_i(t) \in \mathrm{R}^b$ is the control output vector of the i_{th} agent; $i = \{1, 2, \ldots, N\} \triangleq \mathcal{S}_N$; A, F and B are constant matrices with appropriate dimensions.

Assumption 6.1 ([2]) The nonlinear function $h : \mathrm{R}^a \to \mathrm{R}^a$ satisfies the Lipschitz condition, i.e., for $\forall x, z \in \mathrm{R}^a$, there exists a scalar $\epsilon > 0$ such that the following inequality holds.

$$\| h(x) - h(z) \| \leq \epsilon \| x - z \| \tag{6.2}$$

Remark 6.1 In a directed communication graph $\mathbb{G}$, if there exists a directed spanning tree, then its Laplacian matrix L has N eigenvalues which contain a simple eigenvalue 0 and $N - 1$ eigenvalues with positive real parts. According to this fact, one can draw the conclusion that every agent in the MAS (6.1) possesses not only one neighboring agent, which is significant to design the following event-triggered communication mechanism in this chapter.

6.1.3 Design of Event Triggering Scheme

This chapter aims at designing an event-based control strategy such that the system (6.1) under multiple cyber-attacks can reach mean-square exponential consensus. For this purpose, the event-triggered mechanism is adopted for the MAS, the structure of which is exhibited in Fig. 6.1. In this framework, the sensor is time-triggered while the controller, the zero-order-hold (ZOH) and the actuator are event-triggered. The state of each agent i is sampled in a sampling period h, then the sampled-data can be sent out only when the given event-triggering condition is satisfied. Then one can

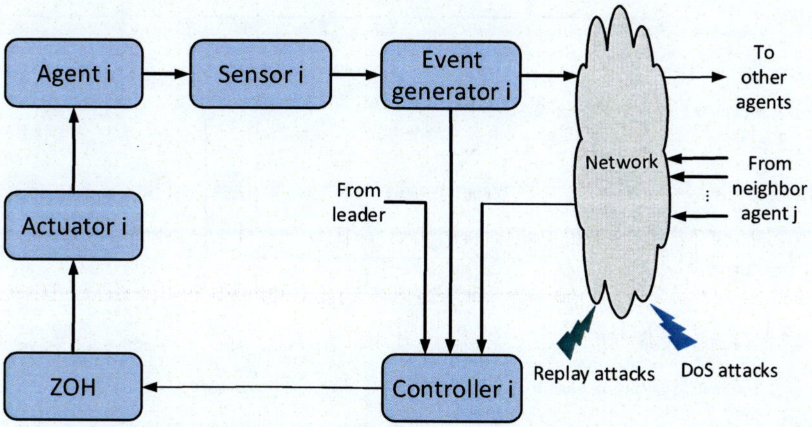

Fig. 6.1 Structure of event-based control for agent i

obtain the following sequence of transmitting instant for agent i.

$$
\begin{aligned}
t_{\lambda+1}^i h =& t_\lambda^i h + \min_{f^i \geq 1} \left\{ f^i h | (\psi^i(t_\lambda^i h))^T \Phi_i \psi^i(t_\lambda^i h) \right. \\
& \left. > \sigma \xi_i^T(t_\lambda^i h + f^i h) \Phi_i \xi_i(t_\lambda^i h + f^i h) \right\}
\end{aligned} \tag{6.3}
$$

where parameter $\sigma \in [0, 1)$, $\Phi_i > 0$, $f^i = 1, 2, \ldots$, and $\psi^i(t_\lambda^i h) = x_i(t_\lambda^i h) - x_i(t_\lambda^i h + f^i h)$, $i \in \mathcal{S}_N$, $h > 0$, $t_\lambda^i h$ denotes the latest transmitting instant, $\xi_i(t_\lambda^i h + f^i h) = \sum_{j=1}^{\mathbb{M}_i} a_{ij} \left[x_i(t_\lambda^i h + f^i h) - x_j(t_\lambda^j h + f^j h) \right]$.

6.1.4 Multiple Cyber Attack Models

Due to the openness of communication network, the networked systems are susceptible to cyber attacks. As exhibited in Fig. 6.1, in this chapter, a class of multiple cyber-attacks is taken into consideration, including replay attacks and DoS attacks. Figure 6.2 shows an example of transmitted signals via the event-triggered mechanism (6.3) under multiple attacks model for agent i, which may be helpful to understand the mechanism of launching a multiple cyber-attack.

The replay attackers launch an attack at time t by executing the following steps [3].

(1) Record transmitted signals via the network from initial instant t_0 to current instant t. The set of the recorded data is represented as $\mathcal{T}_i(t) \triangleq \{\tilde{x}_i(t_0), \tilde{x}_i(t_1), \ldots, \tilde{x}_i(t_m)\}$, $0 \leq t_0 < t_1 < \cdots < t_m < t$.
(2) Select one arbitrary data $\tilde{x}_i(t_r)$ from $\mathcal{T}_i(t)$ for replay.

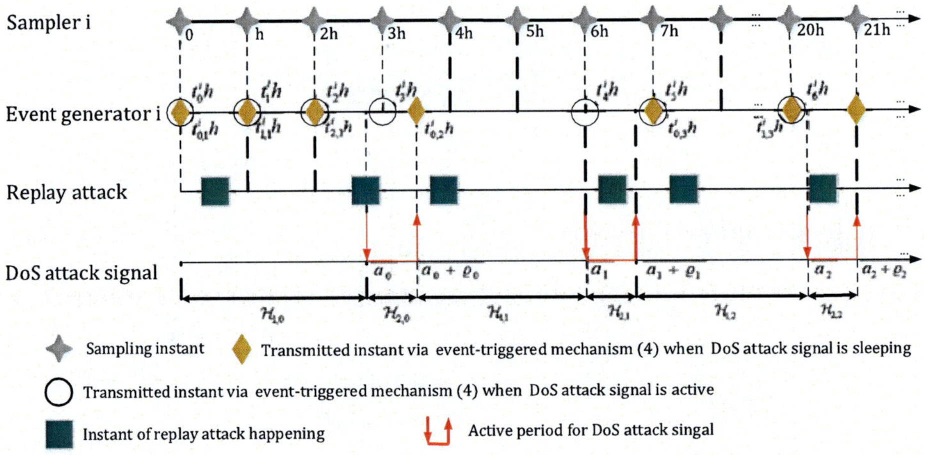

Fig. 6.2 An example of multiple cyber-attacks model

In order to describe replay attacks, a Bernoulli random variable $\alpha_i(t)$ is introduced, whose expectation and mathematical variance are denoted as $\bar{\alpha}_i$ and ρ_i^2, respectively. With the consideration of replay attacks, the i_{th} agent's transmitted signal under replay attacks is presented as

$$\hat{x}_i(t) = \alpha_i(t)\tilde{x}_i(t_r) + (1 - \alpha_i(t))\tilde{x}_i(t) \tag{6.4}$$

where $\alpha_i(t) \in \{0, 1\}$. $\alpha_i(t) = 1$ denotes that replay attacks happen and the normal data is replaced by $\tilde{x}_i(t_r)$; $\alpha_i(t) = 0$ indicates the absence of replay attacks. $\tilde{x}_i(t)$ is the normal signal of agent i, which will be broadcasted over the network; t_r is the previous instant and t is the current instant, obviously, one can get that $0 < t_r < t$, namely, $t_r = t - d(t)$, where $d(t) \in (0, d_M)$, d_M is the upper bound of $d(t)$. It is worth mentioning that the random variables $\alpha_i(t)$, $i \in \mathcal{S}_N$ are independent.

At the same time, the influences of DoS attacks are also taken into account. It needs to be mentioned that a DoS attack calls for a certain amount of energy. Suppose that the energy of DoS jamming signal is limited, then DoS jamming signal is sleeping at certain period to save energy for next attack. Therefore, a reasonable assumption is formalized as follows.

Assumption 6.2 Whether DoS attacks occur or not can be described by a variable $\beta(t)$, which satisfies $\beta(t) = \begin{cases} 1, t \in [a_n, a_n + \varrho_n) \\ 0, t \in [a_n + \varrho_n, a_{n+1}), n \in \mathbb{N} \end{cases}$, where $[a_n, a_n + \varrho_n)$ and $[a_n + \varrho_n, a_{n+1})$ denote the sleeping period and active period of DoS jamming signal, respectively; ϱ_n represents the length of the n_{th} sleeping period. Then we can obtain the starting instants and end instants of DoS sleeping period satisfy $0 \leq a_0 < a_0 + \varrho_0 < a_1 < \cdots < a_n < a_n + \varrho_n < a_{n+1} < \cdots$.

Based on the Assumption 6.2, denote D_M and D_m as a uniform upper bound on the lengths of the DoS active periods $a_{n+1} - a_n - \varrho_n$ and a uniform lower bound on the lengths of the DoS sleeping periods ϱ_n, respectively, then one can acquire the following inequalities:

$$\begin{cases} D_M \geq \sup_{n\in\mathbb{N}}\{a_{n+1} - a_n - \varrho_n\} \\ D_m \leq \inf_{n\in\mathbb{N}}\{\varrho_n\} \end{cases} \tag{6.5}$$

Besides, the number of DoS attacks sleep/active transitions in the interval $[0, t)$ is represented as $o(t)$, then there exist $b_0 \geq 0$ and $f_a \geq h$ so as to satisfy the following condition:

$$o(t) \leq b_0 + \frac{t}{f_a} \tag{6.6}$$

In order to facilitate analysis, denote $\mathcal{H}_{1,n} \triangleq [a_n, a_n + \varrho_n)$, $\mathcal{H}_{2,n} \triangleq [a_n + \varrho_n, a_{n+1})$, $\mathcal{D}^i_{\lambda,n} \triangleq [t^i_{\lambda,n}h, t^i_{\lambda+1,n}h)$ for $\forall n \in \mathbb{N}$, $t_{0,n}h \triangleq a_n$. The corresponding network-induced delay of transmitted instant $t^i_{\lambda,n}h$ is defined as $\delta^i_{\lambda,n}$. Denote $\bar{\delta} = \max\{\delta^i_{\lambda,n}\}$. Inspired by [4, 5], to analyze more easily, $\mathcal{D}^i_{\lambda,n}$ can be divided as $\mathcal{M}^{\chi^i}_{\lambda,n} \cup \mathcal{M}^{p_{\lambda,n}+1}_{\lambda,n}$, where

$$\begin{cases} \mathcal{M}^{\chi^i}_{\lambda,n} = [t^i_{\lambda,n}h + (\chi^i - 1)h + \bar{\delta}, t^i_{\lambda,n}h + \chi^i h + \bar{\delta}), \\ \chi^i \in \{1, 2, \ldots, p_{\lambda,n}\} \\ \mathcal{M}^{p_{\lambda,n}+1}_{\lambda,n} = [t^i_{\lambda,n}h + p_{\lambda,n}h + \bar{\delta}, t^i_{\lambda+1,n}h) \\ p_{\lambda,n} \triangleq \inf\{\chi^i \in \mathbb{N} | t^i_{\lambda,n}h + \chi^i h \geq t^i_{\lambda,n+1}h\} \\ \lambda \in \{0, 1, 2, \ldots, q(n)\} \triangleq \mathcal{U}(n) \\ q(n) = \sup\left\{\lambda \in \mathbb{N} | t^i_{\lambda,n}h \leq a_n + \varrho_n\right\}, n \in \mathbb{N} \end{cases}$$

Then the interval $\mathcal{H}_{1,n}$ can be rewritten as
$\mathcal{H}_{1,n} = \cup_{\lambda=0}^{\mathcal{U}(n)} \cup_{\chi^i=1}^{p_{\lambda,n}+1} \left\{\mathcal{M}^{\chi^i}_{\lambda,n} \cap \mathcal{H}_{1,n}\right\}$. As a result, define

$$\begin{cases} \delta^i_{\lambda,n}(t) = t - t^i_{\lambda,n}h - (\theta^i - 1)h \\ \psi^i_{\lambda,n}(t) = x_i(t^i_{\lambda,n}h) - x_i(t^i_{\lambda,n}h + (\theta^i - 1)h) \end{cases} \tag{6.7}$$

where $t \in \mathcal{M}^{\theta^i}_{\lambda,n} \cap \mathcal{H}_{1,n}$, $\theta^i = 1, 2, \ldots, p_{\lambda,n} + 1$. From (6.7), one can acquire that $0 < \delta^i_{\lambda,n} \leq \delta^i_{\lambda,n}(t) \leq h + \bar{\delta} \triangleq \delta_M$, δ_M is the upper bound of $\delta^i_{\lambda,n}(t)$. Further, the event-triggered sampled state $x_i(t^i_{\lambda,n}h)$ can be given as follows:

$$\tilde{x}_i(t) = x_i(t^i_{\lambda,n}h) = \psi^i_{\lambda,n}(t) + x_i(t - \delta^i_{\lambda,n}(t)) \tag{6.8}$$

with $\psi^i_{\lambda,n}(t)$ violating the following inequality:

$$(\psi^i_{\lambda,n}(t))^T \Phi_i \psi^i_{\lambda,n}(t) - \sigma \xi_i^T(t - \delta^i_{\lambda,n}(t)) \Phi_i \xi_i(t - \delta^i_{\lambda,n}(t)) \leq 0 \tag{6.9}$$

Then by considering the impacts of replay attacks and DoS attacks , the actual control input of agent i received from agent j can be expressed as

$$\bar{x}_j(t) = \begin{cases} \alpha_j(t)\tilde{x}_j(t_r) + (1-\alpha_j(t))\tilde{x}_j(t), \\ t \in \mathcal{D}^j_{\lambda,n} \cap \mathcal{H}_{1,n} \\ 0, t \in \mathcal{H}_{2,n} \end{cases} \tag{6.10}$$

Remark 6.2 In (6.10), when $t \in \mathcal{D}^j_{\lambda,n} \cap \mathcal{H}_{1,n}$ and $\alpha_j(t) = 0$, it means that no cyber-attack occurs in the network, in this case, $\bar{x}_j(t) = \tilde{x}_j(t)$; when $t \in \mathcal{D}^j_{\lambda,n} \cap \mathcal{H}_{1,n}$ and $\alpha_j(t) = 1$, it means that only replay attacks happen, in such situation, $\bar{x}_j(t) = \tilde{x}_j(t_r)$; when $t \in \mathcal{H}_{2,n}$, we can get $\bar{x}_j(t) = 0$, which means that the communication paralysis appears due to the presence of DoS attacks.

6.1.5 Design of Event-Based Control Protocol

Due to the DoS attacks , the communication topologies are no longer always fixed in this chapter. When DoS jamming signal is sleeping in $\cup_{n\in\mathbb{N}}\mathcal{H}_{1,n}$, the communication topology of MASs is connected, where the data transmission is successful. When DoS attacks occur in active periods $\cup_{n\in\mathbb{N}}\mathcal{H}_{2,n}$, consider the effects of DoS attacks on the communication links $\mathbb{E}$ rather than the nodes $\mathbb{V}$. Under such circumstance, the original communication topology with a directed spanning tree is destroyed as a disconnected topological graph where the agents are assumed to give up communications. It should be pointed out that the impacts of DoS attacks will be wiped out at time a_{n+1} by some repairing efforts, then the topological graph is recovered to $\tilde{\mathbb{G}}$ with a directed spanning tree before the next DoS attack. The time spent in repatching the topological graph may be such short that it can be negligible. A piecewise constant function $\tau(t)$: $[0, +\infty) \to \mathrm{M} = \{1, 2, \ldots, m\}$ is utilized to depict the switching among different topology diagrams; $\mathbb{G}^{\tau(t)} \in \tilde{\mathbb{G}}$, for $t \in \mathcal{H}_{1,n}$, $n \in \mathbb{N}$, where $\tilde{\mathbb{G}} = \{\mathbb{G}^1, \mathbb{G}^2, \ldots, \mathbb{G}^m\}$, $m \geq 1$, represents the set of all possible topological graphs when the MASs recover from DoS attacks.

On the basis of the analysis above, the event-based control protocol for system (6.1) is designed as follows:

$$u_i(t) = \begin{cases} -\phi K \left\{ \sum\limits_{j=1}^{\mathbb{M}_i} a_{ij}^{\tau(t)} \left[\tilde{x}_i(t) - \bar{x}_j(t)\right] \right. \\ \left. + d_i^{\tau(t)} \left[\tilde{x}_i(t) - x_0(t)\right] \right\}, t \in \mathcal{D}^i_{\lambda,n} \cap \mathcal{H}_{1,n} \\ 0, t \in \mathcal{H}_{2,n} \end{cases} \tag{6.11}$$

where $\lambda \in \mathcal{U}(n)$, $n \in \mathbb{N}$, $\phi > 0$ denotes the coupling strength of the MASs, K is the controller gain to be determined later; $a_{ij}^{\tau(t)}$ stands for the adjacency element of $\mathbb{G}^{\tau(t)}$; $d_i^{\tau(t)} = 1$ means that follower i can receive the information from the leader.

6.1.6 Modeling of MASs Under Multiple Cyber Attacks

Define error $e_i(t) = x_i(t) - x_0(t)$, by combining (6.1), (6.8), (6.10) and (6.11), and let

$$\begin{aligned}
&x(t) = [x_1^T(t), x_2^T(t), \ldots, x_N^T(t)]^T \\
&e(t) = [e_1^T(t), e_2^T(t), \ldots, e_N^T(t)]^T \\
&\breve{\alpha}(t) = diag\{\alpha_1(t), \alpha_2(t), \ldots, \alpha_N(t)\} \\
&\psi_{\lambda,n}(t) = [(\psi_{\lambda,n}^1)^T(t), (\psi_{\lambda,n}^2)^T(t), \ldots, (\psi_{\lambda,n}^N)^T(t)]^T \\
&h(x(t),t) = [h^T(x_1(t),t), h^T(x_2(t),t), \ldots, h^T(x_N(t),t)]^T \\
&H(x(t),t) = [h^T(x_1(t),t) - h^T(x_0(t),t), h^T(x_2(t),t) \\
&\quad - h^T(x_0(t),t), \ldots, h^T(x_N(t),t) - h^T(x_0(t),t)]^T \\
&x(t - \delta_{\lambda,n}(t)) = [x_1^T(t - \delta_{\lambda,n}^1(t)), x_2^T(t - \delta_{\lambda,n}^2(t)), \ldots, \\
&\quad x_N^T(t - \delta_{\lambda,n}^N(t))]^T
\end{aligned}$$

then we can obtain the consensus tracking error system:

$$\dot{e}(t) = \begin{cases}
(I_N \otimes A)e(t) + (I_N \otimes F)H(x(t),t) \\
-\phi(I_N \otimes B)\left\{[(L^{\tau(t)} \otimes K)e(t - \delta_{\lambda,n}(t)) \right. \\
+(D^{\tau(t)} \otimes K)\psi_{\lambda,n}(t) - (\mathbb{A}^{\tau(t)} \otimes K)\psi_{\lambda,n}(t) \\
-\breve{\alpha}(t)(\mathbb{A}^{\tau(t)} \otimes K)[e(t - r_{\lambda,n}(t)) + \psi_{\lambda,n}(t_r) \\
-e(t - \delta_{\lambda,n}(t)) - \psi_{\lambda,n}(t))] \\
\left. +(D_0^{\tau(t)} \otimes K)\psi_{\lambda,n}(t)\right\}, t \in \mathcal{D}_{\lambda,n}^i \cap \mathcal{H}_{1,n}, \lambda \in \mathcal{U}(n) \\
(I_N \otimes A)e(t) + (I_N \otimes F)H(x(t),t), t \in \mathcal{H}_{2,n}
\end{cases} \tag{6.12}$$

where $r_{\lambda,n}(t) = d(t) + \delta_{\lambda,n}(t - d(t))$, and $r_{\lambda,n}(t) \in [0, r_M)$, r_M is the upper bound of $r_{\lambda,n}(t)$.

For technical convenience, denote

$$\begin{aligned}
&\eta(t) = \begin{bmatrix} x(t) \\ e(t) \end{bmatrix}, \tilde{F}(x(t),t) = \begin{bmatrix} h(x(t),t) \\ H(x(t),t) \end{bmatrix} \\
&\bar{A} = diag\{I_N \otimes A, I_N \otimes A\}, \bar{F} = diag\{I_N \otimes F, I_N \otimes F\} \\
&D_0^{\tau(t)} = diag\left\{d_1^{\tau(t)}, d_2^{\tau(t)}, \ldots, d_N^{\tau(t)}\right\}, \hat{\alpha}(t) = diag\{\breve{\alpha}(t), \breve{\alpha}(t)\}
\end{aligned}$$

$$\bar{B}_1^{\tau(t)} = \begin{bmatrix} 0_{N\times N} & \bar{L}^{\tau(t)} \otimes B \\ 0_{N\times N} & \bar{L}^{\tau(t)} \otimes B \end{bmatrix}, \bar{L}^{\tau(t)} = L^{\tau(t)} + D_0^{\tau(t)}$$

$$\bar{B}_2^{\tau(t)} = \begin{bmatrix} \mathbb{A}^{\tau(t)} \otimes B - D^{\tau(t)} \otimes B - D_0^{\tau(t)} \otimes B \\ \mathbb{A}^{\tau(t)} \otimes B - D^{\tau(t)} \otimes B - D_0^{\tau(t)} \otimes B \end{bmatrix}$$

$$\bar{B}_3^{\tau(t)} = \begin{bmatrix} 0_{N\times N} & \mathbb{A}^{\tau(t)} \otimes B \\ 0_{N\times N} & \mathbb{A}^{\tau(t)} \otimes B \end{bmatrix}, \bar{B}_4^{\tau(t)} = \begin{bmatrix} \mathbb{A}^{\tau(t)} \otimes B \\ \mathbb{A}^{\tau(t)} \otimes B \end{bmatrix}$$

$$H_1 = \begin{bmatrix} I_N & 0_{N\times N} \end{bmatrix}, H_2 = \begin{bmatrix} 0_{N\times N} & I_N \end{bmatrix}$$

where $0_{N\times N}$ represents the N-dimensional zero matrix, which can be abbreviated as 0 sometimes. It needs to point out that in the following, the element 0 in a matrix denotes the zero matrix with appropriate dimension.

Then by combining (6.1) and (6.10), the consensus tracking error system can be written as

$$\dot{\eta}(t) = \begin{cases} \bar{A}\eta(t) + \bar{F}\tilde{F}(x(t), t)) - \phi\bar{B}_1^{\tau(t)}\bar{K}\eta(t - \delta_{\lambda,n}(t)) \\ +\phi\bar{B}_2^{\tau(t)}\bar{K}_1\psi_{\lambda,n}(t) + \phi\hat{\alpha}(t)\bar{B}_3^{\tau(t)}\bar{K}[\eta(t - r_{\lambda,n}(t)) \\ -\eta(t - \delta_{\lambda,n}(t))] + \phi\hat{\alpha}(t)\bar{B}_4^{\tau(t)}\bar{K}_1[\psi_{\lambda,n}(t_r) \\ -\psi_{\lambda,n}(t)], t \in \mathcal{D}_{\lambda,n}^i \cap \mathcal{H}_{1,n}, \lambda \in \mathcal{U}(n) \\ \bar{A}\eta(t) + \bar{F}\tilde{F}(x(t), t)), t \in \mathcal{H}_{2,n} \end{cases} \tag{6.13}$$

where $\bar{K} = I_{2N} \otimes K$, $\bar{K}_1 = I_N \otimes K$; $\zeta(t)$ is the supplemented initial condition of the state $\eta(t)$ with $\zeta(0) \triangleq \zeta_0$.

Remark 6.3 In (6.11), an event-based model of MASs under multiple cyber-attacks has been established. Up to the authors' knowledge, the event-based secure leader-following consensus control issue of MASs with multiple cyber-attacks is firstly addressed in this chapter.

Before we present the main results, the following assumption and definition are given.

Assumption 6.3 For $t \in \cup_{n\in\mathbb{N}}\mathcal{H}_{1,n}$, the topological graph $\mathbb{G}^{\tau(t)}$ has a directed spanning tree where the leader is the root node.

Definition 6.1 ([6]) (Mean-square exponential consensus tracking) The system (6.1) with the event-triggered control strategy $u_i(t)$ can reach mean-square exponential stability under multiple cyber-attacks when studying the problem of leader-following consensus control for MASs, if there exists such a scalar $w > 0$ and a decay rate v that the following inequality holds for any $t \geq 0$.

$$\begin{aligned} &\mathbf{E}\left\{\| x_i(t) - x_0(t) \|^2\right\} \\ &\leq we^{-vt}\mathbf{E}\left\{\| x_i(0) - x_0(0) \|^2\right\}, i \in \mathcal{S}_N \end{aligned} \tag{6.14}$$

6.2 Mean-Square Exponential Consensus Analysis

Theorem 6.1 *The DoS parameters D_M, D_m, b_0, f_a are known. For given matrix K, and scalars $\sigma \in [0, 1)$, $\bar{\alpha}_i \in (0, 1)$, $\delta_M > 0$, $r_M > 0$, $h > 0$, system (6.11) is exponentially stable in mean-square, if for some prescribed positive scalars μ_1, μ_2, $\varsigma = \mu_1 + \mu_2$, ω_1 and ω_2, there exist symmetric positive definite matrices Φ_i ($i \in \mathcal{S}_N$), P_1, P_2, Q_1, Q_2, R_1, R_2, Z_1, Z_2, S_1, S_2, positive scalars ϵ, ϕ, and matrices M_1, M_2, U_1, U_2 with appropriate dimensions such that the following inequalities hold:*

$$\Pi_1 = \begin{bmatrix} \Gamma_{11}^1 & * & * & * \\ \Gamma_{21}^1 & \Gamma_{22}^1 & * & * \\ \Gamma_{31}^1 & 0 & \Gamma_{33}^1 & * \\ \Gamma_{41}^1 & \Gamma_{42}^1 & \Gamma_{43}^1 & \Gamma_{44}^1 \end{bmatrix} < 0 \tag{6.15}$$

$$\Pi_2 = \begin{bmatrix} \Gamma_{11}^2 & * & * \\ \Gamma_{21}^2 & \Gamma_{22}^2 & * \\ \Gamma_{31}^2 & \Gamma_{32}^2 & \Gamma_{33}^2 \end{bmatrix} < 0 \tag{6.16}$$

$$\begin{bmatrix} R_l & * \\ M_l & R_l \end{bmatrix} > 0, \begin{bmatrix} S_l & * \\ U_l & S_l \end{bmatrix} > 0, l = 1, 2 \tag{6.17}$$

$$\begin{cases} (I_{2N} \otimes P_1) \leq \omega_2 (I_{2N} \otimes P_2) \\ (I_{2N} \otimes P_2) \leq \omega_1 e^{2\varsigma h} (I_{2N} \otimes P_1) \end{cases} \tag{6.18}$$

$$Q_l \leq \omega_{3-l} Q_{3-l}, R_l \leq \omega_{3-l} R_{3-l}, l = 1, 2 \tag{6.19}$$

$$\vartheta = \frac{2}{f_a}(\mu_1 D_m - \mu_2 D_M - \varsigma h) - \frac{1}{f_a} \ln(\omega_1 \omega_2) > 0 \tag{6.20}$$

where the elements of Π_1 and Π_2 are given in Appendix I.

Proof See Appendix II. ■

In Theorem 6.1, the sufficient conditions are achieved which can guarantee mean-square exponential consensus of system (6.11). In the following, the controller gain of MASs with event-triggered mechanism and multiple cyber-attacks is derived on the basis of Theorem 6.1.

6.3 Control Gain Design

Theorem 6.2 *For given DoS parameters D_M, D_m, b_0, f_a, and scalars $\sigma \in [0, 1)$, $\bar{\alpha}_i \in (0, 1)$, $\delta_M > 0$, $r_M > 0$, $h > 0$, system (6.11) is exponentially stable in mean-square, if for some prescribed positive scalars μ_1, μ_2, $\varsigma = \mu_1 + \mu_2$, ω_1 and ω_2, there exist symmetric positive definite matrices X_1, X_2, $\hat{\Phi}_i$ ($i \in \mathcal{S}_N$), $\hat{Q}_1$, $\hat{Q}_2$, $\hat{R}_1$, $\hat{R}_2$, $\hat{Z}_1$, $\hat{Z}_2$, $\hat{S}_1$, $\hat{S}_2$, positive scalars ϵ, ϕ, ν_1, ν_2, ν_3, ν_4, ν_{12}, ν_{22}, and matrices Y, $\hat{M}_1$, $\hat{M}_2$, $\hat{U}_1$, $\hat{U}_2$ with appropriate dimensions such that the inequality (6.20) and the* *following linear matrix inequalities hold:*

$$\hat{\Pi}_1 = \begin{bmatrix} \hat{\Gamma}^1_{11} & * & * & * \\ \hat{\Gamma}^1_{21} & \hat{\Gamma}^1_{22} & * & * \\ \hat{\Gamma}^1_{31} & 0 & \hat{\Gamma}^1_{33} & * \\ \hat{\Gamma}^1_{41} & \hat{\Gamma}^1_{42} & \hat{\Gamma}^1_{43} & \hat{\Gamma}^1_{44} \end{bmatrix} < 0 \tag{6.21}$$

$$\hat{\Pi}_2 = \begin{bmatrix} \hat{\Gamma}^2_{11} & * & * \\ \hat{\Gamma}^2_{21} & \hat{\Gamma}^2_{22} & * \\ \hat{\Gamma}^2_{31} & \hat{\Gamma}^2_{32} & \hat{\Gamma}^2_{33} \end{bmatrix} < 0 \tag{6.22}$$

$$\begin{bmatrix} \hat{R}_l & * \\ \hat{M}_l & \hat{R}_l \end{bmatrix} > 0, \begin{bmatrix} \hat{S}_l & * \\ \hat{U}_l & \hat{S}_l \end{bmatrix} > 0, l = 1, 2 \tag{6.23}$$

$$\begin{bmatrix} -\omega_2(I_{2N} \otimes X_2) & * \\ I_{2N} \otimes X_2 & -(I_{2N} \otimes X_1) \end{bmatrix} \leq 0 \tag{6.24}$$

$$\begin{bmatrix} -\omega_1 e^{2\varsigma h}(I_{2N} \otimes X_1) & * \\ I_{2N} \otimes X_1 & -(I_{2N} \otimes X_2) \end{bmatrix} \leq 0 \tag{6.25}$$

$$\begin{bmatrix} -\omega_2 \hat{Q}_2 & * \\ I_{2N} \otimes X_2 & -2\nu_1(I_{2N} \otimes X_1) + \nu_1^2 \hat{Q}_1 \end{bmatrix} \leq 0 \tag{6.26}$$

$$\begin{bmatrix} -\omega_1 \hat{Q}_1 & * \\ I_{2N} \otimes X_1 & -2\nu_3(I_{2N} \otimes X_2) + \nu_3^2 \hat{Q}_2 \end{bmatrix} \leq 0 \tag{6.27}$$

$$\begin{bmatrix} -\omega_2 \hat{R}_2 & * \\ I_{2N} \otimes X_2 & -2\nu_{12}(I_{2N} \otimes X_1) + \nu_{12}^2 \hat{R}_1 \end{bmatrix} \leq 0 \tag{6.28}$$

$$\begin{bmatrix} -\omega_1 \hat{R}_1 & * \\ I_{2N} \otimes X_1 & -2\nu_{22}(I_{2N} \otimes X_2) + \nu_{22}^2 \hat{R}_2 \end{bmatrix} \leq 0 \tag{6.29}$$

where the elements of $\hat{\Pi}_1$ *and* $\hat{\Pi}_2$ *are given in Appendix III.*

Furthermore, the controller gain can be obtained as

$$K = Y X_1^{-1} \tag{6.30}$$

Proof See Appendix IV. ■

6.4 Simulation Example

In this section, a simulation example is provided for validating the feasibility of the event-based leader-following control strategy for MASs, where the dynamic of each agent can be described by an F-18 aircraft model [7, 8]. Then the decoupling linearized longitudinal state dynamical equations of agent i is expressed as

Table 6.1 Meanings of the parameters in Sect. IV

Parameter	Meaning
$x_{i1}(t)$	Angle of attack
$x_{i2}(t)$	Pitch rate
$\dot{x}_{i1}(t)$	Angular velocity of attack
$\dot{x}_{i2}(t)$	Pitch acceleration
$u_{i1}(t)$	Symmetric elevator position
$u_{i2}(t)$	Symmetric pitch thrust velocity nozzle position
A_{Long}	Longitudinal state matrix
B_{Long}	Longitudinal control input matrix

$$\begin{bmatrix} \dot{x}_{i1}(t) \\ \dot{x}_{i2}(t) \end{bmatrix} = A_{Long} \begin{bmatrix} x_{i1}(t) \\ x_{i2}(t) \end{bmatrix} + B_{Long} \begin{bmatrix} u_{i1}(t) \\ u_{i2}(t) \end{bmatrix}$$

where the parameters are given in Table 6.1.

In the following, the velocity of 3 Mach and height of 26 kft are chosen as the flight condition of F-18 aircraft, then one can get the following system matrices A_{Long}, B_{Long} [9]:

$$A_{Long} = \begin{bmatrix} -0.2296 & 0.993 \\ -0.02436 & -0.2406 \end{bmatrix}$$

$$B_{Long} = \begin{bmatrix} -0.01434 & -0.01145 \\ -1.73 & -0.517 \end{bmatrix}$$

$F = I_2$, the state vector of agent i is $x_i(t) = \left[x_{i1}^T(t)\ x_{i2}^T(t)\right]^T$, control output of agent i is $u_i(t) = \left[u_{i1}^T(t)\ u_{i2}^T(t)\right]^T$, then $h(x_i(t), t)=$ $[0.3\sin(0.4x_{i1}(t)),\ -0.2\tan(0.1x_{i2}(t))]^T$. Suppose that there are one leader and eight followers in the agent system, whose communication topology is described as Fig. 6.3(a). Then the Laplacian L has the following form:

$$\begin{bmatrix} 2 & -1 & -1 & 0 & 0 & 0 & 0 & 0 \\ -1 & 3 & 0 & -1 & 0 & -1 & 0 & 0 \\ -1 & 0 & 1 & 0 & 0 & 0 & 0 & 0 \\ 0 & -1 & 0 & 2 & -1 & 0 & 0 & 0 \\ 0 & -1 & 0 & 0 & 2 & -1 & 0 & 0 \\ 0 & 0 & 0 & 0 & -1 & 1 & 0 & 0 \\ 0 & 0 & 0 & 0 & 0 & -1 & 1 & 0 \\ 0 & -1 & 0 & 0 & 0 & 0 & 0 & 1 \end{bmatrix}$$

Remark 6.4 Fig. 6.3(a) satisfying Assumption 6.3 shows one kind of communication topologies for MASs without DoS attacks. When DoS attacks happen, one possible topology graph of MASs is shown in Fig. 6.3(b), where there exists no directed spanning tree and the agents abandon communication. It should be pointed

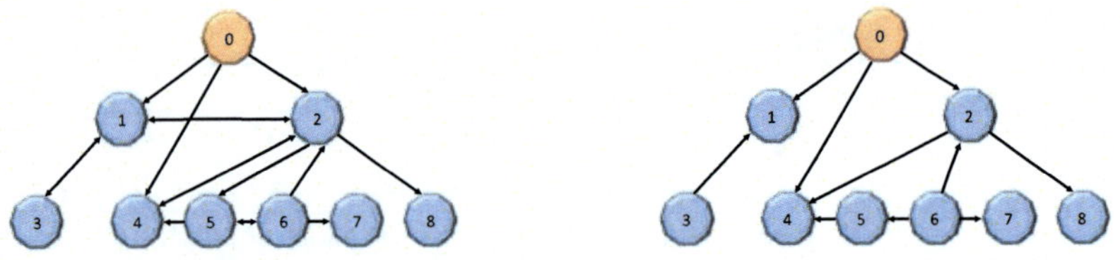

(a) Communication topology of MASs without DoS attacks (b) Communication topology of MASs with DoS attacks

Fig. 6.3 Communication topology of MASs

out that due to page limitation, in this section, suppose that the communication topology of MASs without DoS attacks is exhibited in Fig. 6.3(a). In this case, the controller gain is derived by solving the linear matrix inequalities of Theorem 2 in MATLAB. The other cases of communication topologies for MASs can be discussed and simulated in the same way.

Set $\sigma = 0.2$, $\epsilon = 0.2$, $\phi = 0.45$, $\delta_M = 0.2$, $r_M = 0.5$, $\bar{\alpha}_1 = \bar{\alpha}_5 = 0.035$, $\bar{\alpha}_2 = \bar{\alpha}_6 = \bar{\alpha}_8 = 0.04$, $\bar{\alpha}_3 = \bar{\alpha}_4 = 0.03$, $\bar{\alpha}_7 = 0.02$ and $\nu_1 = \nu_2 = \nu_3 = \nu_4 = 0.1$, $\nu_{12} = \nu_{22} = 0.1$. Let $h = 0.01$s, $D_m = 1.02$s, $D_M = 1.23$s, $\omega_1 = \omega_2 = 1.02$, and $\mu_1 = 0.08$, $\mu_2 = 1.06$. Through solving linear matrix inequalities in MATLAB, it can be obtained that

$$Y = \begin{bmatrix} -19.0258 & 63.8747 \\ 63.8747 & -214.0046 \end{bmatrix}, X = \begin{bmatrix} 49.9773 & 1.2792 \\ 1.2792 & 23.3038 \end{bmatrix}$$

$$\Phi_1 = \begin{bmatrix} 85.8297 & -5.2213 \\ -5.2213 & 65.6375 \end{bmatrix}, \Phi_2 = \begin{bmatrix} 195.4735 & 0.6162 \\ 0.6162 & 78.1812 \end{bmatrix}$$

$$\Phi_3 = \begin{bmatrix} 70.5948 & -12.7171 \\ -12.7171 & 196.1262 \end{bmatrix}, \Phi_4 = \begin{bmatrix} 151.4484 & -17.1506 \\ -17.1506 & 122.6408 \end{bmatrix}$$

$$\Phi_5 = \begin{bmatrix} 391.5966 & -0.0590 \\ -0.0590 & 391.7730 \end{bmatrix}, \Phi_6 = \begin{bmatrix} 391.5907 & -0.0400 \\ -0.0400 & 391.7090 \end{bmatrix}$$

$$\Phi_7 = \begin{bmatrix} 391.5809 & -0.0098 \\ -0.0098 & 391.6106 \end{bmatrix}, \Phi_8 = \begin{bmatrix} 391.5809 & -0.0098 \\ -0.0098 & 391.6105 \end{bmatrix}$$

Then by utilizing equality (6.30) in Theorem 6.2, we can get the controller gain

$$K = \begin{bmatrix} -0.4515 & 2.7657 \\ 1.5153 & -9.2664 \end{bmatrix}$$

The initial conditions are chosen as

$$x_0 = \begin{bmatrix} 0.4 & -0.5 \end{bmatrix}^T, x_1 = \begin{bmatrix} 0.2 & -0.1 \end{bmatrix}^T, x_2 = \begin{bmatrix} -0.2 & 0.3 \end{bmatrix}^T$$
$$x_3 = \begin{bmatrix} -0.8 & 0.5 \end{bmatrix}^T, x_4 = \begin{bmatrix} 0.1 & -0.6 \end{bmatrix}^T, x_5 = \begin{bmatrix} 0.4 & 0.1 \end{bmatrix}^T$$
$$x_6 = \begin{bmatrix} -0.5 & 0.6 \end{bmatrix}^T, x_7 = \begin{bmatrix} -0.4 & 0.9 \end{bmatrix}^T, x_8 = \begin{bmatrix} 0.5 & -0.6 \end{bmatrix}^T$$

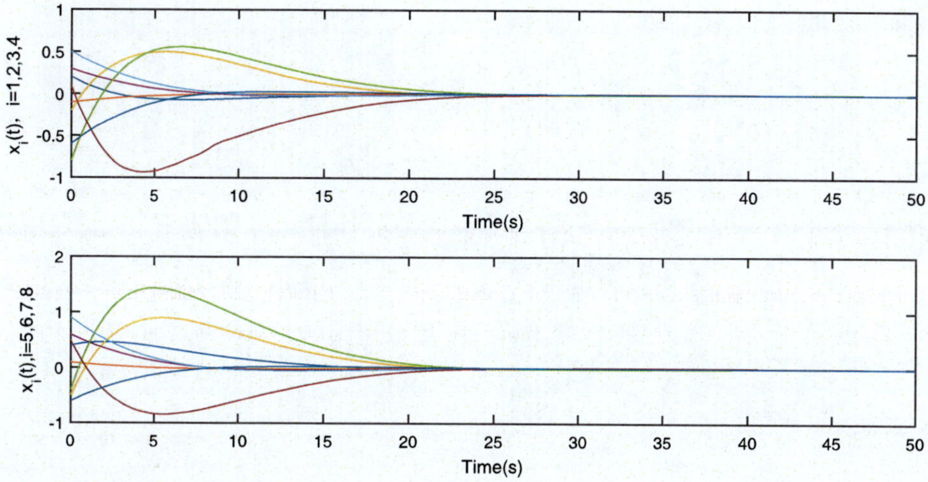

Fig. 6.4 Responses of $x_i(t)$

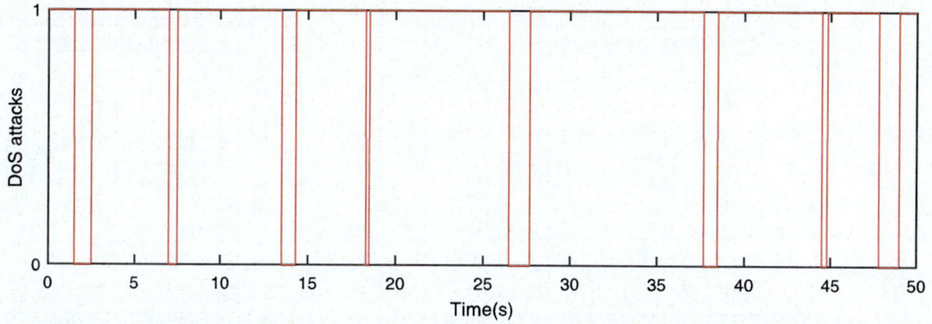

Fig. 6.5 The signal of DoS attacks

Then by using MATLAB, the simulation results are shown in Figs. 6.4, 6.5, 6.6, 6.7, 6.8, 6.9 and 6.10. The state responses of $x_i(t)$ are presented in Fig. 6.4. The DoS jamming signal with $D_M = 1.23$ and $D_m = 1.02$ is described in Fig. 6.5. Figure 6.6 presents the Bernoulli variables for replace attacks. Figures 6.7 and 6.8 exhibit the release time intervals of eight agents, respectively. Besides, the controller inputs are shown in Fig. 6.10. The consensus tracking errors $e_i(t)$ are exhibited in Fig. 6.9, which achieve stable even though the MASs are subjected to multiple cyber-attacks. Based on the above analysis, it can be concluded that the event-triggered mechanism adopted in this chapter can relieve the communication load. The discussed MASs with event-triggered mechanism and multiple cyber-attacks can reach consensus. In other words, the designed controller for MASs under multiple cyber-attacks performs well.

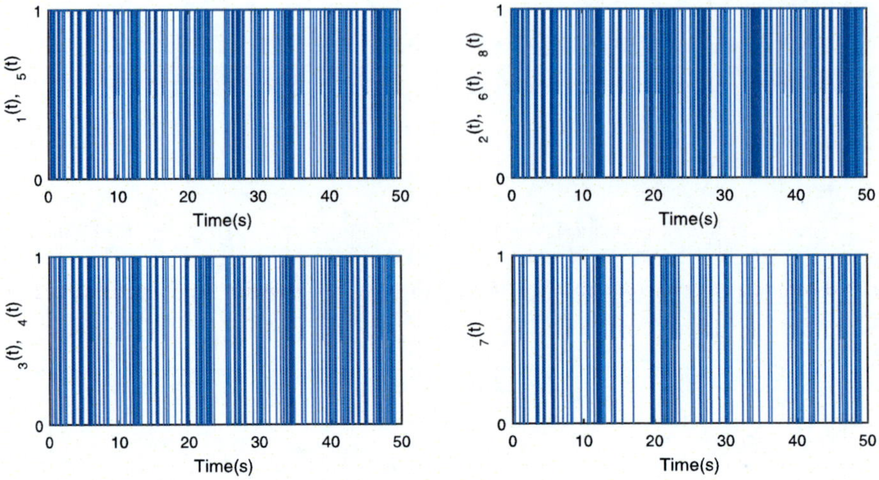

Fig. 6.6 Bernoulli variables for replay attacks

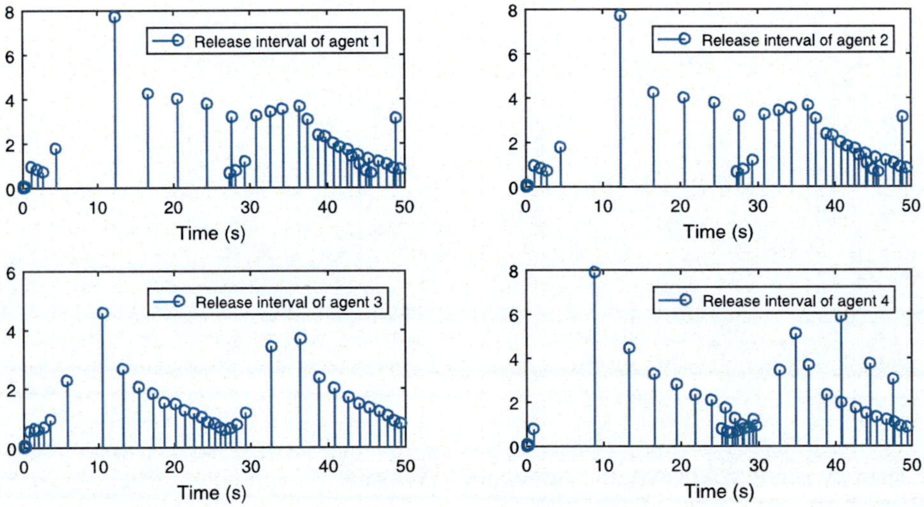

Fig. 6.7 Release instants and release intervals of Agent 1, 2, 3, 4

6.5 Conclusion

In this chapter, the event-based secure leader-following consensus control for MASs with multiple cyber-attacks has been investigated, including replay and DoS attacks. The addressed DoS attacks contribute to the changes of communication topology in MASs. To reduce network bandwidth consumption, the event-triggered mechanism is implemented. Then the event-based consensus control protocol is firstly proposed for MASs under multiple cyber-attacks. Moreover, the sufficient conditions are derived

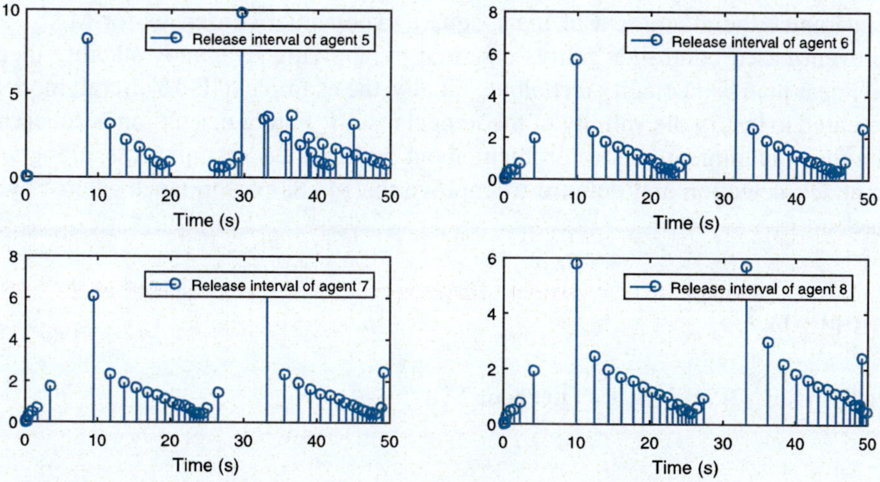

Fig. 6.8 Release instants and release intervals of Agent 5, 6, 7, 8

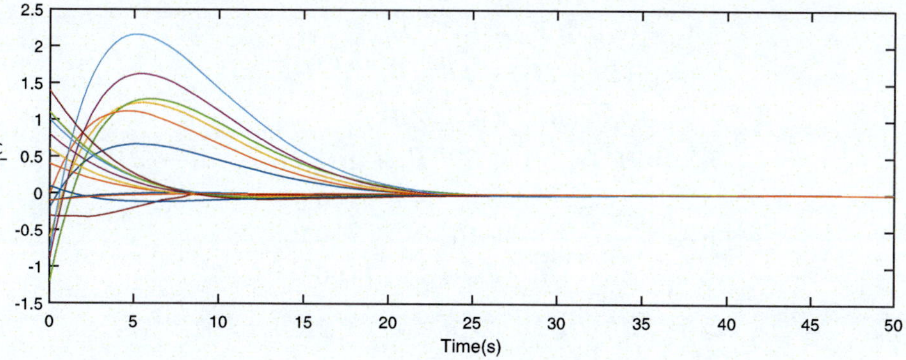

Fig. 6.9 Responses of errors $e_i(t)$

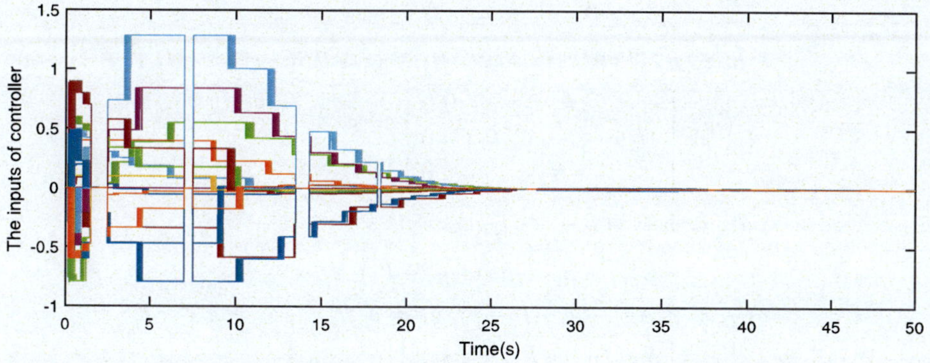

Fig. 6.10 Responses of controller inputs

to guarantee the attainment of mean-square exponential consensus for MASs and the event-based controller gain is acquired by applying Lyapunov stability theory and linear matrix inequality technique. Finally, the example of F-18 aircraft model is presented to testify the validity of theoretical results. Future discussion is concerned with the containment control problem about multiple leaders in MASs, along with the attack detection and defense to improve the MASs performance against cyber attacks.

Appendix I

Elements of Π_1 and Π_2 in Theorem 1

$$\Gamma_{11}^1 = \begin{bmatrix} b_1 & * & * \\ b_2 & b_3 & * \\ -g_1 M_1 & g_1(R_1 + M_1) & -g_1 Q_1 - g_1 R_1 \end{bmatrix}$$

$$\begin{aligned} b_1 &= 2\mu_1(I_{2N} \otimes P_1) + Q_1 + Z_1 - g_1 R_1 - g_2 S_1 \\ &\quad +(I_{2N} \otimes P_1)\bar{A} + \bar{A}^T(I_{2N} \otimes P_1) \\ b_2 &= g_1(R_1 + M_1) - \phi \bar{K}^T(\bar{B}_1^{\tau(t)})^T(I_{2N} \otimes P_1) \\ &\quad -\phi \bar{K}^T(\bar{B}_3^{\tau(t)})^T(I_{2N} \otimes P_1) \\ b_3 &= \sigma H_1^T(\mathbb{A}^{\tau(t)})^T \Phi \mathbb{A}^{\tau(t)} H_1 + g_1(-2R_1 - M_1 - M_1^T) \end{aligned}$$

$$\Gamma_{21}^1 = \begin{bmatrix} b_4 & 0 & 0 \\ -g_2 U_1 & 0 & 0 \\ \phi \bar{K_1}^T(\bar{B}_4^{\tau(t)})^T \bar{\alpha}(I_{2N} \otimes P_1) & 0 & 0 \end{bmatrix}$$

$$\begin{aligned} b_4 &= \phi \bar{K}^T(\bar{B}_3^{\tau(t)})^T \bar{\alpha}(I_{2N} \otimes P_1) + g_2(S_1 + U_1) \\ \Phi &= diag\{\Phi_1, \Phi_2, \ldots, \Phi_N\} \end{aligned}$$

$$\Gamma_{22}^1 = \begin{bmatrix} b_5 & * & * \\ g_2(S_1 + U_1) & -g_2 Z_1 - g_2 S_1 & * \\ 0 & 0 & -\Phi \end{bmatrix}$$

$$b_5 = \sigma H_1^T(\mathbb{A}^{\tau(t)})^T \Phi \mathbb{A}^{\tau(t)} H_1 + g_1(-2S_1 - U_1 - U_1^T)$$

$$\Gamma_{31}^1 = \begin{bmatrix} b_6 & 0 & 0 \\ H_1 \bar{F}^T(I_{2N} \otimes P_1) & 0 & 0 \\ \epsilon H_2 & 0 & 0 \end{bmatrix}$$

$$\begin{aligned} b_6 &= \phi \bar{K_1}^T(\bar{B}_4^{\tau(t)})^T(I_{2N} \otimes P_1) \\ &\quad -\phi \bar{K_1}^T(\bar{B}_2^{\tau(t)})^T \bar{\alpha}(I_{2N} \otimes P_1) \\ \Gamma_{33} &= diag\{-\Phi, -I, -I\}, g_1 = e^{-2\mu_1 \delta_M}, g_2 = e^{-2\mu_1 r_M} \end{aligned}$$

$$\Gamma_{41}^1 = \begin{bmatrix} \delta_M(I_{2N} \otimes P_1)\bar{A} & b_7 & 0 \\ r_M(I_{2N} \otimes P_1)\bar{A} & b_8 & 0 \\ 0 & b_9 & 0 \\ 0 & b_{10} & 0 \end{bmatrix}$$

$$
\begin{aligned}
b_7 &= -\delta_M \phi (I_{2N} \otimes P_1) \bar{B}_1^{\tau(t)} \bar{K} \\
&\quad -\delta_M \phi (I_{2N} \otimes P_1) \bar{\alpha} \bar{B}_3^{\tau(t)} \bar{K} \\
b_8 &= -r_M \phi (I_{2N} \otimes P_1) \bar{B}_1^{\tau(t)} \bar{K} \\
&\quad -r_M \phi (I_{2N} \otimes P_1) \bar{\alpha} \bar{B}_3^{\tau(t)} \bar{K} \\
b_9 &= \delta_M \phi (I_{2N} \otimes P_1) \rho \bar{B}_3^{\tau(t)} \bar{K} \\
b_{10} &= r_M \phi (I_{2N} \otimes P_1) \rho \bar{B}_3^{\tau(t)} \bar{K} \\
\Gamma_{42}^1 &= \begin{bmatrix} \delta_M \phi (I_{2N} \otimes P_1) \bar{\alpha} \bar{B}_3^{\tau(t)} \bar{K} & 0 & \delta_M \phi P_{140} \\ r_M \phi (I_{2N} \otimes P_1) \bar{\alpha} \bar{B}_3^{\tau(t)} \bar{K} & 0 & r_M \phi P_{140} \\ -\delta_M \phi (I_{2N} \otimes P_1) \rho \bar{B}_3^{\tau(t)} \bar{K} & 0 & -\delta_M \phi P_{141} \\ -r_M \phi (I_{2N} \otimes P_1) \rho \bar{B}_3^{\tau(t)} \bar{K} & 0 & -r_M \phi P_{141} \end{bmatrix}
\end{aligned}
$$

$$
\begin{aligned}
P_{140} &= (I_{2N} \otimes P_1) \bar{\alpha} \bar{B}_4^{\tau(t)} \bar{K}_1, \; P_{141} = (I_{2N} \otimes P_1) \rho \bar{B}_4^{\tau(t)} \bar{K}_1 \\
\Gamma_{43}^1 &= \begin{bmatrix} \delta_M \phi b_{11} & \delta_M (I_{2N} \otimes P_1) \bar{F} H_3 & 0 \\ r_M \phi b_{12} & r_M (I_{2N} \otimes P_1) \bar{F} H_3 & 0 \\ \delta_M \phi P_{141} & 0 & 0 \\ r_M \phi P_{141} & 0 & 0 \end{bmatrix} \\
b_{11} &= (I_{2N} \otimes P_1) \bar{B}_2^{\tau(t)} \bar{K}_1 - (I_{2N} \otimes P_1) \bar{\alpha} \bar{B}_4^{\tau(t)} \bar{K}_1 \\
\bar{\alpha} &= diag\{\bar{\alpha}_1, \ldots, \bar{\alpha}_N, \bar{\alpha}_1, \ldots, \bar{\alpha}_N\} \\
\rho &= diag\{\rho_1, \ldots, \rho_N, \rho_1, \ldots, \rho_N\}, \; \rho_i = \sqrt{\bar{\alpha}_i (1 - \bar{\alpha}_i)} \\
\Gamma_{44}^1 &= diag \left\{ (I_{2N} \otimes P_1)^{-1} R_1 (I_{2N} \otimes P_1), \right. \\
&\quad (I_{2N} \otimes P_1)^{-1} S_1 (I_{2N} \otimes P_1), \\
&\quad (I_{2N} \otimes P_1)^{-1} R_1 (I_{2N} \otimes P_1), \\
&\quad \left. (I_{2N} \otimes P_1)^{-1} S_1 (I_{2N} \otimes P_1) \right\} \\
\bar{K} &= I_{2N} \otimes K, \; \bar{K}_1 = I_N \otimes K \\
\Gamma_{11}^2 &= \begin{bmatrix} \mathbf{b}_1 & * & * \\ R_2 + M_2 & -2R_2 - M_2 - M_2^T & * \\ -M_2 & R_2 + M_2 & -g_3 Q_2 - R_2 \end{bmatrix} \\
\mathbf{b}_1 &= -2\mu_2 (I_{2N} \otimes P_2) + Q_2 + Z_2 - R_2 - S_2 \\
&\quad + (I_{2N} \otimes P_2) \bar{A} + \bar{A}^T (I_{2N} \otimes P_2) \\
\Gamma_{21}^2 &= \begin{bmatrix} S_2 + U_2 & 0 & 0 \\ -U_2 & 0 & 0 \\ H_1 \bar{F}^T (I_{2N} \otimes P_2) & 0 & 0 \\ \epsilon H_2 & 0 & 0 \end{bmatrix} \\
\Gamma_{22}^2 &= \begin{bmatrix} -2S_2 - U_2 - U_2^T & * & * & * \\ S_2 + U_2 & \mathbf{b}_2 & * & * \\ 0 & 0 & -I & * \\ 0 & 0 & 0 & -I \end{bmatrix}
\end{aligned}
$$

$$\mathbf{b}_2 = -g_4 Z_2 - S_2, \ g_3 = e^{2\mu_2 \delta_M}, \ g_4 = e^{2\mu_2 r_M}$$
$$\Gamma_{31}^2 = \begin{bmatrix} \delta_M (I_{2N} \otimes P_2)\bar{A} \ 0 \ 0 \\ r_M (I_{2N} \otimes P_2)\bar{A} \ 0 \ 0 \end{bmatrix}$$
$$\Gamma_{32}^2 = \begin{bmatrix} 0 \ 0 \ \delta_M (I_{2N} \otimes P_2)\bar{F} H_3 \ 0 \\ 0 \ 0 \ r_M (I_{2N} \otimes P_2)\bar{F} H_3 \ 0 \end{bmatrix}$$
$$\Gamma_{33}^2 = diag\left\{(I_{2N} \otimes P_2)^{-1} R_2 (I_{2N} \otimes P_2), \right.$$
$$\left. (I_{2N} \otimes P_2)^{-1} S_2 (I_{2N} \otimes P_2)\right\}$$

Appendix II

The proof of Theorem 1 The following time-varying Lyapunov functional is constructed for system (6.11):

$$\begin{aligned} V_{\gamma(t)}(t) =& \eta^T(t)(I_{2N} \otimes P_{\gamma(t)})\eta(t) \\ &+ \int_{t-\delta_M}^{t} \theta(\cdot)\eta^T(v) Q_{\gamma(t)} \eta(v) dv \\ &+ \int_{t-r_M}^{t} \theta(\cdot)\eta^T(v) Z_{\gamma(t)} \eta(v) dv \\ &+ \delta_M \int_{-\delta_M}^{0} \int_{t+s}^{t} \theta(\cdot)\dot{\eta}^T(v) R_{\gamma(t)} \dot{\eta}(v) dv ds \\ &+ r_M \int_{-r_M}^{0} \int_{t+s}^{t} \theta(\cdot)\dot{\eta}^T(v) S_{\gamma(t)} \dot{\eta}(v) dv ds \end{aligned} \tag{6.31}$$

where $P_{\gamma(t)}>0$, $Q_{\gamma(t)}>0$, $Z_{\gamma(t)}>0$, $R_{\gamma(t)}>0$, $S_{\gamma(t)} > 0$, and $\theta(\cdot) \triangleq e^{(-1)^{\gamma(t)} 2\mu_{\gamma(t)}(t-v)}$; $\gamma(t) = 1$ when $t \in \mathcal{H}_{1,n}$ and $\gamma(t) = 2$ when $t \in \mathcal{H}_{2,n}$, $n \in \mathbb{N}$.

In the following, the discussion on two cases of $\gamma(t) = 1$ and $\gamma(t) = 2$ will be given respectively.

When $\gamma(t) = 1$, for $t \in \mathcal{H}_{1,n}$, $n \in \mathbb{N}$, calculate the derivation and mathematical expectation of $V_1(t)$. In addition, it follows from the inequality (6.9) that

$$\begin{aligned} &\sigma \eta^T(t - \delta_{\lambda,n}(t)) H_1^T (\mathbb{A}^{\tau(t)})^T \Phi \mathbb{A}^{\tau(t)} H_1 \eta(t - \delta_{\lambda,n}(t)) \\ &-\psi_{\lambda,n}^T(t) \Phi \psi_{\lambda,n}(t) \geq 0 \end{aligned} \tag{6.32}$$

where $\Phi = diag\{\Phi_1, \Phi_2, \ldots, \Phi_N\}$.

From Assumption 6.1, it is easy to achieve that

$$\epsilon^2 \eta^T(t) H_2^T H_2 \eta(t) - \tilde{F}^T(x(t), t) H_2^T H_2 \tilde{F}(x(t), t) \geq 0 \tag{6.33}$$

Combining (6.32) and (6.33), then by applying Jensen's inequality [?] and Schur complement, we have

$$\mathbf{E}\{\dot{V}_1(t)\} \leq -2\mu_1\mathbf{E}\{V_1(t)\} + \varphi_1^T(t)\Pi_1\varphi_1(t) \tag{6.34}$$

where $\varphi_1(t) = [\eta^T(t) \quad \eta^T(t-\delta_{\lambda,n}(t)) \quad \eta^T(t-\delta_M) \quad \eta^T(t-r_{\lambda,n}(t)) \quad \eta^T(t-r_M) \quad \psi_{\lambda,n}^T(t_r) \; \psi_{\lambda,n}^T(t) \; (H_1^T\tilde{F}(x(t),t))^T \quad I \; I \; I \; I \; I]^T$. Due to $\Pi_1 < 0$, it can be acquired that for $t \in \mathcal{H}_{1,n}$, $n \in \mathbb{N}$, $\mathbf{E}\{\dot{V}_1(t)\} \leq -2\mu_1\mathbf{E}\{V_1(t)\}$ holds.

When $\gamma(t) = 2$, $t \in \mathcal{H}_{2,n}$, $n \in \mathbb{N}$, take the derivation and expectation of $V_2(t)$, and similarly employ the aforementioned method, then we can obtain

$$\mathbf{E}\{\dot{V}_2(t)\} \leq 2\mu_2\mathbf{E}\{V_2(t)\} + \varphi_2^T(t)\Pi_2\varphi_2(t) \tag{6.35}$$

where $\varphi_2(t) = [\eta^T(t) \quad \eta^T(t-\delta_{\lambda,n}(t)) \quad \eta^T(t-\delta_M) \quad \eta^T(t-r_{\lambda,n}(t)) \quad \eta^T(t-r_M) \quad (H_1^T\tilde{F}(x(t),t))^T \quad I \; I \; I]^T$. By applying $\Pi_2 < 0$, it follows that for $t \in \mathcal{H}_{2,n}$, $n \in \mathbb{N}$, $\mathbf{E}\{\dot{V}_2(t)\} \leq 2\mu_2\mathbf{E}\{V_2(t)\}$ holds.

For convenience, define $\ell_{\jmath,n} = \begin{cases} a_n, & \jmath = 1, \\ a_n + \varrho_n, & \jmath = 2 \end{cases}$, then one can get $\mathcal{H}_{\jmath,n} = [\ell_{\jmath,n}, \ell_{3-\jmath,\jmath-1+n})$, then it yields that $\mathbf{E}\{V_\jmath(t)\} \leq e^{2(-1)^\jmath\mu_\jmath(t-\ell_{\jmath,n})}\mathbf{E}\{V_\jmath(\ell_{\jmath,n})\}$ holds for $t \in [\ell_{\jmath,n}, \ell_{3-\jmath,\jmath+n-1})$, $\jmath \in \{1, 2\}$.

Besides, owing to the inequalities in (6.18) holding, then by some simple calculations, we can achieve the following inequalities:

$$\begin{cases} \mathbf{E}\{V_1(\ell_{1,n})\} \leq \omega_2\mathbf{E}\{V_2(\ell_{1,n}^-)\} \\ \mathbf{E}\{V_2(\ell_{2,n})\} \leq \omega_1 e^{2\varsigma h}\mathbf{E}\{V_1(\ell_{2,n}^-)\}, \; \varsigma = \mu_1 + \mu_2 \end{cases} \tag{6.36}$$

If $t \in [\ell_{1,n}, \ell_{2,n})$, according to (6.36), it yields that $\mathbf{E}\{V(t)\} \leq \omega_2 e^{-2\mu_1(t-\ell_{1,n})}\mathbf{E}\{V_2(\ell_{1,n}^-)\}$, then $\mathbf{E}\{V_2(\ell_{1,n}^-)\} \leq e^{2\mu_2(\ell_{1,n}-\ell_{2,n-1})}\mathbf{E}\{V_2(\ell_{2,n-1})\}$ can be derived due to $\ell_{1,n}^- \in [\ell_{2,n-1}, \ell_{1,n})$. By reiterating this process, then we can obtain

$$\mathbf{E}\{V(t)\} \leq e^{o(t)}\mathbf{E}\{V_1(0)\} \tag{6.37}$$

where $o(t) = (b_0 + \frac{t}{f_a})[2(\varsigma h + \mu_2 D_M - 2\mu_1 D_m) + \ln(\omega_1\omega_2)]$. Due to the inequality (6.20) holding, it is easy to acquire that

$$\mathbf{E}\{V(t)\} \leq e^{\xi_1}e^{-\upsilon t}\mathbf{E}\{V_1(0)\} \tag{6.38}$$

where $\xi_1 = b_0[2(\varsigma h + \mu_2 D_M - \mu_1 D_m) + \ln(\omega_1\omega_2)]$, $\upsilon = \frac{1}{f_a}[(-\varsigma h - \mu_2 D_M + \mu_1 D_m) + \frac{1}{2}\ln(\omega_1\omega_2)]$. Then by adopting the similar method and combining inequality (6.20), the following inequality can be obtained:

$$\mathbf{E}\{V(t)\} \leq \frac{e^{\xi_2}}{\omega_2} e^{-\upsilon t} \mathbf{E}\{V_1(0)\} \tag{6.39}$$

where $\xi_2 = (b_0 + 1)[2(\varsigma h + \mu_2 D_M - \mu_1 D_m) + \ln(\omega_1 \omega_2)]$.

Moreover, it yields that

$$\mathbf{E}\{V(t)\} \leq \max\left\{e^{\xi_1}, \frac{e^{\xi_2}}{\omega_2}\right\} e^{-\upsilon t} \mathbf{E}\{V_1(0)\} \tag{6.40}$$

It follows from the definition of $V(t)$ that

$$\begin{cases} \mathbf{E}\{V(t)\} \geq \varphi_{min} \mathbf{E}\left\{\| \eta(t) \|^2\right\} \\ \mathbf{E}\{V_1(0)\} \leq \varphi_{max} \mathbf{E}\left\{\| \zeta_0 \|^2\right\} \end{cases} \tag{6.41}$$

where $\varphi_{min} = \min\{\varpi_n(I_{2N} \otimes P_i)\}$ and $\varphi_{max} = \max\{\varpi_m(I_{2N} \otimes P_i)\} + h\varpi_m(Q_1) + \frac{h^2}{2}\varpi_m(R_1 + S_1)$.

Combining (6.40) and (6.41), it is easy to derive that

$$\forall t \geq 0, \mathbf{E}\left\{\| \eta(t) \|^2\right\} \leq \varepsilon e^{-\frac{\varpi}{2}t} \mathbf{E}\left\{\| \zeta_0 \|^2\right\} \tag{6.42}$$

where $\varepsilon = \frac{\varphi_{max}}{\varphi_{min}} \cdot \max\left\{e^{\xi_1}, \frac{e^{\xi_2}}{\omega_2}\right\}$. According to inequality (6.42) and Definition 6.1, it can be concluded that if the inequalities (6.15)–(6.20) hold, system (12.19) reaches mean-square exponential stability with decay rate $\frac{\varpi}{2}$. That completes the proof.

Appendix III

Elements of $\hat{\Pi}_1$ and $\hat{\Pi}_2$ in Theorem 2

$$\hat{\Gamma}_{11}^1 = \begin{bmatrix} c_1 & * & * \\ c_2 & c_3 & * \\ -g_\varepsilon \hat{M}_1 & g_1(\hat{R}_1 + \hat{M}_1) & -g_1\hat{Q}_1 - g_1\hat{R}_1 \end{bmatrix}$$

$$c_1 = 2\mu_1 \hat{X}_1 + \hat{Q}_1 + \hat{Z}_1 - g_1\hat{R}_1 - g_2\hat{S}_1 + \bar{A}\hat{X}_1 + \hat{X}_1\bar{A}^T$$

$$c_2 = g_1(\hat{R}_1 + \hat{M}_1) - \phi\bar{Y}^T(\bar{B}_1^{\tau(t)})^T - \phi\bar{Y}^T(\bar{B}_3^{\tau(t)})^T$$

$$c_3 = \sigma H_1^T(\mathbb{A}^{\tau(t)})^T \Phi \mathbb{A}^{\tau(t)} H_1 + g_1(-2\hat{R}_1 - \hat{M}_1 - \hat{M}_1^T)$$

$$\hat{\Phi} = diag\{\hat{\Phi}_1, \hat{\Phi}_2, \ldots, \hat{\Phi}_N\}$$

$$\hat{\Gamma}_{21}^1 = \begin{bmatrix} \phi\bar{Y}^T(\bar{B}_3^{\tau(t)})^T\bar{\alpha} + g_2(\hat{S}_1 + \hat{U}_1) & 0 & 0 \\ -g_2\hat{U}_1 & 0 & 0 \\ \phi\bar{Y}_1{}^T(\bar{B}_4^{\tau(t)})^T\bar{\alpha} & 0 & 0 \end{bmatrix}$$

$$\hat{\Gamma}_{22}^1 = \begin{bmatrix} c_4 & * & * \\ g_2(\hat{S}_1 + \hat{U}_1) & -g_2\hat{Z}_1 - g_2\hat{S}_1 & * \\ 0 & 0 & -\hat{\Phi} \end{bmatrix}$$

$$c_4 = \sigma H_1^T (\mathbb{A}^{\tau(t)})^T \Phi \mathbb{A}^{\tau(t)} H_1 + g_1(-2\hat{S}_1 - \hat{U}_1 - \hat{U}_1^T)$$

$$\hat{\Gamma}_{31}^1 = \begin{bmatrix} \phi\bar{Y}_1{}^T(\bar{B}_4^{\tau(t)})^T - \phi\bar{Y}_1{}^T(\bar{B}_2^{\tau(t)})^T\bar{\alpha} & 0 & 0 \\ H_1\bar{F}^T & 0 & 0 \\ \epsilon H_2\hat{X}_1 & 0 & 0 \end{bmatrix}$$

$$\hat{\Gamma}_{33} = diag\{-\hat{\Phi}, -I, -I\},\ g_1 = e^{-2\mu_1\delta_M},\ g_2 = e^{-2\mu_1 r_M}$$

$$\hat{\Gamma}_{41}^1 = \begin{bmatrix} \delta_M\bar{A}\hat{X}_1 & c_5 & 0 \\ r_M\bar{A}\hat{X}_1 & c_6 & 0 \\ 0 & \delta_M\phi\rho\bar{B}_3^{\tau(t)}\bar{Y} & 0 \\ 0 & r_M\phi\rho\bar{B}_3^{\tau(t)}\bar{Y} & 0 \end{bmatrix}$$

$$c_5 = -\delta_M\phi\bar{B}_1^{\tau(t)}\bar{Y} - \delta_M\phi\bar{B}_3^{\tau(t)}\bar{\alpha}\bar{Y}$$

$$c_6 = -r_M\phi\bar{B}_1^{\tau(t)}\bar{Y} - r_M\phi\bar{\alpha}\bar{B}_3^{\tau(t)}\bar{Y}$$

$$\hat{\Gamma}_{42}^1 = \begin{bmatrix} \delta_M\phi\bar{\alpha}\bar{B}_3^{\tau(t)}\bar{Y} & 0 & \delta_M\phi\bar{\alpha}\bar{B}_4^{\tau(t)}\bar{Y}_1 \\ r_M\phi\bar{\alpha}\bar{B}_3^{\tau(t)}\bar{Y} & 0 & r_M\phi\bar{\alpha}\bar{B}_4^{\tau(t)}\bar{Y}_1 \\ -\delta_M\phi\rho\bar{B}_3^{\tau(t)}\bar{Y} & 0 & -\delta_M\phi\rho\bar{B}_4^{\tau(t)}\bar{Y}_1 \\ -r_M\phi\rho\bar{B}_3^{\tau(t)}\bar{Y} & 0 & -r_M\phi\rho\bar{B}_4^{\tau(t)}\bar{Y}_1 \end{bmatrix}$$

$$\hat{\Gamma}_{43}^1 = \begin{bmatrix} \delta_M\phi\bar{B}_2^{\tau(t)}\bar{Y}_1 - \delta_M\phi\bar{\alpha}\bar{B}_4^{\tau(t)}\bar{Y}_1 & \delta_M\bar{F}H_3 & 0 \\ r_M\phi\bar{B}_2^{\tau(t)}\bar{Y}_1 - r_M\phi\bar{\alpha}\bar{B}_4^{\tau(t)}\bar{Y}_1 & r_M\bar{F}H_3 & 0 \\ \delta_M\phi\rho\bar{B}_4^{\tau(t)}\bar{Y}_1 & 0 & 0 \\ r_M\phi\rho\bar{B}_4^{\tau(t)}\bar{Y}_1 & 0 & 0 \end{bmatrix}$$

$$\bar{\alpha} = diag\{\bar{\alpha}_1, \ldots, \bar{\alpha}_N, \bar{\alpha}_1, \ldots, \bar{\alpha}_N\}$$

$$\rho = diag\{\rho_1, \ldots, \rho_N, \rho_1, \ldots, \rho_N\},\ \rho_i = \sqrt{\bar{\alpha}_i(1-\bar{\alpha}_i)}$$

$$\hat{\Gamma}_{44}^1 = diag\Big\{-2\nu_1\hat{X}_1 + \nu_1^2\hat{R}_1, -2\nu_2\hat{X}_1 + \nu_2^2\hat{R}_1, -2\nu_1\hat{X}_1 + \nu_1^2\hat{R}_1, -2\nu_2\hat{X}_1 + \nu_2^2\hat{R}_1\Big\}$$

$$\bar{Y} = I_{2N} \otimes Y,\ \bar{Y}_1 = I_N \otimes Y$$

$$\hat{X}_1 = I_{2N} \otimes X_1,\ \hat{X}_2 = I_{2N} \otimes X_2$$

$$\hat{\Gamma}_{11}^2 = \begin{bmatrix} \mathbf{c}_1 & * & * \\ \hat{R}_2 + \hat{M}_2 & -2\hat{R}_2 - \hat{M}_2 - \hat{M}_2^T & * \\ -\hat{M}_2 & \hat{R}_2 + \hat{M}_2 & -g_3\hat{Q}_2 - \hat{R}_2 \end{bmatrix}$$

$$\mathbf{c}_1 = -2\mu_2\hat{X}_2 + \hat{Q}_2 + \hat{Z}_2 - \hat{R}_2 - \hat{S}_2 + \bar{A}\hat{X}_2 + \hat{X}_2\bar{A}^T$$

$$\hat{\Gamma}_{21}^2 = \begin{bmatrix} \hat{S}_2 + \hat{U}_2 & 0 & 0 \\ -\hat{U}_2 & 0 & 0 \\ H_1 \bar{F}^T & 0 & 0 \\ \epsilon H_2 \hat{X}_2 & 0 & 0 \end{bmatrix}$$

$$\hat{\Gamma}_{22}^2 = \begin{bmatrix} -2\hat{S}_2 - \hat{U}_2 - \hat{U}_2^T & * & * & * \\ \hat{S}_2 + \hat{U}_2 & \mathbf{c}_2 & * & * \\ 0 & 0 & -I & * \\ 0 & 0 & 0 & -I \end{bmatrix}$$

$$\mathbf{c}_2 = -g_4 \hat{Z}_2 - \hat{S}_2,\ g_3 = e^{2\mu_2 \delta_M},\ g_4 = e^{2\mu_2 r_M}$$

$$\hat{\Gamma}_{31}^2 = \begin{bmatrix} \delta_M \bar{A} \hat{X}_2 & 0 & 0 \\ r_M \bar{A} \hat{X}_2 & 0 & 0 \end{bmatrix},\ \hat{\Gamma}_{32}^2 = \begin{bmatrix} 0 & 0 & \delta_M \bar{F} H_3 & 0 \\ 0 & 0 & r_M \bar{F} H_3 & 0 \end{bmatrix}$$

$$\hat{\Gamma}_{33}^2 = diag\left\{-2\nu_3 \hat{X}_2 + \nu_3^2 \hat{R}_2,\ -2\nu_4 \hat{X}_2 + \nu_4^2 \hat{R}_2\right\}$$

Appendix IV

The proof of Theorem 2 Due to $(R_1 - \nu_1^{-1}(I_{2N} \otimes P_1))R_1^{-1}(R_1 - \nu_1^{-1}(I_{2N} \otimes P_1)) \geq 0$, it is easy to obtain that

$$-(I_{2N} \otimes P_1)R_1^{-1}(I_{2N} \otimes P_1) \leq -2\nu_1(I_{2N} \otimes P_1) + \nu_1^2 R_1 \tag{6.43}$$

Similarly, we can get the following inequalities:

$$\begin{cases} -(I_{2N} \otimes P_1)S_1^{-1}(I_{2N} \otimes P_1) \leq -2\nu_2(I_{2N} \otimes P_1) + \nu_2^2 S_1 \\ -(I_{2N} \otimes P_2)R_2^{-1}(I_{2N} \otimes P_2) \leq -2\nu_3(I_{2N} \otimes P_2) + \nu_3^2 R_2 \\ -(I_{2N} \otimes P_2)S_2^{-1}(I_{2N} \otimes P_2) \leq -2\nu_4(I_{2N} \otimes P_2) + \nu_4^2 S_2 \end{cases}$$

Then replacing the terms in Γ_{44}^1 and Γ_{33}^2 with $-2\nu_1(I_{2N} \otimes P_1)+\nu_1^2 R_1, -2\nu_2(I_{2N} \otimes P_1) + \nu_2^2 S_1$ and $-2\nu_3(I_{2N} \otimes P_2) + \nu_3^2 R_2, -2\nu_4(I_{2N} \otimes P_2) + \nu_4^2 S_2$, it follows that

$$\begin{cases} \Gamma_{44}^1 = diag\left\{-2\nu_1(I_{2N} \otimes P_1) + \nu_1^2 R_1,\right. \\ -2\nu_2(I_{2N} \otimes P_1) + \nu_2^2 S_1, -2\nu_1(I_{2N} \otimes P_1) + \nu_1^2 R_1, \\ \left. -2\nu_2(I_{2N} \otimes P_1) + \nu_2^2 S_1\right\} \\ \Gamma_{33}^2 = diag\left\{-2\nu_3(I_{2N} \otimes P_2) + \nu_3^2 R_2,\right. \\ \left. -2\nu_4(I_{2N} \otimes P_2) + \nu_4^2 S_2\right\} \end{cases}$$

Let $X_1 = P_1^{-1}, \hat{X}_1 = I_{2N} \otimes X_1, \bar{X}_1 = I_N \otimes X_1$, $\Delta_1 = diag\{\hat{X}_1, \hat{X}_1, \hat{X}_1, \hat{X}_1, \bar{X}_1, I, I, \hat{X}_1, \hat{X}_1, \hat{X}_1, \hat{X}_1\}$, then pre- and post-multiply Π_1 with Δ_1 and Δ_1^T. Define $Y = KX_1, \bar{Y} = I_{2N} \otimes Y, \bar{Y}_1 = I_N \otimes Y, \hat{Q}_1 = \hat{X}_1 Q_1 \hat{X}_1$,

$\hat{R}_1 = \hat{X}_1 R_1 \hat{X}_1, \hat{Z}_1 = \hat{X}_1 Z_1 \hat{X}_1, \hat{S}_1 = \hat{X}_1 S_1 \hat{X}_1, \hat{M}_1 = \hat{X}_1 M_1 \hat{X}_1, \hat{U}_1 = \hat{X}_1 U_1 \hat{X}_1, \hat{\Phi} = \bar{X}_1 \Phi \bar{X}_1$, then we have $\hat{\Pi}_1$, further, it can be acquired that $\mathbf{E}\{\dot{V}_1(t)\} < 0$.

In addition, denote $X_2 = P_2^{-1}$, $\hat{X}_2 = I_{2N} \otimes X_2$, $\Delta_2 = diag\{\hat{X}_2, \hat{X}_2, \hat{X}_2, \hat{X}_2, \hat{X}_2, I, I, \hat{X}_2, \hat{X}_2\}$, then pre- and post-multiply Π_2 with Δ_2 and Δ_2^T. Define $\hat{Q}_2 = \hat{X}_2 Q_2 \hat{X}_2$, $\hat{R}_2 = \hat{X}_2 R_2 \hat{X}_2$, $\hat{Z}_2 = \hat{X}_2 Z_2 \hat{X}_2$, $\hat{S}_2 = \hat{X}_2 S_2 \hat{X}_2$, $\hat{M}_2 = \hat{X}_2 M_2 \hat{X}_2$, $\hat{U}_2 = \hat{X}_2 U_2 \hat{X}_2$, then we can get $\hat{\Pi}_2$, moreover, one can easily obtain that $\mathbf{E}\{\dot{V}_2(t)\} < 0$.

For the first inequality in (6.18), employing Schur complement, then by multiplying Ω and Ω^T ($\Omega = diag\{\hat{X}_2, \hat{X}_2\}$) on its both sides, we can get the inequality (6.24). Applying the similar method, it follows from the inequalities in (6.18) and (6.19) that inequalities (6.25)–(6.29) hold. Based on the results of Theorem 1, one can obtain that the system (12.19) achieves exponentially stability in mean-square. Owing to $Y = KX_1$, it is easy to derive that the controller gain is $K = YX_1^{-1}$. This completes the proof.

References

1. J. Liu, T. Yin, D. Yue, H.R. Karimi, J. Cao, Event-based secure leader-following consensus control for multiagent systems with multiple cyber attacks. IEEE Trans. Cybern. 1–12 (2020). https://doi.org/10.1109/TCYB.2020.2970556
2. X. Yin, D. Yue, S. Hu, Adaptive periodic event-triggered consensus for multi-agent systems subject to input saturation. Int. J. Control **89**(4), 653–667 (2016)
3. B. Chen, D.W. Ho, G. Hu, L. Yu, Secure fusion estimation for bandwidth constrained cyber-physical systems under replay attacks. IEEE Trans. Cybern. **48**(6), 1862–1876 (2017)
4. J. Liu, Z.-G. Wu, D. Yue, J.H. Park, Stabilization of networked control systems with hybrid-driven mechanism and probabilistic cyber attacks. IEEE Trans. Syst., Man, Cybern.: Syst. (2019). https://doi.org/10.1109/TSMC.2018.2888633
5. S. Hu, D. Yue, Q.-L. Han, X. Xie, X. Chen, C. Dou, Observer-based event-triggered control for networked linear systems subject to denial-of-service attacks. IEEE Trans. Cybern. **50**(5), 1952–1964 (2020)
6. Z. Feng, G. Hu, G. Wen, Distributed consensus tracking for multi-agent systems under two types of attacks. Int. J. Robust Nonlinear Control **26**(5), 896–918 (2016)
7. R.J. Adams, J.M. Buffington, A.G. Sparks, S.S. Banda, *Robust Multivariable Flight Control* (Springer Science & Business Media, Berlin, 2012)
8. X. Ge, Q.L. Han, F. Yang, Event-based set-membership leader-following consensus of networked multi-agent systems subject to limited communication resources and unknown-but-bounded noise. IEEE Trans. Indust. Electron. **64**(6), 5045–5054 (2017)
9. E.M. Jafarov, R. Tasaltin, Robust sliding-mode control for the uncertain mimo aircraft model f-18. IEEE Trans. Aerosp. Electron. Syst. **36**(4), 1127–1141 (2000)

Part III
Active Defense Control of Networked Control System

Chapter 7
Security Control of Networked T-S Fuzzy System Under Nonperiodic DoS Attacks with Event-Based Predictor

This chapter investigates the security issue of networked T-S fuzzy system (NTFS) under intermittent DoS jamming attack (I-DoS-JA). This kind attack often causes the hiatus of control input and output feedback in communication channels. In order to compensate the missing data caused by attacks, a model-based predictive control (MPC) framework is proposed by embedding predictors within the closed-loop NTFS, in which: (1) a T-S fuzzy model is formulated with the consideration of I-DoS-JA. Based on this model, a fuzzy observer is constructed to estimate the unmeasurable system states; (2) the predictors embedded in remote plant and local controller are modeled as a synchronous T-S fuzzy system; and (3) an event-trigger mechanism (ETM) is integrated in the observer-based predictor, which would take great advantages in saving bandwidth sources. Finally, a nonlinear system is given as an example to substantiate the work of this chapter [1].

This chapter is structured as below. The system model and problem formulation are described in Sect. 7.1. The analysis of asymptotically stable is proposed in Sect. 7.2. Section 7.3 provides ETP algorithm. In Sect. 7.4, the simulation example is presented to verify the feasibility of the proposed theory. Finally, Sect. 7.5 concludes the chapter.

7.1 Problem Formulation

7.1.1 System Description

According to the definition of CPS, a networked control system (NCS) framework is proposed as showing in Fig. 7.1 to describe a typical phenomena of CPS under cyber-attack. In this framework, observer is designed to estimate the unmeasurable data, and event-trigger is introduced to make full use of the bandwidth. In view of the complex characteristics of CPS , a nonlinear system is formulated via powerful

D. Yue et al., *Secure Control of Networked Control Systems and Its Applications*,
https://doi.org/10.1007/978-981-33-6730-2_7

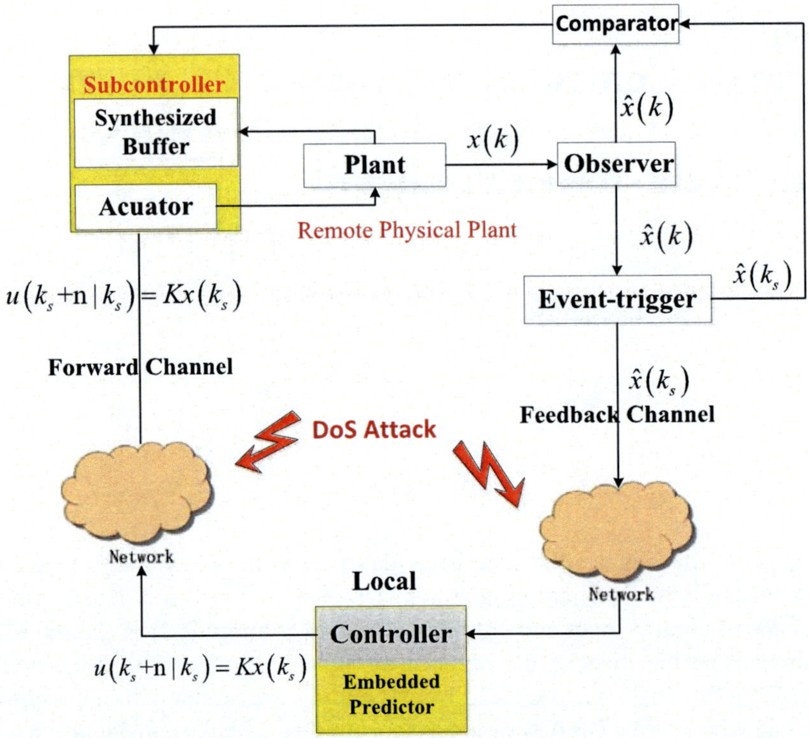

Fig. 7.1 Event-based predictive control for I-DoS-JA schematic

T-S fuzzy model as

$$\begin{cases} x\,(k+1) = A_{\theta(k)}x\,(k) + B_{\theta(k)}u\,(k) \\ y\,(k) \qquad\;\; = C_{\theta(k)}x\,(k) \end{cases} \tag{7.1}$$

where $A_{\theta(k)} = \sum\limits_{i\in\{1,\ldots,r\}} h_i\,(\theta\,(k))\,A_i$, $B_{\theta(k)} = \sum\limits_{i\in\{1,\ldots,r\}} h_i\,(\theta\,(k))\,B_i$ and $C_{\theta(k)} = \sum\limits_{i\in\{1,\ldots,r\}} h_i\,(\theta\,(k))\,C_i$ are the coefficient matrices, and $h_i(\theta(k))$ is the membership function, $\theta(k)$ is the premise variable. In addition, it satisfies the features $h_i(\theta(k)) > 0$ and $\sum_{i=1}^{r} h_i(\theta(k)) = 1$. $x(k) \in R^{n_x}$ is plant state, $u(k) \in R^{n_u}$ is the control input and $y(k) \in R^{n_y}$ is the measurement of sensors.

In order to obtain the unmeasurable states, an observer approach is adopted to estimate the states. According to the powerful T-S fuzzy IF-THEN rules, the nonlinear observer system is presented as

$$\begin{cases} \hat{x}(k+1) = A_{\hat{\theta}(k)}\hat{x}(k) + B_{\hat{\theta}(k)}u(k) + L_{\hat{\theta}(k)}\left[y(k) - \hat{y}(k)\right] \\ \hat{y}\,(k) \qquad\;\; = C_{\hat{\theta}(k)}\hat{x}(k) \end{cases} \tag{7.2}$$

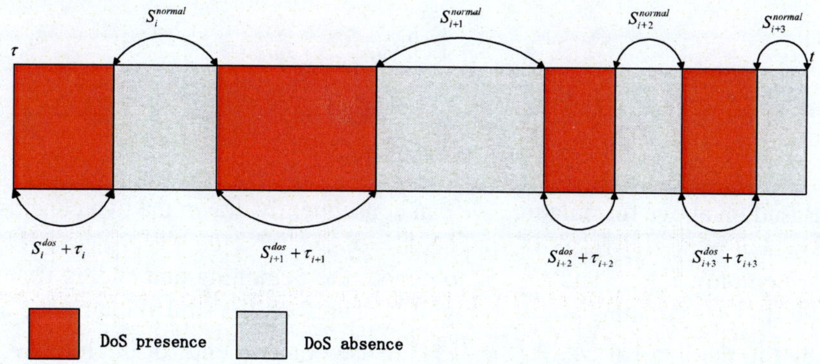

Fig. 7.2 DoS attack schematic

where $A_{\hat{\theta}(k)} = \sum_{j\in\{1,\ldots,r\}} h_j(\hat{\theta}(k))A_j$, $B_{\hat{\theta}(k)} = \sum_{j\in\{1,\ldots,r\}} h_j(\hat{\theta}(k))B_j$, $C_{\hat{\theta}(k)} = \sum_{j\in\{1,\ldots,r\}} h_j(\hat{\theta}(k))C_j$ and $L_{\hat{\theta}(k)} = \sum_{j\in\{1,\ldots,r\}} h_j(\hat{\theta}(k))L_j$. The notation $h_i(\hat{\theta}(k))$ is the membership function, and $\hat{\theta}(k)$ is the premise variable. In addition, it satisfies the features $h_i(\hat{\theta}(k)) > 0$ and $\sum_{i=1}^{r} h_i(\hat{\theta}(k)) = 1$. $\hat{x}(k) \in R^{n_x}$ and $\hat{y}(k) \in R^{n_y}$ denote the estimated state and output of the observer, respectively.

Remark 7.1 According to the standard T-S fuzzy process, we realize that the plant and the observer in the fuzzy system have the same premise variables. In order to distinguish the two, different symbols are used in this chapter. Thus, if the subscripts $i = j$, we have $A_{\theta(k)} = A_{\hat{\theta}(k)}$, $B_{\theta(k)} = B_{\hat{\theta}(k)}$, $C_{\theta(k)} = C_{\hat{\theta}(k)}$ and $L_{\theta(k)} = L_{\hat{\theta}(k)}$.

The remote controller with feedback control law is presented as

$$u(k) = \mathcal{K}_{h(k)}x(k) \tag{7.3}$$

where $\mathcal{K}_{h(k)} = \sum_{l\in\{1,\ldots,r\}} \eta_l(h(k))\mathcal{K}_l$ denotes the feedback gain set, which contains $\{\mathcal{K}_1 \ldots \mathcal{K}_r\}$. The notation $\eta_l(h(k))$ is the membership function, and $h(k)$ is the premise variable. In addition, it satisfies the features $\eta_l(h(k)) > 0$ and $\sum_{l=1}^{r} \eta_l(h(k)) = 1$.

Considering the event-based scheme and the feedback original signal is from observer, thus the controller output in this chapter is described as

$$u(k) = \mathcal{K}_{h(k)}\hat{x}(k_s)$$

The periodic DoS jamming attack process to be considered is illustrated in Fig. 7.2.

From Fig. 7.2, we can get that the DoS attack process is intermittent and stochastic during the interval $[k_0, k_p]$. Thus, it can be formulated with the follow subsections as

$$S_{nor} := \bigcup_{i=1}^{\infty} [s_{att}^{i} + \tau_i, s_{att}^{i+1}) \tag{7.4}$$

$$S_{att} := \{s_{att}^{i}\} \cup [s_{att}^{i}, s_{att}^{i} + \tau_i), \ \ i \in [0, \infty) \tag{7.5}$$

Based on above formulation, we can consequently drawn the two conclusions as (i) $[k_0, k_p] = S_{att} \bigcup S_{nor}$; (ii) $S_{att} \bigcap S_{nor} = \phi$.

We employ $k_{off/on}^{i}$ and $k_{on/off}^{i}$ to denote the beginning and ending time instant of the ith DoS jamming attack in Fig. 7.2. Therefore, the ith DoS jamming attack is located in the interval $\left[k_{off/on}^{i}, k_{on/off}^{i}\right]$. Based on two ends of the interval, the ith DoS attack duration is drawn as $\Pi_i = |k_{on/off}^{i} - k_{off/on}^{i}|$ and then we can describe the union set of DoS attack as

$$\Pi_{DoS} = \bigcup_{i=1}^{\infty} \left[k_{off/on}^{i}, k_{on/off}^{i}\right] \tag{7.6}$$

According to Assumptions 7.1 and 7.2 in [2], two important definitions of DoS duration time and frequency are presented.

Definition 7.1 ([2, 3]) Given scalars $\varepsilon \geq 0$ and $\tau_D \in \mathbb{R}_{\geq 0}$, the I-DoS-JA frequency can be described as

$$\mathcal{N}_{(k_0, k_p)} \leq \varepsilon + \frac{k_p - k_0}{\tau_D} \tag{7.7}$$

where k_0 and k_p denote the beginning and ending time of the DoS attack interval, and τ_D stands for the average number of DoS attack per unit time.

Definition 7.2 ([2, 3]) Given scalars $\epsilon \geq 0$ and $\mathcal{T} \in \mathbb{R}_{\geq 0}$, the I-DoS-JA duration can be described as

$$|\Pi(k_0, k_p)| \leq \epsilon + \frac{k_p - k_0}{\mathcal{T}} \tag{7.8}$$

where k_0 and k_p denote the beginning and ending time of the DoS attack interval, and $\frac{1}{\mathcal{T}}$ stands for the upper limit of average dwell-on time of intermittent DoS attack within a unit time.

For the sake of pointing out the differences between packet loss and intermittent DoS jamming attack, we define τ_{PC} and τ_R to denote the predict span and the upper bound of packet loss, respectively. The probability of packet loss is often fixed and not greater than a certain supremum for a fixed bandwidth and hardware conditions, thus, the packet loss span τ_R is bounded. While the data missing span τ_{PC} (or called DoS duration time) caused by DoS attack is changeable and stochastic. Such that, the following relationships can be summerized.

1. $0 < \Pi_i < \tau_R$;
2. $\tau_R < \Pi_i < \tau_R + \tau_{PC}$;

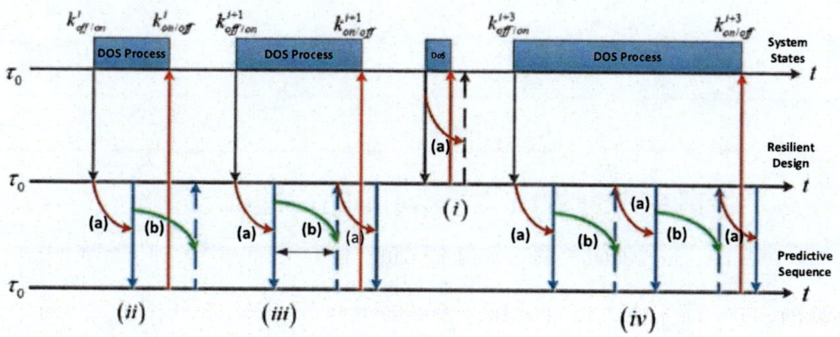

Fig. 7.3 Different DoS attack duration time schematic

3. $n * (\tau_R + \tau_{PC}) < \Pi_i < n * (\tau_R + \tau_{PC}) + \tau_R, i = 1, 2, \ldots, n\ (\tau_0, t)$;
4. $n * (\tau_R + \tau_{PC}) + \tau_R < \Pi_i < (n + 1) * (\tau_R + \tau_{PC}), i = 1, 2, \ldots, \mathcal{N}_{(k_0, k_p)}$.

Based on above relationships, an strategy of predictive control is presented for stability and security control when the DoS duration time is longer than the upper bound of packet loss (UBPL). The detailed processes are presented in Fig. 7.3.

In Fig.7.3, the interval (a) denotes the process of DoS belongs to Π_i, which indicates the length of attack the system can withstand. The interval (b) denotes a complete predictive period for missing date compensation. The processes (i), (ii), (iii) and (iv) denote the different data compensation cases. In addition, the red line indicates the ending of data compensation. In practical process, the time instant described by the solid blue line with downward arrow is not existing. It is designed for explaining the predictive control scenario.

According to above analysis, an event-based predictive control scenario will be designed in the next section. In addition, some meaningful assumptions, which will play important roles in the rest parts, are shown as follows:

Assumption 7.1 If the DoS attack duration time Π_i is shorter than UBPL τ_R, system is assumed to be stable.

Assumption 7.2 In the NCS framework as presented in Fig. 7.1, the controller and actuator are assumed to be event-driven, while the sensor is clock-driven.

Assumption 7.3 The forward and backward communication channels are not attacked simultaneously. If both of the two channels are under intermittent DoS attack, the processes are intermittent and alternate.

7.1.2 Modeling of Event-Based Fuzzy Predictive System

Intermittent DoS attack is a kind of the attack that the communications between plant and controller are fitful, a part of the communication is impossible. Due to this malicious attack, many important information will be lost, and the performances

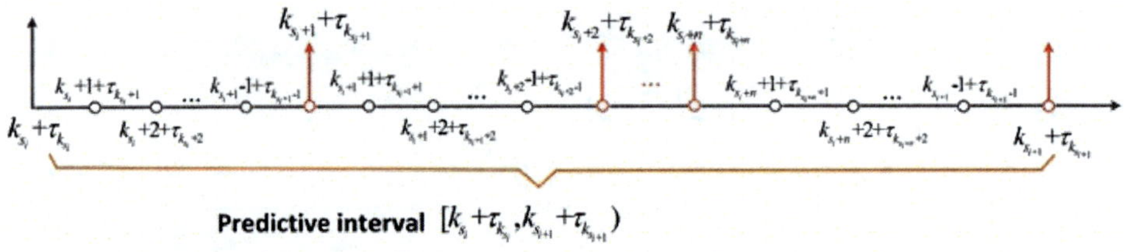

Fig. 7.4 Event-based predictive instants schematic

of the communication will decline as well. For the communication limitations, the inevitable cyber attack often do harm to the availability and authenticity of the data, such that event-based predictor is designed in this note to mitigate effect of malicious attack.

Considering the advantage of saving communication resources, ETS is adopted to weaken the effect of intermittent DoS attack [4].

Since the states of the plant can not be measured directly, thus, the observer-based approach is employed. Motivated by the design philosophy of dynamic ETS [4–6], the transmission instants can be determined via the following rule

$$k_{s+1} = \inf_{k} \left\{ k > k_s : e_s^T(k) W e_s(k) > \mu \hat{x}^T(k_s) W \hat{x}(k_s) \right\} \tag{7.9}$$

where $e_s(k) = \hat{x}(k) - \hat{x}(k_s)$ is the errors between observer and predictor, $\mu \in [0, 1]$ is a parameter for ETS modulation and $W > 0$ is a weighting matrix, which will be determined later.

To model the DoS attack, we consider the time interval $[k_{s_i}, k_{s_{i+1}})$, with $k_{s_i+m} \in [k_{s_i}, k_{s_{i+1}})$ and $m \in \mathcal{S}_1 := \{0, 1, \ldots, s_{i+1} - s_i - 1\}$, $i = 1, 2, \ldots, \mathcal{N}_{(k_0, k_p)}$, are the data transmitted instants generated by event-trigger. Then we can formulate the interval sets as

$$\Gamma = \bigcup_{m=0}^{s_{i+1}-s_i-1} [k_{s_i+m}, k_{s_i+m+1}), m \in \mathcal{S}_1$$

as the subintervals, $k \in \Gamma$.

The interval $[k_{s_i}, k_{s_{i+1}})$ is taken as a completed predictive period, k_{s_n+m} are the trigger instants without DoS attack, with $m \in S_1$. Therefore, the lost triggered data caused by DoS attack can be produced by the algorithm

$$\tilde{x}\left(\tilde{k}+1|k_{s_n}\right) = \tilde{A}\tilde{x}\left(\tilde{k}|k_{s_n}\right) + \tilde{B}\tilde{u}\left(\tilde{k}|k_{s_n}\right), \ \tilde{k} \in \left[\tilde{k}_{s_n}, \tilde{k}_{s_{n+1}}\right) \tag{7.10}$$

Thus, within the predict intervals $[k_{s_i}, k_{s_i+1})$, the triggered instants can be described as

In Fig. 7.4, the instants k_{s_i+m} denote the instants of successful data transmission, if we ignore the effects of time delay, we can get $\tilde{k} \in \left[\tilde{k}_{s_n}, \tilde{k}_{s_{n+1}}\right)$ equals to $\tilde{k} \in$

$[k_{s_n}, k_{s_{n+1}})$, based on this condition, we have

$$\tilde{x}\left(\tilde{k}_{s_n+m} + j_m + 1|k_{s_n}\right) = \tilde{A}\tilde{x}\left(\tilde{k}_{s_n+m} + j_m|k_{s_n}\right) + \tilde{B}\tilde{u}\left(\tilde{k}_{s_n+m}|k_{s_n}\right) \tag{7.11}$$

where $j_m \in \mathcal{S}_2 := \left\{0, 1, \ldots, \tilde{k}_{s_n+m+1} - \tilde{k}_{s_n+m} - 1\right\}$, $\tilde{A}$, $\tilde{B}$ are the coefficient matrix, $\tilde{u}\left(\tilde{k}_{s_n+m}|k_{s_n}\right)$ is the signal transmitted from controller to actuator, k_{s_n+m} indicates the time when data transmission is triggered.

For a complete prediction process, all of event-based predictive time can be formulated as

$$\Gamma_{mn} = \bigcup_{n=1}^{s_{n+1}-s_n-1} \bigcup_{j_m=0}^{\tilde{k}_{s_n+m+1}-\tilde{k}_{s_n+m}-1} [\tilde{k}_{s_n+m} + j_m, \tilde{k}_{s_n+m} + j_m + 1) \tag{7.12}$$

where $m \in \mathcal{S}_1, \tilde{k} \in \Gamma_{mn}$.

From the model-based predictive (MPC) scenario design, the corresponding trigger condition is updated as

$$\begin{aligned} &\left[\tilde{x}\left(k\right) - \tilde{x}\left(k_{s_i+m}|k_{s_i}\right)\right]^T W \left[\tilde{x}\left(k\right) - \tilde{x}\left(k_{s_i+m}|k_{s_i}\right)\right] \geq \\ &\quad \mu\tilde{x}^T\left(k_{s_i+m} + j_m|k_{s_i}\right) W\tilde{x}\left(k_{s_i+m} + j_m|k_{s_i}\right) \end{aligned} \tag{7.13}$$

where $\tilde{x}(k) = \tilde{x}(k_{s_i+m} + j_m|k_{s_i})$, with $j_m \in \mathcal{S}_2, m \in \mathcal{S}_1$.

Motivated by [7], the event-trigger judgement function is defined as

$$\begin{aligned} f\left(k|k_{s_i}\right) = &\left[\tilde{x}(k) - \tilde{x}(k_{s_i+m}|k_{s_i})\right]^T W \left[\tilde{x}(k) - \tilde{x}(k_{s_i+m}|k_{s_i})\right] \\ &-\mu\tilde{x}^T(k_{s_i+m} + j_m|k_{s_i}) W\tilde{x}(k_{s_i+m} + j_m|k_{s_i}) \end{aligned} \tag{7.14}$$

where $f(\cdot)$ is the function of event-trigger scheme, if $f(\cdot) > 0$, the transmission is allowed, otherwise, the transmission is infeasible.

According to the predictive scenario, the system model is updated as

$$\begin{aligned} x(k+1) &= A_{\theta(k)}x(k) + B_{\theta(k)}\tilde{u}(k_s) \\ &= A_{\theta(k)}x(k) + B_{\theta(k)}\mathcal{K}_{h(\tilde{k})}\hat{x}(k_s) \end{aligned} \tag{7.15}$$

where $\tilde{u}(k_s)$ is the control signal produced by predictor, and $\mathcal{K}_{h(\tilde{k})}$ is the feedback gain.

In the framework in Fig. 7.2, an observer is presented as

$$\hat{x}(k+1) = \left(A_{\theta(k)} - \hat{L}_{\theta(k)}C_{\theta(k)}\right)\hat{x}\left(k\right) + B_{\theta(k)}\mathcal{K}_{\eta(\hat{k})}\hat{x}\left(k_s\right) + L_{\theta(k)}y\left(k\right) \tag{7.16}$$

In the controller side, the state of the predictor is formulated as

$$\tilde{x}(k+1) = \tilde{A}_{\theta(k)}x(k) + B_{\theta(k)}\tilde{u}(k_s) \tag{7.17}$$

where $\tilde{A}_{\theta(k)} = \sum\limits_{i\in\{1,\ldots,r\}} \tilde{A}_i$ and $\tilde{B}_{\theta(k)} = \sum\limits_{i\in\{1,\ldots,r\}} \tilde{B}_i$ are the coefficient matrices.

Some important errors are presented $e(k) = x(k) - \hat{x}(k)$, $\tilde{e}(k) = \hat{x}(k) - \tilde{x}(k)$, $e_{im}(k) = \tilde{x}(k) - \hat{x}(k_s)$ and $e_s(k) = \hat{x}(k) - \hat{x}(k_s)$ with $k \in \Gamma_{mn}$ to denote the state errors between plant and observer , observer and predictor, predictor and predictive event-trigger, respectively.

According to above definitions of the errors, we have $\hat{x}\,(k_s) = \hat{x}\,(k_s) - \hat{x}\,(k) + \hat{x}\,(k) - x\,(k) + x\,(k) = x\,(k) - e\,(k) - e_s\,(k)$ and $\tilde{x}\left(k_{s_n+m}|k_{s_n}\right) = \tilde{x}\left(k_{s_n+m}|k_{s_n}\right) - \tilde{x}\,(k) + \tilde{x}\,(k) - \hat{x}\,(k) + \hat{x}\,(k) - x\,(k) + x\,(k) = x\,(k) - e\,(k) - \tilde{e}\,(k) - e_{nm}\,(k)$.

A stacking vector is defined as $z\,(k) = \left[x^T\,(k)\; e^T\,(k)\; \tilde{e}^T\,(k)\right]^T$, then the synthesis system model is formulated as

$$z\,(k+1) = \Psi z\,(k) + \Upsilon e_{nm}\,(k) \tag{7.18}$$

where

$$\Psi = \sum_{i=1}^{r}\sum_{j=1}^{r}\sum_{l=1}^{r}\begin{bmatrix} \Pi_{11}^{ijl} & Pi_{12}^{ijl} & \Pi_{13}^{ijl} \\ \Pi_{21}^{ijl} & \Pi_{22}^{ijl} & \Pi_{23}^{ijl} \\ \Pi_{31}^{ijl} & \Pi_{32}^{ijl} & \Pi_{33}^{ijl} \end{bmatrix}$$

and

$$\Upsilon = \sum_{i=1}^{r}\sum_{j=1}^{r}\sum_{l=1}^{r}\begin{bmatrix} -B_i\tilde{K}_l \\ \left(B_j - B_i\right)\tilde{K}_l \\ -\left(B_j - \tilde{B}_i\right)\tilde{K}_l \end{bmatrix},$$

with $\Pi_{11}^{ijl} = A_i + B_i\tilde{K}_l, \Pi_{12}^{ijl} = -B_i\tilde{K}_l, \Pi_{12}^{ijl} =, \Pi_{13}^{ijl} = -B_i\tilde{K}_l, \Pi_{21}^{ijl} = \left(A_i - A_j\right) + \left(B_i - B_j\right)\tilde{K}_l + L_j\left(C_j - C_i\right), \Pi_{22}^{ijl} = A_j - L_jC_j + \left(B_j - B_i\right)\tilde{K}_l, \Pi_{23}^{ijl} = \left(B_j - B_i\right)$ $\tilde{K}_l$, $\Pi_{31}^{ijl} = \left(B_j - \tilde{B}_i\right)\tilde{K}_l + A_j + L_j\left(C_i - C_j\right) - \tilde{A}_i$, $\Pi_{32}^{ijl} = \tilde{A}_i - A_j + L_jC_j - \left(B_j - \tilde{B}_i\right)\tilde{K}_l$, $\Pi_{33}^{ijl} = \tilde{A}_i - \left(B_j - \tilde{B}_i\right)\tilde{K}_l$.

For further, we have

$$z\,(k+1) = \Psi_1 z\,(k) + \Psi_2\tilde{K}z\,(k) + \Upsilon_1 K e_{nm}\,(k) \tag{7.19}$$

where $\Psi_1 = \sum\limits_{i=1}^{r}\sum\limits_{j=1}^{r}\begin{bmatrix} A_i & 0 & 0 \\ \left(A_i - A_j\right) + L_j\left(C_j - C_i\right) & A_j - L_jC_j & 0 \\ A_j - \tilde{A}_i + L_j\left(C_i - C_j\right) & \tilde{A}_i - A_j + L_jC_j & \tilde{A}_i \end{bmatrix}$,

$\Psi_2 = \sum\limits_{i=1}^{r}\sum\limits_{j=1}^{r}\begin{bmatrix} B_i & -B_i & -B_i \\ B_i - B_j & B_j - B_i & B_j - B_i \\ B_j - \tilde{B}_i & B_j - \tilde{B}_i & \tilde{B}_i - B_j \end{bmatrix}$, $\Upsilon_1 = \sum\limits_{i=1}^{r}\sum\limits_{j=1}^{r}\begin{bmatrix} -B_i \\ B_j - B_i \\ \tilde{B}_i - B_j \end{bmatrix}$,

$\tilde{K} = \sum\limits_{l=1}^{r}\begin{bmatrix} K_l & 0 & 0 \\ 0 & K_l & 0 \\ 0 & 0 & K_l \end{bmatrix}$, $K_{h(k)} = \sum\limits_{l=1}^{r} K_l$.

According to the definitions in the event-trigger scheme design part, we can derive

$$e_{nm}^{T}(k)\,We_{nm}(k) \le \mu z^{T}(k)\,\Gamma z(k) \tag{7.20}$$

where $\Gamma = W \times \begin{bmatrix} 1 & -1 & -1 \\ -1 & 1 & 1 \\ -1 & 1 & 1 \end{bmatrix}$.

In order to obtain the event-trigger condition of observer, we define an error vector $e_s(k) = \hat{x}(k) - \hat{x}(k_s)$, then we have

$$\left[\hat{x}(k) - \hat{x}(k_s)\right]^{T} W \left[\hat{x}(k) - \hat{x}(k_s)\right] \le \mu \hat{x}^{T}(k)\,W\hat{x}(k) \tag{7.21}$$

According to above definitions of error vectors, we have made before, we can get $\hat{x}(k) = \hat{x}(k) - x(k) + x(k) = x(k) - e(k)$, thus, a further useful event-trigger condition in () is derived as

$$e_{s}^{T}(k)\,We_{s}(k) \le \mu \eta^{T}(k)\,\bar{W}\eta(k) \tag{7.22}$$

where $\eta(k) = \begin{bmatrix} x(k) \\ e(k) \end{bmatrix}$, $\tilde{W} = \begin{bmatrix} W & -W \\ -W & W \end{bmatrix}$, $\mu \in [0, 1]$.

7.2 Stability Analysis

Lemma 7.1 *Given real matrices H and E with appropriate dimensions, for all $F(t)$ satisfying $F^{T}(t)\,F(t) \le I$, there exists inequality*

$$HF(t)\,E + E^{T}F^{T}(t)\,H^{T} < 0$$

holds, if and only if there exists a scalar $\varepsilon > 0$, such that

$$\varepsilon HH^{T} + \varepsilon^{-1}E^{T}E < 0$$

Based on above design and **Lemma** 7.1, stability analysis results are derived below.

Theorem 7.1 *Given positive parameters $\mu \in (0, 1)$, $\alpha > 0$, $\beta > 0$, system (7.17)–(7.18) are asymptotically stable (AS), if there exists $\bar{R} > 0$, $\bar{W} > 0$ and appropriate dimensions matrix Y_j, $j \in \{1, 2, \ldots, r\}$, which satisfy the in equations*

$$\begin{bmatrix} \Omega_{11} & * & * & * \\ 0 & \Omega_{22} & * & * \\ 0 & 0 & \Omega_{33} & * \\ \Psi & \Upsilon & 0 & \Omega_{44} \end{bmatrix} < 0 \tag{7.23}$$

where $\Omega_{11} = \begin{bmatrix} 2\mu\bar{W} - \bar{R} & -2\mu\alpha\bar{W} & -\mu\beta\bar{W} \\ -2\mu\alpha\bar{W} & 2\mu\alpha^2\bar{W} - \alpha\bar{R} & \mu\alpha\beta\bar{W} \\ -\mu\beta\bar{W} & \mu\alpha\beta\bar{W} & \mu\beta^2\bar{W} - \beta\bar{R} \end{bmatrix}$, $\Omega_{22} = -\delta_1^2\bar{W}$, $\Omega_{33} = -\delta_2^2\bar{W}$, $\bar{W} = \bar{R}W\bar{R}$, $\Omega_{44} = -diag\{\bar{R}\ \alpha\bar{R}\ \beta\bar{R}\}$, $Y_j = K_j\bar{R}$.

Proof A discrete-time Lyapunov candidate function is selected as

$$V(z(k), k) = z^T(k) Pz(k) \tag{7.24}$$

P is a positive matrix with appropriate dimension. Thus, we can obtain

$$V(z(k+1), k+1) = z^T(k+1) Pz(k+1) \tag{7.25}$$

According to above definition, the difference between Lyapunov function candidates for two consecutive time instants yields

$$\begin{aligned}
&\Delta V(z(k), k) = V(z(k+1), k+1) - V(z(k), k) \\
&= z^T(k+1) Pz(k+1) - z^T(k) Pz(k) \\
&\le [\Psi z(k) + \Upsilon e_{nm}(k)]^T P [\Psi z(k) + \Upsilon e_{nm}(k)] \\
&-z^T(k) Pz(k) + \mu z^T(k) \Gamma z(k) \\
&-e_{nm}^T(k) We_{nm}(k) + \mu\eta^T(k) \tilde{W}\eta(k) - e_s^T(k) We_s(k) \\
&= \begin{bmatrix} z(k) \\ e_{nm}(k) \\ e_s(k) \end{bmatrix}^T \begin{bmatrix} \Psi^T P\Psi + \mu\bar{\Gamma} - P & \Psi^T P\Upsilon & 0 \\ \Upsilon^T P\Psi & \Upsilon^T P\Upsilon - W & 0 \\ 0 & 0 & -W \end{bmatrix} \begin{bmatrix} z(k) \\ e_{nm}(k) \\ e_s(k) \end{bmatrix}
\end{aligned}$$

where $\bar{\Gamma} = \Gamma + \tilde{W} = \begin{bmatrix} 2W & -2W & -W \\ -2W & 2W & W \\ -W & W & W \end{bmatrix}$, and Ψ, Υ have been defined in the previous article.

If the system (7.17)–(7.18) are asymptotically stable, the condition $\Delta V(z(k), k) < 0$ should be satisfied, which equals to

$$\begin{bmatrix} \Psi^T P\Psi + \mu\bar{\Gamma} - P & \Psi^T P\Upsilon & 0 \\ \Upsilon^T P\Psi & \Upsilon^T P\Upsilon - W & 0 \\ 0 & 0 & -W \end{bmatrix} < 0 \tag{7.26}$$

Applying Shur complement,

$$\begin{bmatrix} \mu\bar{\Gamma} - P & * & * & * \\ 0 & -W & * & * \\ 0 & 0 & -W & * \\ \Psi & \Upsilon & 0 & -P^{-1} \end{bmatrix} < 0 \tag{7.27}$$

To facilitate analysis, we define matrix P as $P = diag\left\{R\ \alpha^{-1}R\ \beta^{-1}R\right\}$

$$diag\left\{ P^{-1}\ \delta_1 R^{-1}\ \delta_2 R^{-1}\ I \right\}$$

where α, β, δ_1 and δ_2 are the positive parameters that can be adjusted.

$$\begin{bmatrix} \mu P^{-1}\bar{\Gamma} P^{-1} - P^{-1} & * & * & * \\ 0 & -\delta_1^2 R^{-1} W R^{-1} & * & * \\ 0 & 0 & -\delta_2^2 R^{-1} W R^{-1} & * \\ \Psi & \Upsilon & 0 & -P^{-1} \end{bmatrix} < 0$$

We denote $\bar{R} = R^{-1}$, $Y_j = K_j R^{-1} = K_j \bar{R}$, $\bar{W} = R^{-1} W R^{-1} = \bar{R} W \bar{R}$. Therefore, we can get $K_j = Y_j \bar{R}^{-1}$ and $W = \bar{R}^{-1} \bar{W} \bar{R}^{-1}$

Substituting the above defined parameters, we obtain matrix blocks as follows

$$\Omega_{11} = \mu P^{-1}\bar{\Gamma} P^{-1} - P^{-1} = \begin{bmatrix} 2\mu\bar{W} - \bar{R} & -2\mu\alpha\bar{W} & -\mu\beta\bar{W} \\ -2\mu\alpha\bar{W} & 2\mu\alpha^2\bar{W} - \alpha\bar{R} & \mu\alpha\beta\bar{W} \\ -\mu\beta\bar{W} & \mu\alpha\beta\bar{W} & \mu\beta^2\bar{W} - \beta\bar{R} \end{bmatrix},$$

$\Omega_{22} = -\delta_1^2 R^{-1} W R^{-1} = -\delta_1^2 \bar{W}$, $\Omega_{33} = -\delta_2^2 R^{-1} W R^{-1}$,
$\Omega_{44} = -diag\{ R^{-1}\ \alpha R^{-1}\ \beta R^{-1} \}$.

This completes the proof.

Assumption 7.4 In this note, an assumption that $\Delta A_j(k) = A_j - \tilde{A}_j$ and $\Delta B_j(k) = B_j - \tilde{B}_j$ are given. For the uncertainties, we have $\Delta A_j(k) = H_{A_j} F_{A_j}(k) E_{A_j}$ and $\Delta B_j(k) = H_{B_j} F_{B_j}(k) E_{B_j}$, where H_{A_j}, E_{A_j}, H_{B_j} and E_{B_j} are known matrices. $F_{A_j}(k)$ and $F_{B_j}(k)$ are unknown matrices, which are varying with time and satisfying the conditions $F_{A_j}^T(k) F_{A_j}(k) \le I$ and $F_{B_j}^T(k) F_{B_j}(k) \le I$.

7.3 Robust Security Controller Design

Theorem 7.2 *For given known matrices A_j, B_j, scalars $\mu \in (0, 1)$, $\alpha, \beta, \delta > 0$, and positive parameters $\varepsilon_1, \varepsilon_2$, system (7.17)–(7.18) are asymmetrically stable, if there exist matrices $\bar{R} > 0$, $\bar{W} > 0$ and matrices Y_l, $l \in \{1, 2, ...r\}$ satisfying the LMIs*

$$\bar{\Lambda} = \begin{bmatrix} \Omega_{11} & * & * & * & * & * \\ 0 & \Omega_{22} & * & * & * & * \\ 0 & 0 & \Omega_{33} & * & * & * \\ \Omega_{41}^1 & \Omega_{42}^1 & 0 & \bar{\Omega}_{44} & * & * \\ \Omega_{51} & 0 & 0 & 0 & -\varepsilon_1 I & * \\ \Omega_{61} & \Omega_{62} & 0 & 0 & 0 & -\varepsilon_2 I \end{bmatrix} < 0 \tag{7.28}$$

where $\bar{\Omega}_{44} = diag\{-\bar{R} - \alpha\bar{R} - \beta\bar{R} + \bar{\Sigma}\}$, $\bar{\Sigma} = \varepsilon_1 H_{A_j} H_{A_j}^T + \varepsilon_2 H_{B_j} H_{B_j}^T$,
$\Omega_{51} = \sum_{j=1}^{r} \eta_j [E_{A_j}\bar{R} - \alpha E_{A_j}\bar{R} - \beta E_{A_j}\bar{R}]$, $\Omega_{62} = -\delta_1 \sum_{j=1}^{r}\sum_{l=1}^{r} \eta_j \varpi_l E_{B_j} Y_l$,
$\Omega_{61} = \sum_{j=1}^{r}\sum_{l=1}^{r} \eta_j \varpi_l [E_{B_j} Y_l - \alpha E_{B_j} Y_l - \beta E_{B_j} Y_l]$.

Proof From Theorem 7.1, we know that the uncertainties are concluded in Ω_{41} and Ω_{42}. Thus, we decompose (7.23) as

$$\Lambda = \begin{bmatrix} \Omega_{11} & * & * & * \\ 0 & \Omega_{22} & * & * \\ 0 & 0 & \Omega_{33} & * \\ \Omega_{41} & \Omega_{42} & 0 & \Omega_{44} \end{bmatrix} = \begin{bmatrix} \Omega_{11} & * & * & * \\ 0 & \Omega_{22} & * & * \\ 0 & 0 & \Omega_{33} & * \\ \Omega_{41}^1 & \Omega_{42}^1 & 0 & \Omega_{44} \end{bmatrix} + \begin{bmatrix} 0 & * & * & * \\ 0 & 0 & * & * \\ 0 & 0 & 0 & * \\ \Omega_{41}^2 & \Omega_{42}^2 & 0 & 0 \end{bmatrix} < 0 \tag{7.29}$$

to ensure the stability of system (7.17)–(7.18), where Ω_{11}, Ω_{22}, Ω_{33}, Ω_{44} have been presented in Theorem 7.1. And $\Omega_{41}=\Psi$, $\Omega_{42}=\Upsilon$,

$$\Omega_{41}^1 = \sum_{i=1}^{r}\sum_{j=1}^{r}\sum_{l=1}^{r} \eta_i \eta_j \varpi_l \begin{bmatrix} A_i\bar{R} + B_i Y_l & -\alpha B_i Y_l & -\beta B_i Y_l \\ \nabla_{21} & \nabla_{22} & -\beta(\Delta B_{ij}) Y_l \\ L_j(\Delta C_{ij})\bar{R} & \alpha L_j C_j \bar{R} & \beta A_j \bar{R} \end{bmatrix},$$

with $\Delta A_{ij} = A_i - A_j$, $\Delta B_{ij} = B_i - B_j$, $\Delta C_{ij} = C_i - C_j$, $\nabla_{21} = (\Delta A_{ij})\bar{R} + (\Delta B_{ij}) Y_l - L_j(\Delta C_{ij})\bar{R}$, $\nabla_{22} = \alpha A_j \bar{R} - \alpha L_j C_j \bar{R} - \alpha(\Delta B_{ij}) Y_l$,

$$\Omega_{41}^2 = \sum_{j=1}^{r}\sum_{l=1}^{r} \eta_j \varpi_l \begin{bmatrix} 0 & 0 & 0 \\ 0 & 0 & 0 \\ \Delta A_j \bar{R} + \Delta B_j Y_l & -\alpha\Delta A_j\bar{R} - \alpha\Delta B_j Y_l & -\beta\Delta A_j\bar{R} - \beta\Delta B_j Y_l \end{bmatrix},$$

$$\Omega_{42}^1 = \sum_{i=1}^{r}\sum_{j=1}^{r}\sum_{l=1}^{r} \eta_i \eta_j \varpi_l \begin{bmatrix} -\delta_1 B_i Y_l \\ -\delta_1 (B_i - B_j) Y_l \\ 0 \end{bmatrix},$$

$$\Omega_{42}^2 = \sum_{j=1}^{r}\sum_{l=1}^{r} \eta_j \varpi_l \begin{bmatrix} 0 \\ 0 \\ -\delta_1 \Delta B_j Y_l \end{bmatrix}, \Lambda = \begin{bmatrix} \Omega_{11} & * & * & * \\ 0 & \Omega_{22} & * & * \\ 0 & 0 & \Omega_{33} & * \\ \Omega_{41}^1 & \Omega_{42}^1 & 0 & \Omega_{44} \end{bmatrix} + \sum_{j=1}^{r}\sum_{l=1}^{r}$$
$$\left\{\pi_j^1 F_{A_j}(k)\pi_j^2 + (\pi_j^2)^T F_{A_j}^T(k)(\pi_j^1)^T + \pi_j^3 F_{B_j}(k)\pi_{jl}^4 + (\pi_{jl}^4)^T F_{B_j}^T(k)(\pi_j^3)^T\right\}.$$

where $\pi_j^1 = [0\,0\,0\,0\,0\,0\,0\; H_{A_j}^T]^T$, $\pi_j^2 = [E_{A_j}\bar{R} \; -\alpha E_{A_j}\bar{R} \; -\beta E_{A_j}\bar{R} \; 0\,0\,0\,0\,0]$,
$\pi_j^3 = [0\,0\,0\,0\,0\,0\,0\; H_{B_j}^T]^T$,
$\pi_{jl}^4 = [E_{B_j} Y_l \; -\alpha E_{B_j} Y_l \; -\beta E_{B_j} Y_l \; -\delta_1 E_{B_j} Y_l \; 0\,0\,0\,0]$.

Using Lemma 7.1, under the condition $F_{A_j}(k) F_{A_j}^T(k) \leq I$ and $F_{B_j}(k) F_{B_j}^T(k) \leq I$, there exists $\varepsilon_n > 0, n = 1, 2$, such that

$$\pi_j^1 F_{A_j}(k)\pi_j^2 + (\pi_j^2)^T F_{A_j}^T(k)(\pi_j^1)^T \leq \varepsilon_1 \pi_j^1 (\pi_j^1)^T + \varepsilon_1^{-1} (\pi_j^2)^T \pi_j^2 \tag{7.30}$$

$$\pi_j^3 F_{B_j}(k)\pi_{jl}^4 + \left(\pi_{jl}^4\right)^T F_{B_j}^T(k)\left(\pi_j^3\right)^T \le \varepsilon_2\pi_j^3\left(\pi_j^3\right)^T + \varepsilon_2^{-1}\left(\pi_{jl}^4\right)^T\pi_{jl}^4 \tag{7.31}$$

Based on above process, we have

$$\Sigma_{ijls} = \Sigma_1 + \Sigma_2 \le \Sigma_1 + \varepsilon_1\pi_j^1\left(\pi_j^1\right)^T + \varepsilon_1^{-1}\left(\pi_j^2\right)^T\pi_j^2 + \varepsilon_2\pi_j^3\left(\pi_j^3\right)^T + \varepsilon_2^{-1}\left(\pi_{jl}^4\right)^T\pi_{jl}^4 \tag{7.32}$$

where $\Sigma_1 = \begin{bmatrix} \Omega_{11} & * & * & * \\ 0 & \Omega_{22} & * & * \\ 0 & 0 & \Omega_{33} & * \\ \Omega_{41}^1 & \Omega_{42}^1 & 0 & \Omega_{44} \end{bmatrix}$, $\Sigma_2 = \begin{bmatrix} 0 & * & * & * \\ 0 & 0 & * & * \\ 0 & 0 & 0 & * \\ \Omega_{41}^2 & \Omega_{42}^2 & 0 & 0 \end{bmatrix}$.

Applying schur complement for $\varepsilon_1^{-1}\left(\pi_{ij}^2\right)^T\pi_{ij}^2$ and $\varepsilon_2^{-1}\left(\pi_{ij}^4\right)^T\pi_{ij}^4$ respectively, and then we can get

$$\bar{\Lambda} = \begin{bmatrix} \Omega_{11} & * & * & * & * & * \\ 0 & \Omega_{22} & * & * & * & * \\ 0 & 0 & \Omega_{33} & * & * & * \\ \Omega_{41}^1 & \Omega_{42}^1 & 0 & \bar{\Omega}_{44} & * & * \\ \Omega_{51} & 0 & 0 & 0 & -\varepsilon_1 I & * \\ \Omega_{61} & \Omega_{62} & 0 & 0 & 0 & -\varepsilon_2 I \end{bmatrix}$$

where $\bar{\Omega}_{44} = \begin{bmatrix} -\bar{R} & 0 & 0 \\ 0 & -\alpha\bar{R} & 0 \\ 0 & 0 & -\beta\bar{R} + \bar{\Sigma} \end{bmatrix}$, $\bar{\Sigma} = \varepsilon_1 H_{A_j} H_{A_j}^T + \varepsilon_2 H_{B_j} H_{B_j}^T$,

$\Omega_{51} = \sum_{j=1}^{r} \eta_j \left[\, E_{A_j}\bar{R} \;\; -\alpha E_{A_j}\bar{R} \;\; -\beta E_{A_j}\bar{R} \,\right]$,

$\Omega_{61} = \sum_{j=1}^{r}\sum_{l=1}^{r} \eta_j\varpi_l \left[\, E_{B_j}Y_l \;\; -\alpha E_{B_j}Y_l \;\; -\beta E_{B_j}Y_l \,\right]$ and

$\Omega_{62} = -\delta_1 \sum_{j=1}^{r}\sum_{l=1}^{r} \eta_j\varpi_l E_{B_j}Y_l$.

This is the end of proof. Although the results in Theorem 7.2 are universal, they are more complex and should not be directly applied in engineering practice. Therefore, we consider the ideal case, that is $A_j = \tilde{A}_s$ and $B_j = \tilde{B}_s$, $j, s \in \{1, \ldots, r\}$.

Theorem 7.3 *For given positive parameter α, β, δ_1, δ_2 and $\mu \in (0, 1)$, system (7.17)–(7.18) are AS, if there exist $\bar{R} > 0$, $\bar{W} > 0$ and the matrices Y_l, $l \in \{1, \ldots, r\}$, such that the follow LMI holds*

$$\begin{bmatrix} \Omega_{11} & * & * & * \\ 0 & \Omega_{22} & * & * \\ 0 & 0 & \Omega_{33} & * \\ \tilde{\Omega}_{41} & \tilde{\Omega}_{42} & 0 & \Omega_{44} \end{bmatrix} < 0$$

where $\Omega_{11} = \mu P^{-1}\bar{\Gamma}P^{-1} - P^{-1} = \begin{bmatrix} 2\mu\bar{W} - \bar{R} & -2\mu\alpha\bar{W} & -\mu\beta\bar{W} \\ -2\mu\alpha\bar{W} & 2\mu\alpha^2\bar{W} - \alpha\bar{R} & \mu\alpha\beta\bar{W} \\ -\mu\beta\bar{W} & \mu\alpha\beta\bar{W} & \mu\beta^2\bar{W} - \beta\bar{R} \end{bmatrix}$,
$\Omega_{22} = -\delta_1^2 R^{-1}WR^{-1} = -\delta_1^2\bar{W}$, $\Omega_{33} = -\delta_2^2 R^{-1}WR^{-1}$,
$\Omega_{44} = -diag\{R^{-1}\ \alpha R^{-1}\ \beta R^{-1}\}$,

$$\tilde{\Omega}_{41} = \sum_{i=1}^{r}\sum_{j=1}^{r}\sum_{l=1}^{r}\eta_i\eta_j\varpi_l \begin{bmatrix} A_i\bar{R} + B_iY_l & -\alpha B_iY_l & -\beta B_iY_l \\ \Delta A_{ij}\bar{R} + \Delta B_{ij}Y_l - L_j\Delta C_{ij}\bar{R} & \alpha A_j\bar{R} - \alpha L_jC_j\bar{R} - \alpha\Delta B_{ij}Y_l & -\beta\Delta B_{ij}Y_l \\ L_j\Delta C_{ij}\bar{R} & \alpha L_jC_j\bar{R} & \beta A_j\bar{R} \end{bmatrix}$$

$\tilde{\Omega}_{42} = \sum_{i=1}^{r}\sum_{j=1}^{r}\sum_{l=1}^{r}\eta_i\eta_j\varpi_l \begin{bmatrix} -\delta_1 B_iY_l \\ -\delta_1\left(B_i - B_j\right)Y_l \\ 0 \end{bmatrix}$ with $\Delta A_{ij} = A_i - A_j$, $\Delta B_{ij} = B_i - B_j$, $\Delta C_{ij} = C_i - C_j$.

Proof Considering the Lyapunov candidate function

$$V(\eta(k), k) = \eta^T(k)P\eta(k)$$

where $P = P^T > 0$, for any $k \in \Gamma_{mn}$, along the trajectory of system (7.17)–(7.18) to calculate the difference for $V(\eta(k), k)$, observe the inequality

$$\tilde{\Lambda} = \begin{bmatrix} \Omega_{11} & * & * & * \\ 0 & \Omega_{22} & * & * \\ 0 & 0 & \Omega_{33} & * \\ \tilde{\Omega}_{41} & \tilde{\Omega}_{42} & 0 & \Omega_{44} \end{bmatrix} < 0$$

meanwhile.

Ideally, $A_j = \tilde{A}_s$ and $B_j = \tilde{B}_s$, with $j, s \in \{1, \ldots, r\}$, which implies $\tilde{x}\left(\tilde{k}_{s_n+m}|k_{s_n}\right) = \hat{x}\left(k_{s_n+m}|k_{s_n}\right)$. Then, in (31), we have $\Sigma_2 = 0$, such that $\Omega_{11} = \mu P^{-1}\bar{\Gamma}P^{-1} - P^{-1} = \begin{bmatrix} 2\mu\bar{W} - \bar{R} & -2\mu\alpha\bar{W} & -\mu\beta\bar{W} \\ -2\mu\alpha\bar{W} & 2\mu\alpha^2\bar{W} - \alpha\bar{R} & \mu\alpha\beta\bar{W} \\ -\mu\beta\bar{W} & \mu\alpha\beta\bar{W} & \mu\beta^2\bar{W} - \beta\bar{R} \end{bmatrix}$, $\Omega_{22} = -\delta_1^2 R^{-1}WR^{-1} = -\delta_1^2\bar{W}$, $\Omega_{33} = -\delta_2^2 R^{-1}WR^{-1}$,
$\Omega_{44} = -diag\{R^{-1}\ \alpha R^{-1}\ \beta R^{-1}\}$, $\tilde{\Omega}_{41} = \Omega_{41}^1$ and $\tilde{\Omega}_{42} = \Omega_{42}^1$.

7.4 Event-Triggered Security Predictive Control Algorithm

In order to complement the missing data caused by intermittent DoS jamming attack, an effective algorithm is designed based on ETPC strategy, the algorithm is presented

Table 7.1 The algorithm of predictive process for attack caused data missing

Algorithm 7.1: Event-Based Predictive Security Control Scenario under DoS Attacks	
Begin for $i = 1$	
1	**if** $x\left(k^i_{off/on}\right) = 1$
2	$x\left(k_{s_n}\right) = x\left(k^i_{off/on} + k_R\right)$;
3	**if** $x\left(k^i_{off/on} + k_R\right) < x\left(k^i_{on/off}\right)$
4	**Predictive Sequence** $\tilde{x}\left(\tilde{k}_{s_n+m} + j_m \vert k_{s_n}\right)$;
5	**if** $\left(\tilde{k}_{s_n+m} + j_m \vert k_{s_n}\right) < \left(k^i_{on/off}\right)$
6	**return to step.4**
7	**else if** $m = s_{n+1} - s_n$;
8	$\left(\tilde{k}_{s_{n+1}} + k_R \vert k_{s_n}\right) < \left(k^i_{on/off}\right)$
9	**else return back to step 4;**
10	**if** $\left(\tilde{k}_{s_{n+1}} + k_R \vert k_{s_n}\right) < \left(k^i_{on/off}\right)$
11	$\tilde{x}\left(\tilde{k}_{s_{n+1}}\right) = \tilde{x}\left(k_{s_{n+1}} + k_R \vert k_{s_n}\right)$;
12	$n = n + 1$;
13	return to step.4;
14	**else** $i = i + 1$;
15	return to begin;
16	**else return to begin;**
17	**else return to begin;**
18	**end**

in Table 7.1 And at the same time, an flow chart to explain the algorithm is presented as well.

The event-based predictive control scenario for I-DoS-JA is showing in the flow chart as presented in Fig. 7.5.

7.5 Simulation Example

In this section, the system is considered as

$$x(k+1) = \sum_{i=1}^{2} h_i(x(k))\left[A_i x(k) + B_i u(k)\right] \tag{7.33}$$

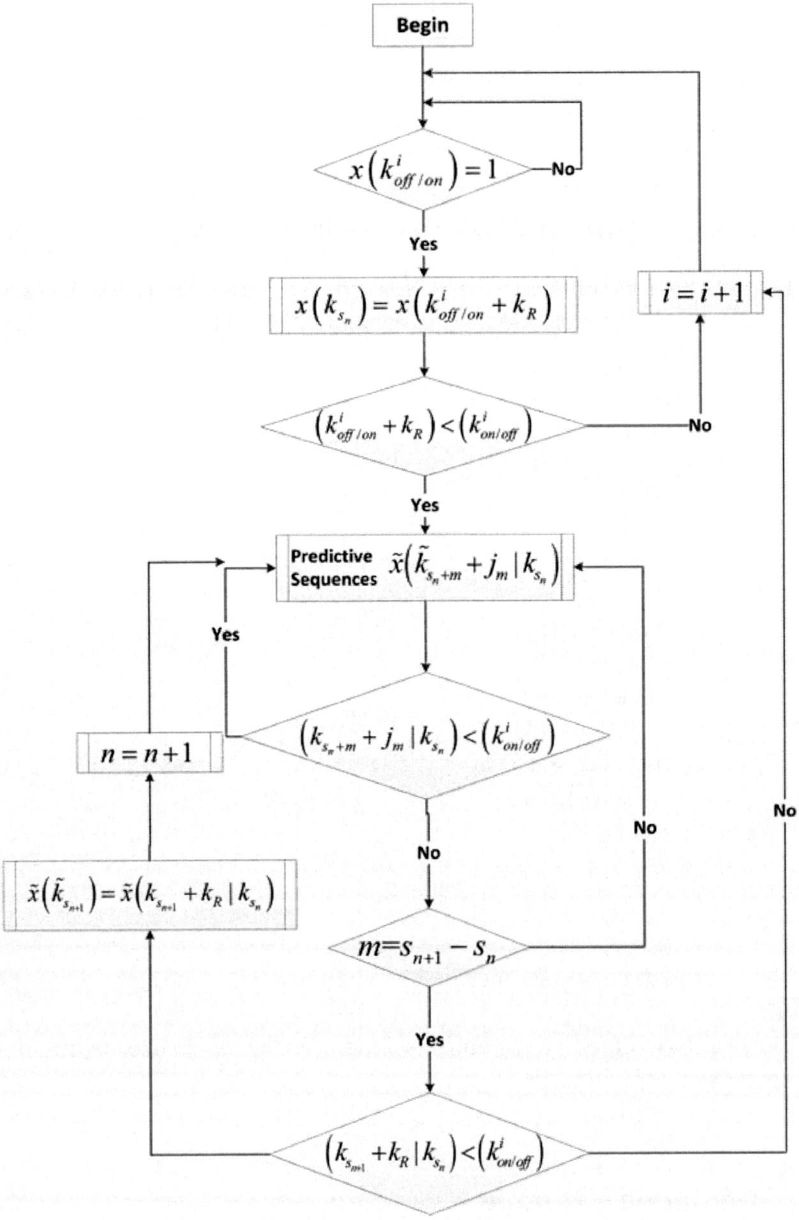

Fig. 7.5 The flow chart of event-based predictive sequence making schematic

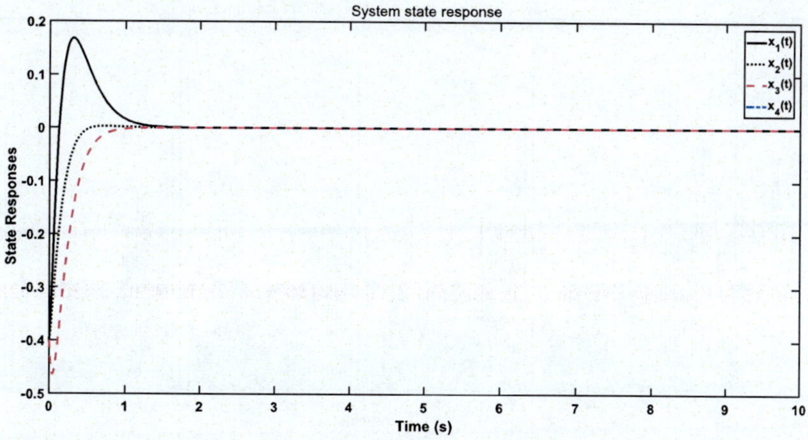

Fig. 7.6 The output of system states with event-based predictive approach

where $A_1 = \begin{bmatrix} 0.9000 & -0.0200 & -0.1000 & 0 \\ 0 & 0.9000 & -0.0200 & 0 \\ 0 & 0 & 0.9000 & 0.2000 \\ 0 & 0 & 0 & 0.7000 \end{bmatrix}$,

$$A_2 = \begin{bmatrix} 0.7000 & -0.2000 & -0.1000 & 0 \\ 0 & 0.8000 & -0.2000 & 0 \\ 0 & 0 & 0.7800 & 0.2000 \\ 0 & 0 & 0 & 0.7500 \end{bmatrix}, B_1 = B_2 = \begin{bmatrix} -0.0000 \\ 0.0000 \\ 0.0020 \\ 0.0200 \end{bmatrix},$$

$$L_1 = L_2 = \begin{bmatrix} 0.0291 \\ -0.0207 \\ 0.0291 \\ -0.0207 \end{bmatrix},$$

$C_1 = C_2 = \begin{bmatrix} 0.0291 & 0.1945 & -0.1013 & -0.0207 \end{bmatrix}$. The fuzzy membership function is chosen as $\theta_1(x(k)) = \frac{1}{1-\exp(-2x_1(k))}$ and $\theta_2(x(k)) = 1 - \theta_1(x(k))$.

According to **Theorem** 7.3, the event-triggering weighting matrix W and feedback gain K subjected to control scenario are as follows:

$$W_1 = \begin{bmatrix} 9.0770 & 6.9279 & -4.3485 & -1.6288 \\ 6.9279 & 5.3328 & -3.0645 & -1.0377 \\ -4.3485 & -3.0645 & 3.5276 & 1.9420 \\ -1.6288 & -1.0377 & 1.9420 & 1.2734 \end{bmatrix},$$

$$W_2 = \begin{bmatrix} 1.4191 & 0.0205 & 1.9708 & 0.3409 \\ 0.0205 & 3.0108 & 3.0007 & 0.5136 \\ 1.9708 & 3.0007 & 3.2545 & 0.3834 \\ 0.3409 & 0.5136 & 0.3834 & 0.3808 \end{bmatrix}, K_1 = \begin{bmatrix} 0.2269 & 0.1603 & -0.1752 & -0.1444 \end{bmatrix}$$

and $K_2 = \begin{bmatrix} 0.1019 & 0.1440 & 0.2673 & -0.1890 \end{bmatrix}$. Substitute the results, the system state and event-triggered release instants under intermittent DoS attack with event-based predictor are obtained (Figs. 7.6 and 7.7).

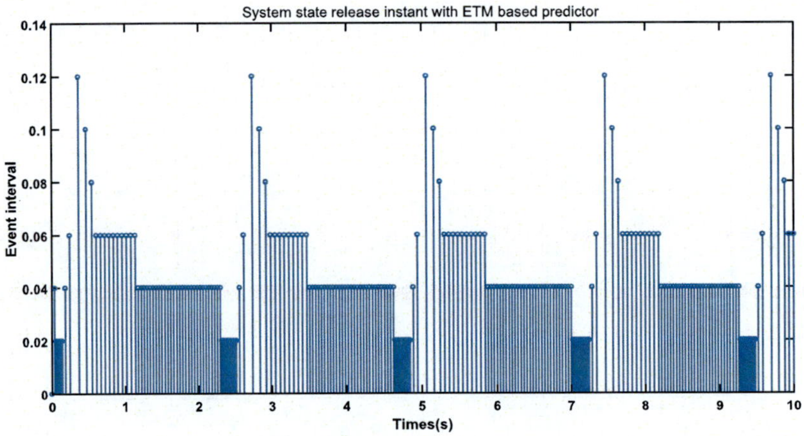

Fig. 7.7 Event-trigger transmitted instants

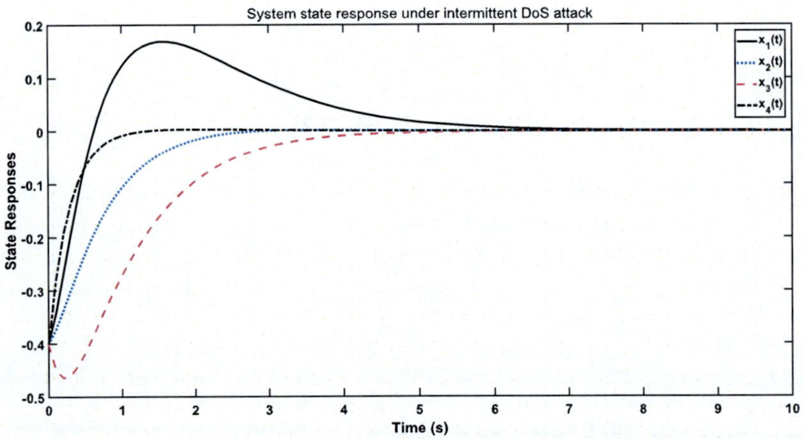

Fig. 7.8 The output of system states under intermittent DoS attack

In order to demonstrate the problem reasonably, we assume that the intermittent DoS duration time equals to 4 times of the sampling time. Thus, the system state and event-triggered release instants are obtained as follows (Figs. 7.8 and 7.9).

The results of simulations can be summarized from the following four indexes in Table 7.2. Under assumed attack, we can find that from Table 7.2, without predictive data compensation, the system can maintain stability performance, but the transition time is is nearly 5 times longer comparing with the case with predictive data compensation. However, the event-based release number is greatly reduced, which reduces the communication burden of the channels as well. Meanwhile, the average trigger intervals and maximum trigger intervals have enhanced dramatically. In other words, despite the large amount of data transmission and the smaller data transmission intervals, the stable response speed of the system has been greatly improved for the predictive data compensation case.

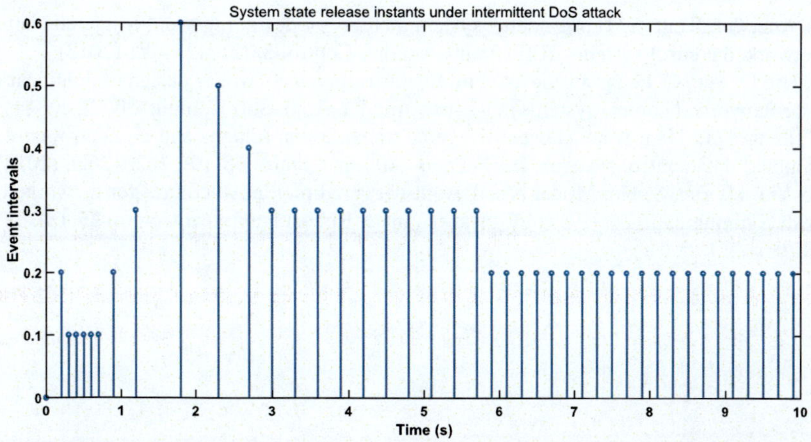

Fig. 7.9 System state release instants under intermittent DoS attack based on ETM

Table 7.2 Comparison of before and after attack compensation with event-based predictive control

Different cases	Without compensation	With compensation
Up to stability time	$t = 7$ s	$t = 1.5$
Number of ETM release	42	230
Average trigger intervals	0.2381 s	0.0435 s
Maximum trigger intervals	0.6 s	0.12 s

7.6 Conclusion

In this chapter, the effects of intermittent DoS jamming attack have been discussed. Based on the analysis, an event-based predictor is designed to defense the periodic DoS jamming attack. A parallel framework of the embedded predictor in controller side is proposed, the missing data caused by DoS attack is compensated by the predictor. Then the security control of NCSs under DoS attack is achieved. In addition, the performance of system resilience is enhance via this control scheme as well.

References

1. H. Ge, D. Yue, X. Xie, C. Dou, S. Wang, Security control of cyber-physical system based on switching approach for intermittent denial-of-service jamming attack. ISA Trans. **20**, 700–714 (2019)
2. C.D. Persis, P. Tesi, Input-to-state stabilizing control under denial-of-service. IEEE Trans. Autom. Control **60**(11), 2930–2944 (2015)
3. J.P. Hespanha, A.S. Morse, Stability of switched systems with average dwell-time, in *Proceedings of the IEEE Conference on Decision and Control, 1999*, vol. 3 (2002), pp. 2655–2660

4. D. Yue, Q.L. Han, E. Tian, A delay system method for designing event-triggered controllers of networked control systems. IEEE Trans. Autom. Control **58**(2), 475–481 (2013)
5. S. Hu, D. Yue, C. Peng, X. Xie, X. Yin, Event-triggered controller design of nonlinear discrete-time networked control systems in t-s fuzzy model. Appl. Soft Comput. **30**(C), 400–411 (2015)
6. C. Peng, Q.L. Han, A novel event-triggered transmission scheme and $\mathcal{L}_2$ control co-design for sampled-data control systems. IEEE Trans. Autom. Control **58**(10), 2620–2626 (2013)
7. X. Yin, D. Yue, S. Hu, Model based event-triggered predictive control for networked systems with communication delays compensation. Int. J. Robust Nonlinear Control **25**(18), 3572–3595 (2016)

Chapter 8
Bandwidth Allocation-Based Switched Dynamic Triggering Control Against Nonperiodic DoS Attacks

The chapter is organized as follows. The primary-redundancy control structure and system modeling are presented in Sect. 8.1. Section 8.2 proposes the design method of defense switching law. Stability analysis and switching control design are given by Sects. 8.3 and 8.4, respectively. A simulation example is illustrated in Sect. 8.5. Finally, the conclusion is stated in Sect. 8.6.

8.1 Problem Formulation

8.1.1 System Description

A class of system is described to represent networked continue system with redundant channel based switching control under DoS attacks as shown in Fig. 8.1.

$$\dot{x}(t) = Ax(t) + Bu_{s(t)}(t) + Bu_{(1-s(t))\rho(t)}(t), \tag{8.1}$$

where A, B are known matrices, $u_{s(t)}$ represents control input transmitted over primary channel while $u_{(1-s(t))\rho(t)}(t)$ control input over redundancy channel. The DoS attacks signal $s(t)$ is subjected to attack strategy. To defend DoS attacks, the switching signal $\rho(t)$ is to be designed by defender. The communication switching between PR channels depends on the switch signals $s(t)$ and $\rho(t)$.

$$s(t) = \begin{cases} 1 & sleep, \\ 0 & attack, \end{cases} \qquad \rho(t) = \begin{cases} 2 & defense, \\ 0 & conceal. \end{cases}$$

Then, a composite switching signal is introduced by

D. Yue et al., *Secure Control of Networked Control Systems and Its Applications*,
https://doi.org/10.1007/978-981-33-6730-2_8

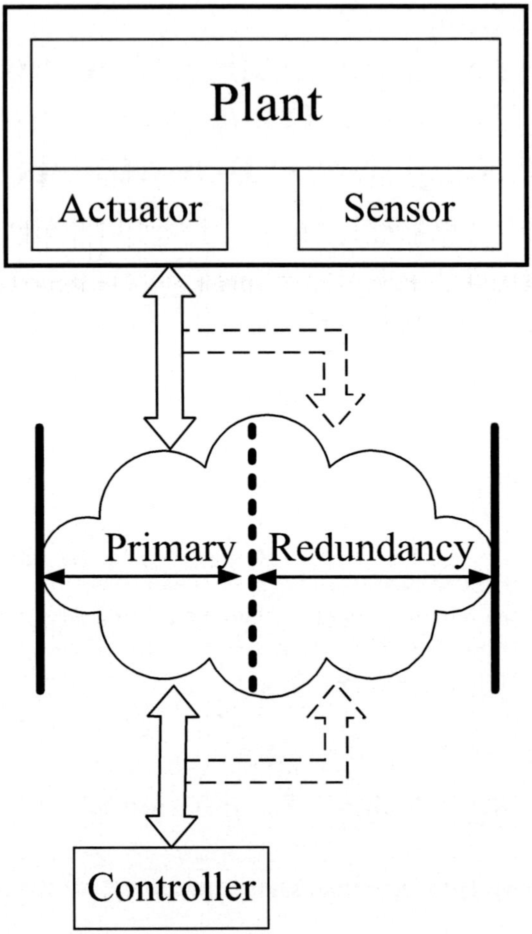

Fig. 8.1 Primary-redundancy control structure

$$\sigma(t) = s(t) + (1 - s(t))\rho(t). \tag{8.2}$$

where $\sigma(t) : [0, \infty) \rightarrow M = \{0, 1, 2\}$. Accordingly, the switching intervals can be defined by

$$\begin{cases} \mathcal{D}_n \triangleq [d_n, d_n + l_n), & \sigma(t) = 0, \\ \mathcal{H}_n \triangleq [h_n, h_n + g_n), & \sigma(t) = 1, \\ \mathcal{R}_n \triangleq [r_n, r_n + f_n), & \sigma(t) = 2. \end{cases}$$

where d_n, h_n and r_n are the beginning time of DoS attacks intervals $\mathcal{D}_n$, primary control intervals $\mathcal{H}_n$ and redundancy control intervals $\mathcal{R}_n$, respectively and l_n, g_n and f_n are the corresponding intervals' duration.

Both primary control loop and redundancy control loop adopt state feedback control law.

$$u(t) = \begin{cases} 0, & t \in \mathcal{D}_n, \\ K_1 x(t), & t \in \mathcal{H}_n, \\ K_2 x(t), & t \in \mathcal{R}_n, \end{cases} \tag{8.3}$$

where the control gains K_1 and K_2 are to be designed.

When control inputs are transmitted over the primary-redundancy channels, network induced time delay may cause system unstable. In this study, a new delay problem is explored to be called correlated delay to represent the delay of PR channels allocated by limited bandwidth. First, introduce a virtual delay $\tau_0(t)$ in case of DoS attacks; denote $\tau_1(t)$ and $\tau_2(t)$ as the transmission delay of the primary channel and the redundancy channel, respectively.

Assumption 8.1 The delay margin and delay derivative satisfy $0 \le \tau_0(t) \le \bar{\tau}$ and $\dot{\tau}_0 = 1$; $0 \le \tau_1(t) \le \bar{\tau}_1$ and $\dot{\tau}_1 = 1$; $0 \le \tau_2(t) \le \bar{\tau}_2$ and $\dot{\tau}_2 = 1$. The correlated characteristic between $\tau_1(t)$ and $\tau_2(t)$ is represented by

$$\begin{aligned} \bar{\tau} &\le \bar{\tau}_1 \le e^{(\frac{1}{\lambda}-1)\phi}\bar{\tau}, \\ \bar{\tau} &\le \bar{\tau}_2 \le e^{(\frac{1}{1-\lambda}-1)\phi}\bar{\tau}, \end{aligned} \tag{8.4}$$

where $\lambda \in [0, 1]$ and $\phi \in [0, 1]$.

Remark 8.1 To defense DoS attacks , we allocate part of bandwidth to redundant channel for reconstructing closed loop of control system when the primary channel breaks. We here assume both primary control channel and redundant control channel are homogeneous communication network. Thus, it is reasonable that there exists the correlation of induced delay in PR channels under the allocation of limited bandwidth. $\bar{\tau}$ denote the delay bound of the communication channel equipped with the total bandwidth. λ represents the ratio of bandwidth we configure to primary transmission. ϕ reflects the network load in system working mode. The large ϕ reflects the heavy transmission load.

8.1.2 Design of Dynamic Event-Triggered Scheme

To improve the communication efficiency of the PR channels, a dynamic event-triggered scheme or mechanism is introduced which can be used to reduce the communication load ϕ.

The principle of the dynamic event-triggered mechanism is same in PR channels. Without loss of generality, the triggering process of primary channel will be discussed in detail.

Denote $\{t_{k,n}h\}$ as the triggered instants

$$t_{k,n}h = \{t_{k_j}h | t_{k_j}h \; satisfying \; (8.6) \; and \; t_{k_j}h \in \mathcal{H}_n\} \cup \{h_n\} \tag{8.5}$$

where $n,\ j,k_j \in N$. The relationship (8.6) is represented by

$$\begin{aligned}&[x(t_{k_j}h)-x(t_{k,n}h)]^T\Omega[x(t_{k_j}h)-x(t_{k,n}h)]\\&>\varepsilon x^T(t_{k,n}h)\Omega x(t_{k,n}h)+\frac{1}{\varphi}\eta(t_{k_j}h)\end{aligned}\tag{8.6}$$

where $t_{k_j}h \triangleq t_k h + jh, \varepsilon > 0, \varphi > 0$ and positive triggering matrix Ω to be designed. The internal dynamic function $\eta(t)$ is defined as

$$\begin{cases}\dfrac{d\eta(t)}{dt} = -\mu\eta(t)+\varepsilon x^T(t_{k,n}h)\Omega x(t_{k,n}h)\\ \qquad\quad -[x(t_{k_j}h)-x(t_{k,n}h)]^T\Omega[x(t_{k_j}h)-x(t_{k,n}h)],\\ \eta(0)=\eta_0,\end{cases}\tag{8.7}$$

where $\eta_0 > 0$.

Remark 8.2 Note that $\eta(t) > 0, \forall t$ with $\varphi \neq 0$. Substituting (8.6) to (8.7), for $t \in [t_{k_j}h, t_{k_{j+1}}h)$, one has

$$\begin{aligned}\frac{d\eta(t)}{dt} &\geq -\mu\eta(t)-\frac{1}{\varphi}\eta(t_{k_j}h),\\ \eta(t) &\geq \frac{(\mu\varphi+1)e^{-\mu(t-t_{k_j}h)}-1}{\mu\varphi}\eta(t_{k_j}h)\\ &\geq \frac{(\mu\varphi+1)e^{-\mu h}-1}{\mu\varphi}\eta(t_{k_j}h)\\ &\geq \left[\frac{(\mu\varphi+1)e^{-\mu h}-1}{\mu\varphi}\right]^N\eta(0).\end{aligned}$$

where $N \geq 0$. Then, it satisfies $\eta(t) > 0$ if $ln(1+\mu\varphi) > \mu h$.

Remark 8.3 The continue function $\eta(t)$ is similar with the discrete one in [2]. Both of them are positive decay function. However, the work in [2] did not consider the influence of DoS attacks. Here the function $\eta(t)$ is piecewise continued function due to the intermittent operation of dynamic ETM. At the beginning time of intervals $\mathcal{H}_n, \mathcal{R}_n$, the dynamic ETM will be forced to trigger communication.

8.1.3 Modeling of Switched Control Systems

According to the triggered sequence $\{t_{k,n}h\}$ with delay, the primary control interval $\mathcal{H}_n$ can be divided by

$$\mathcal{H}_n = \cup_{k=0}^{k(n)} \mathcal{H}_{k,n},$$

where

$$\mathcal{H}_{k,n} = \begin{cases} [h_n,\ t_{1,n}h + \tau_1(t_{1,n}h)),\ k = 0, \\ [t_{k,n}h + \tau_1(t_{k,n}h),\ t_{k+1,n}h + \tau_1(t_{k+1,n}h)), \\ \qquad\qquad k = 1, ..., k(n) - 1, \\ [t_{k(n),n}h + \tau_1(t_{k(n),n}h),\ h_n + g_n),\ k = k(n), \end{cases}$$

and $k(n) = sup\{k \in N | t_{k,n}h + \tau_1(t_{k,n}h) < h_n + g_n\}$.

For each $\mathcal{H}_{k,n}$, it can be divided by

$$\mathcal{H}_{k,n} = \cup_{j=0}^{j(k)} \mathcal{H}_{k,n}^j,$$

where

$$\mathcal{H}_{k,n}^j = \begin{cases} [t_{k,n}h + \tau_1(t_{k,n}h),\ t_{k,n}h + \bar{\tau}_1 + h),\ j = 0, \\ [t_{k,n}h + \bar{\tau}_1 + jh,\ t_{k,n}h + \bar{\tau}_1 + (j+1)h), \\ \qquad\qquad j = 1, ..., j(k) - 1, \\ [t_{k,n}h + \bar{\tau}_1 + j(k)h,\ t_{k+1,n}h + \tau_1(t_{k+1,n}h)), \\ \qquad\qquad j = j(k), \end{cases}$$

and $j(k) = sup\{j \in N | t_{k,n}h + \bar{\tau}_1 + jh < t_{k+1,n}h + \tau_1(t_{k+1,n}h)\}$. Then, a virtual input delay is introduced

$$\tau_1(t) = \begin{cases} t - t_{k,n}h, & t \in \mathcal{H}_{k,n}^0, \\ t - t_{k,n}h - jh, & t \in \mathcal{H}_{k,n}^j,\ j = 1, ..., j(k) - 1 \\ t - t_{k,n}h - j(k)h, & t \in \mathcal{H}_{k,n}^{j(k)}. \end{cases} \tag{8.8}$$

Thus, one can has

$$\tau_1(t_{k,n}h) \le \tau_1(t) \le \tilde{\tau}_1,\ t \in \mathcal{H}_{k,n}.$$

where $\tilde{\tau}_1 = \bar{\tau}_1 + h$.

Furthermore, the triggered error can be represented by

$$e_1(t) = \begin{cases} x(t_{k,n}h) - x(t_{k,n}h), & t \in \mathcal{H}_{k,n}^0, \\ x(t_{k,n}h) - x(t_{k,n}h + jh), & t \in \mathcal{H}_{k,n}^j,\ j = \\ & 1, ..., j(k) - 1, \\ x(t_{k,n}h) - x(t_{k,n}h + j(k)h), & t \in \mathcal{H}_{k,n}^{j(k)}. \end{cases} \tag{8.9}$$

When the redundancy channel is active, similarly, the triggering instants and intervals will be generated as same as the above process. Denote triggering instants $\{\hat{t}_{k,n}h\}$ generated by $\{r_n\}$ and the triggered condition (8.6) in which the $t_{k_j}h$ and $t_{k,n}h$ are replaced by $\hat{t}_{k_j}h$ and $\hat{t}_{k,n}h$, where $\hat{t}_{k_j}h \in \mathcal{R}_n$, $\hat{t}_{k_j}h \triangleq \hat{t}_k h + jh$. Then, we can obtain that

$$\tau_2(\hat{t}_{k,n}h) \leq \tau_2(t) \leq \tilde{\tau}_2, \ t \in \mathcal{R}_{k,n}, \tag{8.10}$$

where $\tilde{\tau}_2 = \bar{\tau}_2 + h$ and the triggered error $e_2(t) = x(\hat{t}_{k,n}h) - x(\hat{t}_{k,n}h + jh), t \in \mathcal{R}^j_{k,n}, \mathcal{R}^j_{k,n} = [\hat{t}_{k,n}h + jh + \bar{\tau}_2, \hat{t}_{k,n}h + (j+1)h + \bar{\tau}_2)$.

According to (8.1), (8.2), (8.8), (8.9), (8.10), under DoS attacks, networked continued system (8.1) with redundant channel is represented by a class of switched delay system consisting of three subsystems $\Sigma_{\sigma(t)}$:

$$\dot{x}(t) = \begin{cases} Ax(t) + BK_1[x(t - \tau_1(t)) + e_1(t)], & t \in \mathcal{H}^j_{k,n}, \\ Ax(t) + BK_2[x(t - \tau_2(t)) + e_2(t)], & t \in \mathcal{R}^j_{k,n}, \\ Ax(t), & t \in \mathcal{D}_n, \end{cases} \tag{8.11}$$

where $\tau_1(t) \in [0, \tilde{\tau}_1]$ and $\tau_2(t) \in [0, \tilde{\tau}_2]$ and the triggering error $e_i(t)$ $(i = 1, 2)$ satisfies

$$\begin{aligned} e_i^T(t)\Omega e_i(t) \leq & \varepsilon[x(t - \tau_i(t)) + e_i(t)]^T \Omega [x(t - \tau_i(t)) + e_i(t)] \\ & + \frac{1}{\varphi}\eta(t - \tau_i(t)), \end{aligned} \tag{8.12}$$

with

$$\begin{aligned} \frac{d\eta(t)}{dt} = & -\mu\eta(t) + \varepsilon[x(t - \tau_i(t)) + e_i(t)]^T \Omega \\ & [x(t - \tau_i(t)) + e_i(t)] - e_i(t)^T \Omega e_i(t). \end{aligned}$$

In this study, our control objective is to design feedback gains K_1, K_2 and dynamic ETM parameter Ω to stabilize the system (8.11).

Remark 8.4 The system model (8.11) includes three subsystems dynamic. As a special case of our model, the study on resilient control problem in [3, 4] only considered two subsystem dynamic, namely Σ_0 and Σ_1. The subsystem Σ_2 is described individually because its control channel has different and correlated communication environment with relative to primary channel, as reflected by the correlated delay $\tau_1(t)$, $\tau_2(t)$. This study case is also different from the multi-channel system with the assumption of independent delay or packet drop [5, 6]. Although the system we established has only two channels, it can represent a class of multi-channel system from a viewpoint that one part of channels is for primary control while the other part of channels is for redundancy control to prevent DoS attacks.

8.2 Design of Defense Switching Law

Although the strategy of attackers is priori unknown, our defenders should keep the system in an invincible state by well using the redundant resources. We aim to design a defense switching law $\rho(t)$ for the redundant control channel. Before that, an assumption is that DoS attacks can be detected by a real-time on-line detection mechanism.

Assumption 8.2 DoS attacks on the primary channel can be detected by intrusion detection system of cyber layer.

Remark 8.5 Although the strategy of DoS attacks is unknown for us, it can be detected in real-time [7, 8]. In this study, we focus on studying defensive strategy on the basis of the given detection results which identifies whether the primary channel is under attacking at the current moment or not. As DoS attacks is detected, we will generate a switching law to determine when to switch to redundancy control channel.

To eliminate the influence of DoS attacks on system performance, the switching law of redundant control channel is very important to assistant primary control loop to achieve system security. In following, two kinds of defense switching laws are discussed here.

(1) Seamless switching law: Order $\rho(t) = 2$ for all time. Both $u_1(t)$ and $u_2(t)$ are transmitted by primary channel and redundant channel respectively. But actuators will only choose $u_1(t)$ and drop $u_2(t)$ when primary channel is secure without DoS attacks. And system (8.11) will switch to subsystem Σ_2 immediately DoS attacks disrupt the control channel transmitting signal $u_1(t)$. This is perfect to defense DoS attacks. However, seamless switching law requires the redundant channel is always on-line. Thus, this switching law not only wastes much resources and energies but also increases the risk to be attacked. The redundancy channel once invaded by DoS attacks will betray our original intention to achieve active defense.

(2) Concealing switching law: It is not guaranteed that the redundancy control channel would not be found and broken by DoS attacks. Thus, we should design one kind of concealing switching law in order that the redundant channel is to be switched as possible as less while the system performance is compensated under DoS attacks.

In this study, according to Hurwitz convex combination method [9, 10], a concealing switching law is generated. First, we assume that there exist a Hurwizt convex combination

$$E = \sum_{i=1}^{2} \theta_i (A_i + BK_i), \tag{8.13}$$

which satisfies

$$E^T P + PE = -Q.$$

where $\theta_1 + \theta_2 = 1$ and P, Q are positive definite matrices.

Based on P and Q, the switching region is constructed by

$$\begin{aligned}\Omega_i = \{x \in R^n | x^T(t)[(A_i + BK_i)^T P + P(A_i + BK_i)]x(t) \\ \leq -x^T(t)Qx(t)\}\end{aligned} \tag{8.14}$$

According to the obtained Ω_i, a set of switching regions with no overlapping is given by

$$\begin{aligned}\tilde{\Omega}_1 &= \Omega_1, \\ \tilde{\Omega}_2 &= \Omega_2 - (\Omega_2 \cap \tilde{\Omega}_1),\end{aligned} \tag{8.15}$$

where $\cup_{i=1}^{2} \tilde{\Omega}_i = R^n$ and $\tilde{\Omega}_1 \cap \tilde{\Omega}_2 = \varnothing$.

The concealing switching law is constructed by

$$\begin{cases}\rho(t) = 0, \quad when \ \ x(t) \in \tilde{\Omega}_1 \quad or \quad s(t) = 1, \\ \rho(t) = 2, \quad when \ \ x(t) \in \tilde{\Omega}_2 \ \ and \ \ s(t) = 0.\end{cases} \tag{8.16}$$

Remark 8.6 The switching law design problem in this study is different from conventional one [9, 10]. First, we use $A + BK_1$ but A to represent the subsystem Σ_0 in Hurwizt convex combination. On the one hand, if the subsystem Σ_0 with unstable A were in Hurwizt convex combination, all states $x(t)$ will always be fallen into the region $\tilde{\Omega}_2$ and then we only obtain a seamless switching law. On the other hand, considering that system under DoS attacks has some kinds of resilience, we choose subsystem $A + BK_1$ to construct Hurwizt convex combination (8.13). Second, because the switching law to be designed is under the comprehensive constraint of DoS attacks strategy and defend strategy. In convention study [9, 10], the switching law only between subsystem Σ_1 and subsystem Σ_2 is generated by

$$\sigma(t) = i, \quad x(t) \in \tilde{\Omega}_i, \quad (i = 1, 2). \tag{8.17}$$

According to [10], the system with subsystem Σ_1, Σ_2 is stable under the switching law (8.17). However, the switching law (8.17) will be disturbed by attacks signal and another unstable subsystem Σ_0 caused by attacks will be introduced to damage the stability of system (8.11). It can not be guaranteed that system (8.11) is stable under the disturbed switching law.

8.3 Stability Analysis

In this section, we will demonstrate that the concealing switching law (8.16) can stabilize system (8.11) under DoS attacks. First, the switching signals $\sigma(t)$ is generated by the superposition of the concealing switching law (8.16) and DoS attacks signal.

$$\sigma(t)=\begin{cases}0, & s(t)=0, \ \ and \ \ \rho(t)=0,\\ 1, & s(t)=1, \ \ and \ \ \rho(t)=0,\\ 2, & s(t)=0, \ \ and \ \ \rho(t)=2.\end{cases} \tag{8.18}$$

According to the resulting switching signal $\sigma(t)$ (8.18), the switchings between subsystems have several situations as explanation in follows Table 8.1.

And the subsystem switching logic is depicted in Fig. 8.2. The stability of system driven by the switching logic is developed in following Theorem 1.

Theorem 8.1 *Given λ, ϕ, $\bar{\tau}$, h, α_i, $(i=0,1,2)$, θ_i, $(i=1,2)$, $\varepsilon(>\alpha_1,\alpha_2)$, φ, μ and δ, system (8.11) is exponential stable under switching law (8.18) if there exists positive definite matrices P, R, Z, W and matrices K_1, K_2, X with appropriate dimensions satisfying*

$$\Xi_i<0, (i=0,1,2), \tag{8.19}$$

Table 8.1 Explanation of subsystem switching

(1) $\sigma(t^-)=0\rightarrow\sigma(t^+)=1$	DoS attacks ending
(2) $\sigma(t^-)=0\rightarrow\sigma(t^+)=2$	Defense during DoS attacks
(3) $\sigma(t^-)=1\rightarrow\sigma(t^+)=0$	DoS attacks starting without defense
(4) $\sigma(t^-)=1\rightarrow\sigma(t^+)=2$	DoS attacks starting while defense starting
(5) $\sigma(t^-)=2\rightarrow\sigma(t^+)=0$	Defense ending during DoS attacks
(6) $\sigma(t^-)=2\rightarrow\sigma(t^+)=1$	DoS attacks ending while defense ending

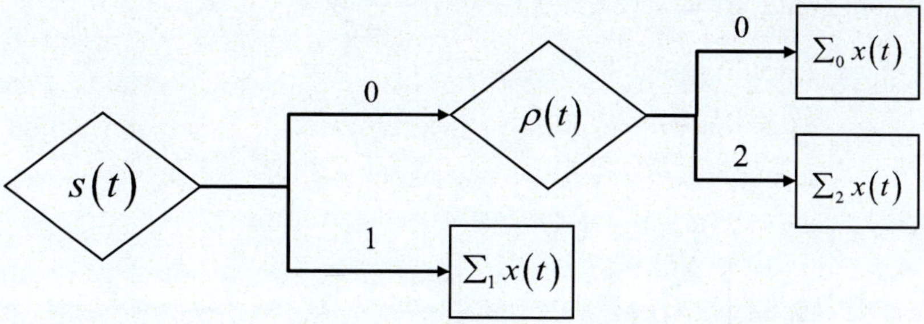

Fig. 8.2 Switching subsystem under attacks signal $s(t)$ and defense law $\rho(t)$

where

$$\Xi_1 = \begin{bmatrix} \Phi_0 & * & * \\ \tilde{\tau}_1\Phi_1 & -Z^{-1} & * \\ \tilde{\tau}_1\Phi_1 & 0 & -W^{-1} \end{bmatrix},$$

$$\Phi_0 = \begin{bmatrix} \Phi_0(1,1) & \Phi_0(1,2) & -\rho_1 X & PBK_1 \\ * & \Phi_0(2,2) & \rho_1(X-Z) & \psi_1\varepsilon\Omega \\ * & * & \Phi_0(3,3) & 0 \\ * & * & * & \psi_1(\varepsilon-1)\Omega \end{bmatrix},$$

$$\Phi_0(1,1) = He(PA) + \alpha_1 P + R + \rho_1 Z - \frac{\pi^2}{4}W,$$

$$\Phi_0(1,2) = PBK_1 - \rho_1 Z + \rho_1 X + \frac{\pi^2}{4}W,$$

$$\Phi_0(2,2) = \rho_1(2Z - X - X^T) - \frac{\pi^2}{4}W + \psi_1\varepsilon\Omega,$$

$$\Phi_0(3,3) = -e^{-\alpha_1\tilde{\tau}_1}R + \rho_1 Z,$$

$$\Phi_1 = [A, BK_1, 0, BK_1],$$

$$\tilde{\tau}_1 = e^{(\frac{1}{\lambda}-1)\phi}\bar{\tau} + h,\ \rho_1 = \alpha_1\tilde{\tau}_1/(1-e^{\alpha_1\tilde{\tau}_1}),$$

$$\psi_1 = \varphi(\mu - \alpha_1) + 1.$$

$$\Xi_2 = \begin{bmatrix} \Gamma_0 & * & * \\ \tilde{\tau}_2\Gamma_1 & -Z^{-1} & * \\ \tilde{\tau}_2\Gamma_1 & 0 & -W^{-1} \end{bmatrix},$$

$$\Gamma_0 = \begin{bmatrix} \Gamma_0(1,1) & \Gamma_0(1,2) & -\rho_2 X & PBK_2 \\ * & \Gamma_0(2,2) & \rho_2(X-Z) & \psi_2\varepsilon\Omega \\ * & * & \Gamma_0(3,3) & 0 \\ * & * & * & \psi_2(\varepsilon-1)\Omega \end{bmatrix},$$

$$\Gamma_0(1,1) = He(P(E - B_2K_2)) + \alpha_2 P + R + \rho_2 Z - \frac{\pi^2}{4}W,$$

$$\Gamma_0(1,2) = PBK_2 - \rho_2 Z + \rho_2 X + \frac{\pi^2}{4}W,$$

$$\Gamma_0(2,2) = \rho_2(2Z - X - X^T) - \frac{\pi^2}{4}W + \psi_2\varepsilon\Omega,$$

$$\Gamma_0(3,3) = -e^{-\alpha_2\tilde{\tau}_2}R + \rho_2 Z,$$

$$\Gamma_1 = [A, BK_2, 0, BK_2],$$

$$\tilde{\tau}_2 = e^{(\frac{1}{1-\lambda}-1)\phi}\bar{\tau} + h,\ \rho_2 = \alpha_2\tilde{\tau}_2/(1-e^{\alpha_2\tilde{\tau}_2}),$$

$$\psi_2 = \varphi(\mu - \alpha_2) + 1,$$

$$\Xi_0 = \begin{bmatrix} \Lambda_0 & * & * \\ \bar{\tau}\Lambda_1 & -Z^{-1} & * \\ \bar{\tau}\Lambda_1 & 0 & -W^{-1} \end{bmatrix},$$

$$\Lambda_0 = \begin{bmatrix} \Lambda_0(1,1) & \Lambda_0(1,2) & -\rho_0 X \\ * & \Lambda_0(2,2) & \rho_0(X-Z) \\ * & * & -e^{-\alpha_0 \bar{\tau}} R + \rho_0 Z \end{bmatrix},$$

$$\Lambda_0(1,1) = He(P(E - B_1 K_1)) + \alpha_0 P + R + \rho_0 Z - \frac{\pi^2}{4} W,$$

$$\Lambda_0(1,2) = -\rho_0 Z + \rho_0 X + \frac{\pi^2}{4} W,$$

$$\Lambda_0(2,2) = \rho_0(2Z - X - X^T) - \frac{\pi^2}{4} W,$$

$$\Lambda_1 = [A, 0, 0], \ \rho_0 = \alpha_0 \bar{\tau}/(1 - e^{\alpha_0 \bar{\tau}}).$$

Proof According to Remark 8.6, the switched system that is consisted of the subsystem $\sigma(t) = 1$ and $\sigma(t) = 2$ can be stabilized by the switching law (8.17) without of considering the DoS attacks. Comparing with the conventional switching law (8.17), we have a variation of switching law (8.18) which is the superposition of the concealing switching control law (8.16) and DoS attacks signal $s(t)$.

Further, we will show that the variation of switching law (8.18) can also stabilize the switched system (8.11).

For each subsystem, choose a Lyapunov–Krasovskii functional as followings.

$$\begin{aligned} V_i(t) &= x^T(t) P x(t) + \int_{t-\bar{\tau}_i}^{t} x^T(s) exp(\star) R x(s) ds \\ &+ \bar{\tau}_i \int_{-\bar{\tau}_i}^{0} \int_{t+\theta}^{t} \dot{x}^T(s) exp(\star) Z \dot{x}(s) ds d\theta, \\ &+ \bar{\tau}_i^2 \int_{t-\tau_i(t)}^{t} \dot{x}^T(s) exp(\star) W \dot{x}(s) ds, \\ &- \frac{\pi^2}{4} \int_{t-\tau_i(t)}^{t} [x(s) - x(t-\tau_i(t))]^T \\ &\quad exp(\star) W [x(s) - x(t-\tau_i(t))] ds. \end{aligned} \tag{8.20}$$

where $i = 0, 1, 2$ and $exp(\star) = e^{\alpha_i(s-t)}$.

Case I: when $\sigma(t) = 1$, the subsystem Σ_1 is active. Based on (8.20), choose a Lyapunov functional for $\Sigma_1, t \in \mathcal{H}_{k,n}^j$

$$V(t) = \eta(t)|_{t=t_{k_j}h^+} + V_1(t). \tag{8.21}$$

Taking derivation of Lyapunov–Krasovskii functional (8.21) along subsystem $\Sigma_1, t \in \mathcal{H}_{k,n}^j$, it has

$$
\begin{aligned}
\dot{V}(t) \leq & -\alpha_1 V(t) + \alpha_1 \eta(t_{k_j}h^+) + \alpha_1 x^T(t)Px(t) \\
& - \mu\eta(t_{k_j}h^+) - e_1^T(t)\Omega e_1(t) \\
& + \varepsilon[x(t-\tau_1(t)) + e_1(t)]^T \Omega [x(t-\tau_1(t)) + e_1(t)] \\
& + 2x^T(t)P[Ax(t) + BK_1x(t-\tau_1(t)) + BK_1e_1(t)] \\
& + x^T(t)Rx(t) - e^{-\alpha_1\tilde{\tau}_1}x^T(t-\tilde{\tau}_1)Rx(t-\tilde{\tau}_1) \\
& + \tilde{\tau}_1^2\xi_1^T(t)\Phi_1^T(Z+W)\Phi_1\xi_1(t) \\
& - \tilde{\tau}_1 \int_{t-\tilde{\tau}_1}^{t} e^{\alpha_1(s-t)}\dot{x}^T(s)Z\dot{x}(s)ds \\
& - \frac{\pi^2}{4}[x(t) - x(t-\tau_1(t))]^T W [x(t) - x(t-\tau_1(t))]
\end{aligned}
\tag{8.22}
$$

where $\xi_1(t) = col[x(t), x(t-\tau_1(t)), x(t-\tilde{\tau}_1), e_1(t)]$.

Recalling $\eta(t_{k_j}h^+) = \eta(t-\tau_i(t))$ and (8.12), one has

$$
\begin{aligned}
& -(\mu - \alpha_1)\eta(t_{k_j}h^+) \leq -\varphi(\mu - \alpha_1)e_1^T(t)\Omega e_1(t) + \varepsilon\varphi \\
& (\mu - \alpha_1)[x(t-\tau_1(t)) + e_1(t)]^T \Omega [x(t-\tau_1(t)) + e_1(t)].
\end{aligned}
$$

According to Lemmas 1 and 2 in [11], we have

$$
\begin{aligned}
& -\tilde{\tau}_1 \int_{t-\tilde{\tau}_1}^{t} e^{\alpha_1(s-t)}\dot{x}^T(s)Z\dot{x}(s)ds \\
& \leq \rho_1 \begin{bmatrix} \eta_1^T & \eta_2^T \end{bmatrix} \begin{bmatrix} Z & X \\ X^T & Z \end{bmatrix} \begin{bmatrix} \eta_1 \\ \eta_2 \end{bmatrix}
\end{aligned}
$$

where $\rho_1 = \alpha_1\tilde{\tau}_1/(1 - e^{\alpha_1\tilde{\tau}_1})$, $\eta_1 = x(t) - x(t-\tau_1(t))$, $\eta_2 = x(t-\tau_1(t)) - x(t-\tilde{\tau}_1)$.

By using Shur Complement lemma, it is further obtained

$$
\dot{V}(t) \leq -\alpha_1 V(t) + \xi_1^T(t)\Xi_1\xi_1(t). \tag{8.23}
$$

According to (8.19), one has

$$
\dot{V}(t) \leq -\alpha_1 V(t). \tag{8.24}
$$

Case II: When $\sigma(t) = 2$, the subsystem Σ_2 is active. Based on (8.20), choose a Lyapunov functional for Σ_2, $t \in \mathcal{R}_{k,n}^j$

$$
V(t) = \eta(t)|_{t=\hat{t}_{k_j}h^+} + V_2(t). \tag{8.25}
$$

Because $x(t) \in \tilde{\Omega}_2$, according to (8.13) and (8.14), one has

$$0 \leq -x^T(t)[(A + B_2K_2)^T P + P(A + B_2K_2)]x(t) \\ + x^T(t)(E^T P + PE)x(t) \tag{8.26}$$

Taking derivation of Lyapunov–Krasovskii functional (8.25) along subsystem Σ_2, and adding (8.26) to the right side of the derivation of Lyapunov–Krasovskii functional (8.25), we obtain

$$\dot{V}(t) \leq -\alpha_2 V(t) + \xi_2^T(t)\Xi_2\xi_2(t). \tag{8.27}$$

where $\xi_2(t) = col[x(t), x(t - \tau_2(t)), x(t - \tilde{\tau}_2), e_2(t)]$ and

According to (8.19), one has

$$\dot{V}(t) \leq -\alpha_2 V(t). \tag{8.28}$$

Case III: when $\sigma(t) = 0$, the subsystem Σ_0 is active. Based on (8.20), choose a Lyapunov functional as follows

$$V(t) = V_0(t), \tag{8.29}$$

where $\tilde{\tau}_0 = \bar{\tau}$. Because $x(t) \in \tilde{\Omega}_1$, according to (8.13) and (8.14), the constraint should be satisfied

$$0 \leq -x^T(t)[(A + B_1K_1)^T P + P(A + B_1K_1)]x(t) \\ + x^T(t)(E^T P + PE)x(t). \tag{8.30}$$

Taking derivation of Lyapunov–Krasovskii functional (8.29) along subsystem Σ_0 and adding (8.30) to the right side of the derivation of Lyapunov–Krasovskii functional (8.29), we obtain

$$\dot{V}(t) \leq -\alpha_0 V(t) + \xi_0^T(t)\Xi_0\xi_0(t), \tag{8.31}$$

where $\xi_0(t) = col[x(t), x(t - \tau_0(t)), x(t - \bar{\tau})]$.

According to (8.19), one has

$$\dot{V}(t) \leq -\alpha_0 V(t). \tag{8.32}$$

Denote $\alpha = min\{\alpha_0, \alpha_1, \alpha_2\}$. According to (8.24), (8.28) and (8.32), one has $\dot{V}(t) \leq -\alpha V(t)$. By denoting $\varpi_1 = \lambda_{min}(P)$, $\tau_{max} = max(\tau_1, \tau_2)$, $\pi_0 = \lambda_{max}(P) + \tau_{max}\lambda_{max}(R) + \tau_{max}^3/2\lambda_{max}(Z) + \tau_{max}^3\lambda_{max}(W)$, $\|\chi_0\| = max\{\|x(t)\|, \|\dot{x}(t)\|\}$, $t \in [-\tau_{max}, 0]$, one has

$$\|x(t)\| \leq \sqrt{\frac{\varpi_0}{\varpi_1}} e^{-\frac{\alpha t}{2}} \|\chi_0\|$$

which indicates that the switched delay system (8.11) is exponential stable.

8.4 Switching Controller Design

In this section, we use LMIs technique to provide a co-design method of feedback control gains K_1, K_2 and dynamic triggering parameter Ω.

Theorem 8.2 *Given λ, ϕ, $\bar{\tau}$, h, α_i, $(i = 0, 1, 2)$, θ_i, $(i = 1, 2)$, $\varepsilon(> \alpha_1, \alpha_2)$, φ, μ and δ, system (8.11) is exponential stable under the concealing switching law (8.18) if there exists positive definite matrices $\tilde{P}$, $\tilde{R}$, $\tilde{Z}$, $\tilde{W}$, $\tilde{\Omega}$ and matrices $\tilde{K}_1$, $\tilde{K}_2$, $\tilde{X}$ with appropriate dimensions satisfying*

$$\tilde{\Xi}_i < 0, (i = 0, 1, 2), \tag{8.33}$$

$$\begin{bmatrix} \tilde{Z} & \tilde{X} \\ \tilde{X}^T & \tilde{Z} \end{bmatrix} \geq 0, \tag{8.34}$$

$$E\tilde{P} + \tilde{P}E^T \leq 0. \tag{8.35}$$

where

$$\tilde{\Xi}_1 = \begin{bmatrix} \tilde{\Phi}_0 & * & * \\ \bar{\tau}_1\tilde{\Phi}_1 & \delta^2\tilde{Z} - 2\delta\tilde{P} & * \\ \bar{\tau}_1\tilde{\Phi}_1 & 0 & \delta^2\tilde{W} - 2\delta\tilde{P} \end{bmatrix},$$

$$\tilde{\Phi}_0 = \begin{bmatrix} \tilde{\Phi}_0(1,1) & \tilde{\Phi}_0(1,2) & -\rho_1\tilde{X} & B\tilde{K}_1 \\ * & \tilde{\Phi}_0(2,2) & \rho_1(\tilde{X} - \tilde{Z}) & \psi_1\varepsilon\tilde{\Omega} \\ * & * & \tilde{\Phi}_0(3,3) & 0 \\ * & * & * & \psi_1(\varepsilon - 1)\tilde{\Omega} \end{bmatrix},$$

$$\begin{aligned}
\tilde{\Phi}_0(1,1) &= He(A\tilde{P}) + \alpha_1\tilde{P} + \tilde{R} + \rho_1\tilde{Z} - \frac{\pi^2}{4}\tilde{W}, \\
\tilde{\Phi}_0(1,2) &= B\tilde{K}_1 - \rho_1\tilde{Z} + \rho_1\tilde{X} + \frac{\pi^2}{4}\tilde{W}, \\
\tilde{\Phi}_0(2,2) &= \rho_1(2\tilde{Z} - \tilde{X} - \tilde{X}^T) - \frac{\pi^2}{4}\tilde{W} + \psi_1\varepsilon\tilde{\Omega}, \\
\tilde{\Phi}_0(3,3) &= -e^{-\alpha_1\bar{\tau}_1}\tilde{R} + \rho_1\tilde{Z}, \\
\tilde{\Phi}_1 &= [A\tilde{P}, B\tilde{K}_1, 0, B\tilde{K}_1], \\
\bar{\tau}_1 &= e^{(\frac{1}{\lambda}-1)\phi}\bar{\tau} + h, \ \rho_1 = \alpha_1\bar{\tau}_1/(1 - e^{\alpha_1\bar{\tau}_1}), \\
\psi_1 &= \varphi(\mu - \alpha_1) + 1,
\end{aligned}$$

$$\tilde{\Xi}_2 = \begin{bmatrix} \tilde{\Gamma}_0 & * & * \\ \bar{\tau}_2\tilde{\Gamma}_1 & \delta^2\tilde{Z} - 2\delta\tilde{P} & * \\ \bar{\tau}_2\tilde{\Gamma}_1 & 0 & \delta^2\tilde{W} - 2\delta\tilde{P} \end{bmatrix},$$

$$\tilde{\Gamma}_0 = \begin{bmatrix} \tilde{\Gamma}_0(1,1) & \tilde{\Gamma}_0(1,2) & -\rho_2\tilde{X} & B\tilde{K}_2 \\ * & \tilde{\Gamma}_0(2,2) & \rho_2(\tilde{X}-\tilde{Z}) & \psi_2\varepsilon\tilde{\Omega} \\ * & * & \tilde{\Gamma}_0(3,3) & 0 \\ * & * & * & \psi_2(\varepsilon-1)\tilde{\Omega} \end{bmatrix},$$

$$\begin{aligned}
\tilde{\Gamma}_0(1,1) =& He(\sum_{i=1}^{2}\theta_i(A\tilde{P}+B\tilde{K}_i) - B_2\tilde{K}_2) + \alpha_2\tilde{P} \\
&+ \tilde{R} + \rho_2\tilde{Z} - \frac{\pi^2}{4}\tilde{W}, \\
\tilde{\Gamma}_0(1,2) =& B\tilde{K}_2 - \rho_2\tilde{Z} + \rho_2\tilde{X} + \frac{\pi^2}{4}\tilde{W}, \\
\tilde{\Gamma}_0(2,2) =& \rho_2(2\tilde{Z} - \tilde{X} - \tilde{X}^T) - \frac{\pi^2}{4}\tilde{W} + \psi_2\varepsilon\tilde{\Omega}, \\
\tilde{\Gamma}_0(3,3) =& -e^{-\alpha_2\bar{\tau}_2}\tilde{R} + \rho_2\tilde{Z}, \\
\tilde{\Gamma}_1 =& [A\tilde{P}, B\tilde{K}_2, 0, B\tilde{K}_2], \\
\bar{\tau}_2 =& e^{(\frac{1}{1-\lambda}-1)\phi}\bar{\tau} + h,\ \rho_2 = \alpha_2\bar{\tau}_2/(1-e^{\alpha_2\bar{\tau}_2}), \\
\psi_2 =& \varphi(\mu-\alpha_2)+1,
\end{aligned}$$

$$\tilde{\Xi}_0 = \begin{bmatrix} \tilde{\Lambda}_0 & * & * \\ \bar{\tau}\tilde{\Lambda}_1 & \delta^2\tilde{Z} - 2\delta\tilde{P} & * \\ \bar{\tau}\tilde{\Lambda}_1 & 0 & \delta^2\tilde{W} - 2\delta\tilde{P} \end{bmatrix},$$

$$\tilde{\Lambda}_0 = \begin{bmatrix} \tilde{\Lambda}_0(1,1) & \tilde{\Lambda}_0(1,2) & -\rho_0\tilde{X} \\ * & \tilde{\Lambda}_0(2,2) & \rho_0(\tilde{X}-\tilde{Z}) \\ * & * & -e^{-\alpha_0\bar{\tau}}\tilde{R} + \rho_0\tilde{Z} \end{bmatrix},$$

$$\begin{aligned}
\tilde{\Lambda}_0(1,1) =& He(\sum_{i=1}^{2}\theta_i(A\tilde{P}+B\tilde{K}_i) - B_1\tilde{K}_1) \\
&+ \alpha_0\tilde{P} + \tilde{R} + \rho_0\tilde{Z} - \frac{\pi^2}{4}\tilde{W}, \\
\tilde{\Lambda}_0(1,2) =& -\rho_0\tilde{Z} + \rho_0\tilde{X} + \frac{\pi^2}{4}\tilde{W}, \\
\tilde{\Lambda}_0(2,2) =& \rho_0(2\tilde{Z} - \tilde{X} - \tilde{X}^T) - \frac{\pi^2}{4}\tilde{W}, \\
\tilde{\Lambda}_1 =& [A\tilde{P}, 0, 0],\ \rho_0 = \alpha_0\bar{\tau}/(1-e^{\alpha_0\bar{\tau}}).
\end{aligned}$$

The state feedback control gains can be obtained by $K_i = \tilde{K}_i\tilde{P}^{-1}$.

Proof Denote $\tilde{P} = P^{-1}, \tilde{K} = K\tilde{P}, \tilde{Q} = \tilde{P}Q\tilde{P}, \tilde{R} = \tilde{P}R\tilde{P}, \tilde{Z} = \tilde{P}Z\tilde{P}, \tilde{W} = \tilde{P}W\tilde{P}$ and $\tilde{X} = \tilde{P}X\tilde{P}$. Pre-and pro-multiply $diag\{\tilde{P}, \tilde{P}, \tilde{P}, \tilde{P}, I, I\}$ and its transpose on both sides of $\Xi_1 < 0$ and pre-and pro-multiply $diag\{\tilde{P}, \tilde{P}, \tilde{P}, I, I\}$ and its transpose on both sides of $\Xi_0 < 0$ By using inequality technique $-Z^{-1} < \delta^2\tilde{Z} - 2\delta\tilde{P}$, LMI

(8.33) $\hat{\Xi}_i < 0$, $(i = 0, 1, 2)$ is derived. With the same process, (8.34) and (8.35) can be obtained.

Remark 8.7 The feasibility of LMIs (8.33)–(8.35) depends on the following tuning parameters: referring to sample periodic h; delay bound λ, ϕ, $\bar{\tau}$; DETM ε, φ, μ; Hurwitz convex combination θ_i, $(i = 1, 2)$; exponential decay rates α_i, $(i = 0, 1, 2)$ and tight inequality δ. The above parameters are coupling with each other to influence the feasibility of LMIs and the triggering control performance. Especially, the sample periodic h and the tight inequality δ can be viewed as the fundamental parameters for the feasibility of LMIs. And it is better to choose small value h and large value δ. For the feasible LMIs, other parameters refer to exponential stability, triggering efficiency, switching concealment and tolerable delay bound, respectively. It could be right for better exponential stability with large α_i, $(i = 0, 1, 2)$, for larger triggering interval with small value φ, μ and large value ε, for better switching concealment with large θ_1 $(\theta_2 = 1 - \theta_1)$, for larger tolerable delay bound with large ϕ, $\bar{\tau}$. However, due to the coupling characteristic of these parameters, the mentioned performances will happen to be a tradeoff by choosing suitable corresponding parameters.

8.5 Simulation Example

In this section, we simulate networked control systems under DoS attacks to verify that system can be exponential stable under redundant control regulated by the designed concealing switching law. Both numerical example and practice control system simulation will be carried out.

8.5.1 A Numerical Example

The system matrices are given by

$$A = \begin{bmatrix} 3 & 2 \\ -1 & 1 \end{bmatrix}, B = \begin{bmatrix} -1 & 0 \\ 1 & -1 \end{bmatrix}.$$

Other parameters are given by $\alpha_0 = \alpha_1 = \alpha_2 = 0.1$, $\theta_1 = \theta_2 = 0.5$, $\delta = 0.3$. $\lambda = 0.7$, $\phi = 0.2$, $\bar{\tau} = 0.1s$, $h = 0.01s$, $ts = 0.001s$ and $\varepsilon = 0.1$, $\varphi = 0.3$, $\mu = 0.2$. According to Theorem 8.2, we obtain that

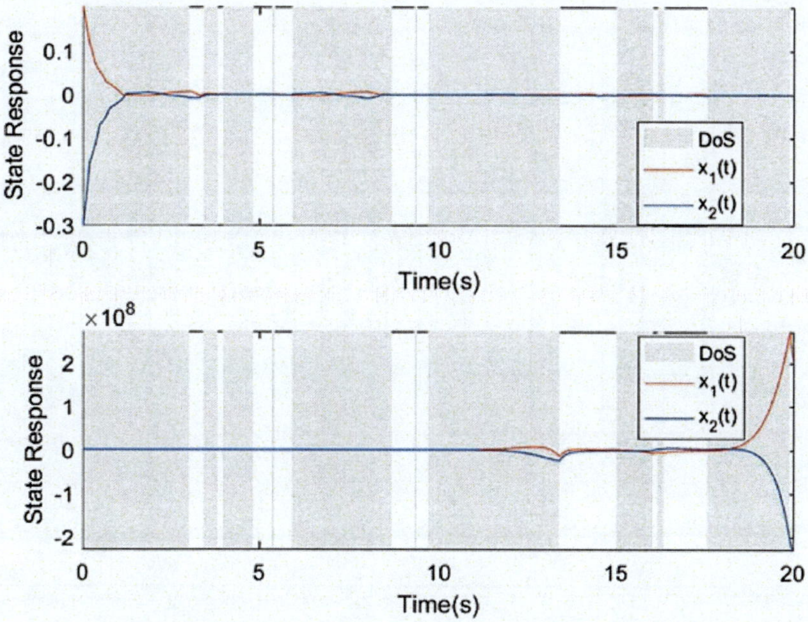

Fig. 8.3 System responses under DoS with and without redundancy control

$$K_1 = \begin{bmatrix} 8.0044 & 3.9747 \\ 11.1930 & 8.2087 \end{bmatrix}, K_2 = \begin{bmatrix} 5.7209 & 2.2924 \\ 16.3787 & 13.1491 \end{bmatrix},$$

$$\Omega = 10^{11} \begin{bmatrix} 5.5411 & 3.0286 \\ 3.0286 & 1.7841 \end{bmatrix}, P = 10^{9} \begin{bmatrix} 7.0905 & 3.5045 \\ 3.5045 & 2.5818 \end{bmatrix},$$

$$Q = -(E^T P + PE) = 10^{11} \begin{bmatrix} 1.1031 & 0.6497 \\ 0.6497 & 0.4174 \end{bmatrix}.$$

Figure 8.3 shows the dynamic trajectories of the simulated system under DoS attacks with adopting our proposed method and conventional control, respectively. The grey areas are the attack intervals. In the first subgraph of Fig. 8.3, intuitively, the states are stable under dynamic triggering control as well as switching law. It illustrates the validness of our defense method by comparing with the conventional control having unstable trajectories in the second subgraph of Fig. 8.3.

As the main contribution in our study, the concept of concealing switching law is well illustrated in Fig. 8.4. First, it can be observed that during sleeping intervals the switching signal is always taking $\sigma(t) = 1$ representing the online of primary channel while the offline of redundancy channel to avoid being detected by attacker before attack implement. Second, during DoS attacks some intervals with $\sigma(t) = 2$ represents the redundancy channel in active while others with $\sigma(t) = 0$ represents the concealing of redundancy channel to conceal the defense aim. Even though the use time of redundancy channel is not completely overlapping with attack intervals, the resulting stable system as shown in Fig. 8.3 verifies the validness of the concealing

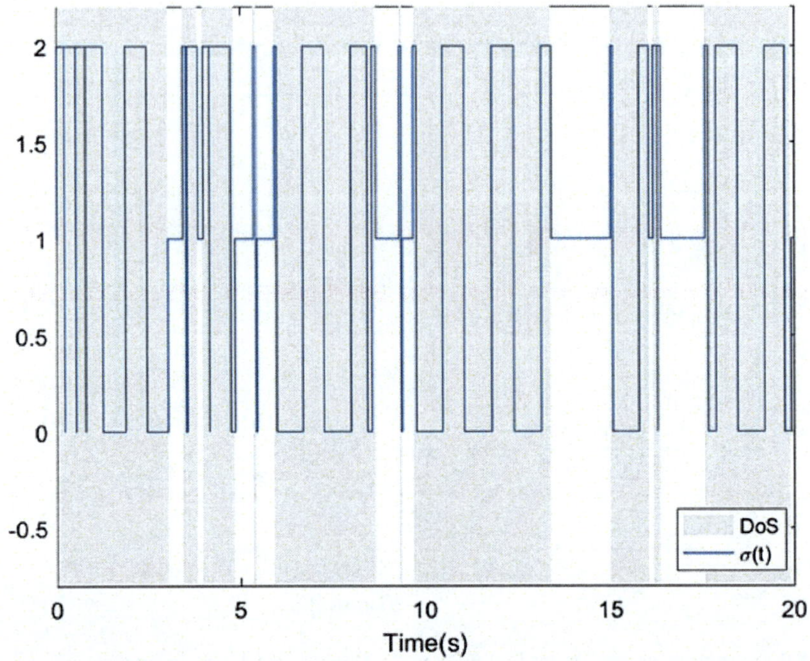

Fig. 8.4 Switching law

Fig. 8.5 The triggered instants and release intervals in PR channels

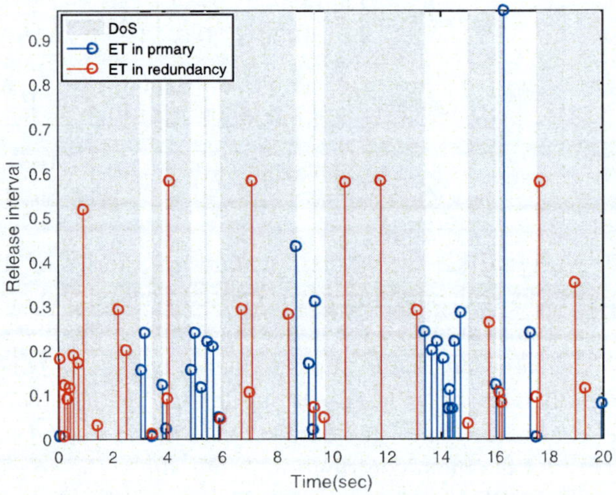

switch law especially for system with the limited communication resources and energy.

To have efficient use of limited communication resources, dynamic event-triggered mechanism is presented. The triggering instants and release intervals are depicted in Fig. 8.5. The blue stems launch in primary channel while the red stems in redundancy channel. Comparing with the sample periodic $h = 0.01s$, the sparse release

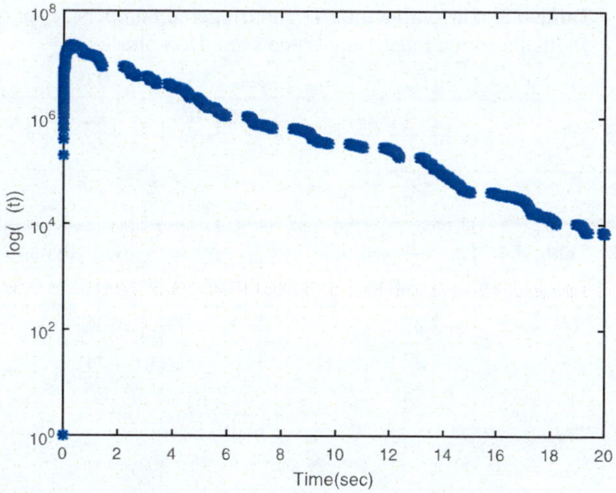

Fig. 8.6 $log(\eta(t))$ with $\eta(t_0) = 1$

Table 8.2 The delay upper bound $\tilde{\tau}_1$, $\tilde{\tau}_2$ for various resources allocation rates λ

λ	0.1	0.3	0.5	0.7	0.9
$\tilde{\tau}_1$	0.6150	0.1695	0.1321	0.1189	0.1122
$\tilde{\tau}_2$	0.1122	0.1189	0.1321	0.1695	0.6150

intervals proof that our designed dynamic event-triggered mechanism can save more communication bandwidth.

Figure 8.6 shows the trajectory of the positive piecewise function $\eta(t)$ due to the existence of DoS attacks when $t \in \mathcal{D}_n$. It appears to decay fast with decay rate $\mu = 0.2$. The positive character of $\eta(t)$ will delay the average triggering periodic while the decay character will have the dynamic tuning ability of ETM as well as guarantee the necessary triggering control to stabilize system.

Then, we explore the relationship between the resources allocation rates λ and the communication situation like delay bound. Table 8.2 can be observed that with the increase of λ the delay upper bound $\tilde{\tau}_1$ of the primary channel is decreased while the delay upper bound $\tilde{\tau}_2$ of the redundancy channel is increased. Given that λ is the resources allocation rate for primary channel, it is obvious that the large value of λ represents the more communication resource bandwidth will be configured in primary channel. Then, the delay bound $\tilde{\tau}_1$ become small with the large communication bandwidth under the specified transmitted load ϕ. It is also reasonable that the delay bound $\tilde{\tau}_2$ become large with the equipped small communication bandwidth.

From Table 8.3, the online time T_2 of the redundancy channel has the tendency of from reducing to raising. The minimum value of T_2 appears with $\lambda = 0.7$. It is suggested that it is better to obtain the best concealing defense strategy by choosing the allocation ratio nearby 0.7. The triggering numbers N_2, however, are positive correlation to the online time T_2. Under the dynamic ETM, the average triggering

Table 8.3 The online time T_2 and triggered numbers N_2 of redundancy channel for various bandwidth allocation rates λ under the same DoS attacks

λ	0.2	0.4	0.6	0.7	0.8
T_2	14.5060	13.6090	12.7020	10.1060	14.5060
N_2	80	59	55	51	70

Table 8.4 The released numbers N_1, N_2 for various dynamic triggering parameters φ

φ	0.2	0.4	0.6	0.8	1.2
N_1	27	27	26	27	29
N_2	27	31	34	35	37
N	54	58	60	62	66

Table 8.5 The total released numbers N and the total online time T of primary-redundancy (PR) channels and the average triggering periodic T/N for various trigger parameters μ

μ	0.3	0.4	0.5	0.6	0.7
N	33	32	23	38	36
T	14.7160	13.5520	14.5850	14.8020	14.0550
T/N	0.3003	0.2337	0.2917	0.2552	0.2130

periodic is relative average under different bandwidth allocation to preserve system stable within the limited control time of redundancy channel.

Table 8.4 shows the function of the trigger tuning parameters φ. As the increase of φ, the triggered numbers N_1 and N_2 are becoming more and more large. According to (8.6), one can understand that the right hand of the triggering condition will become small to reduce the triggering intervals as well as increase the triggering numbers.

Table 8.5 will try to illustrate the function of tuning parameter μ. According to (8.7), μ is the decay rate of differential equation $\eta(t)$. It can be inferred that the large μ will lead to the fast reduction of the function value $\eta(t)$ to reduce the triggering periodic T/N. Table 8.5, however, disagree the above analysis according to the decreasing trend but with the wave at $\mu = 0.5$. Because the wave might be subjected to the term $\varepsilon x^T(t_k h)\Omega x(t_k h) - [x(t_{k_j} h) - x(t_k h)]^T \Omega [x(t_{k_j} h) - x(t_k h)]$.

8.5.2 A Practice Example

In this section, the proposed dynamic triggering control method is employed to a batch reactor system (8.1). The matrices of linearized system model are given as follows.

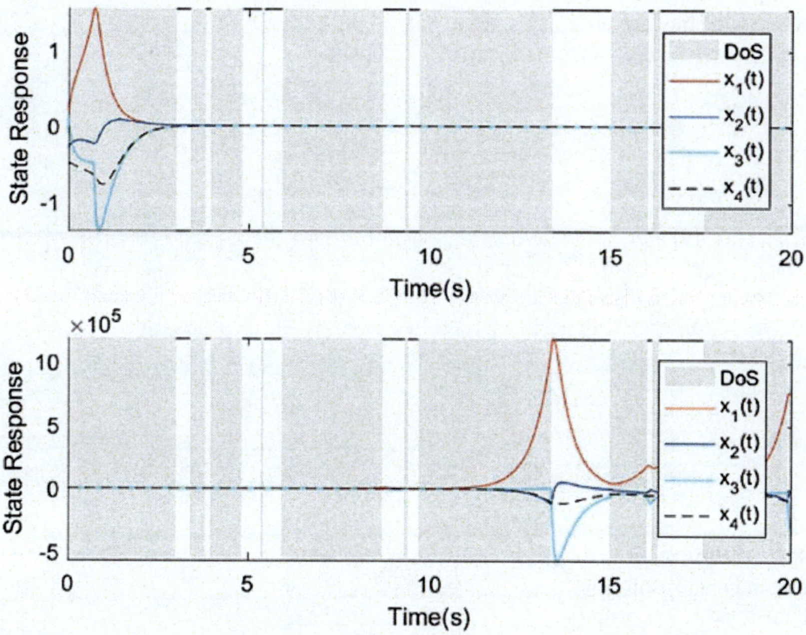

Fig. 8.7 System responses under DoS with redundancy control (upper sub-figure) and without (lower sub-figure)

$$A = \begin{bmatrix} 1.38 & -0.2077 & 6.715 & -5.676 \\ -0.5814 & -4.29 & 0 & 0.675 \\ 1.067 & 4.273 & -6.654 & 5.893 \\ 0.048 & 4.273 & 1.343 & -2.104 \end{bmatrix},$$

$$B = \begin{bmatrix} 0 \; 5.679 & 1.136 & 1.136 \\ 0 \quad 0 & -3.146 & 0 \end{bmatrix}^T.$$

Other parameters are given by $\alpha_0 = \alpha_1 = \alpha_2 = 1$, $\theta_1 = \theta_2 = 0.5$, $\delta = 0.3$. $\lambda = 0.712$, $\phi = 0.2$, $\bar{\tau} = 0.2s$, $h = 0.01s$, $ts = 0.001s$ and $\varepsilon = 0.1$, $\varphi = 0.3$, $\mu = 5$.

According to Theorem 8.2, the states trajectories are depicted in Fig. 8.7. As can be seen, the states are approaching zeros under DoS attacks with our designed redundancy channel based dynamic triggering law. However, the convention control can not stabilize system. Thus, the validness of our proposed method is verified.

The switching law is depicted in Fig. 8.8. A interesting phenomenon can be observed that the redundancy channel has not been stimulated at the neighborhood of the initial time. Because the primary channel at initial time is responsible for the transmission of the feedback control inputs. Even though DoS attacks occurs after the initial time immediately, system has some kind of resilience performance which can be view as the nature of system ability of tolerating DoS attacks.. This kind of resilience is the important evidence to achieve the concealing switch control. That

Fig. 8.8 Switching law

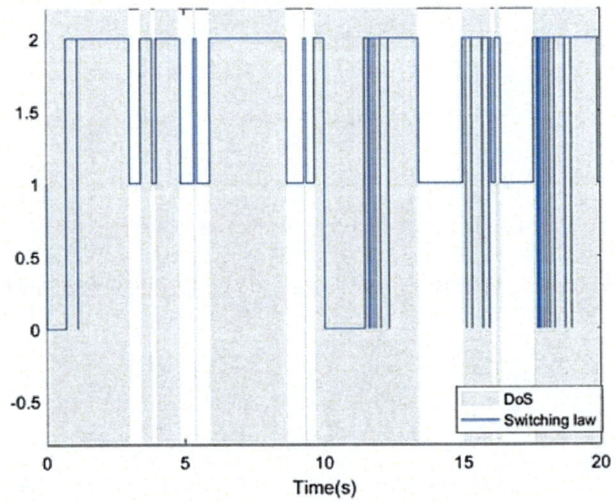

Fig. 8.9 The triggered instants and release intervals in PR channels

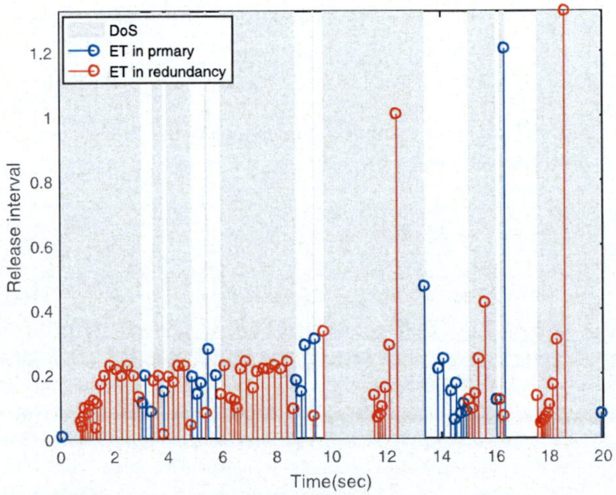

is why we use the subsystem Σ_1 to replace the subsystem Σ_0 to construct Hurwizt convex combination as mentioned in Remark 8.6.

Figure 8.9 shows the triggering instants and release intervals of dynamic event-triggered mechanism. As can be seen, the largest triggering intervals is about $1.2s$ which illustrate the dynamic ETM having better efficiency of saving communication resources.

Further, set $\phi = 0.01, \bar{\tau} = 0.05s, \theta_1 \in [0.1, 0.9]$ and $\lambda \in [0.1, 0.9]$. Other parameters are same as before. To study concealment capability of the redundancy channel and the event-trigger efficiency, we introduce two index: concealed time of redundancy channel (CR) and event-triggering communication rate (ETR), respectively.

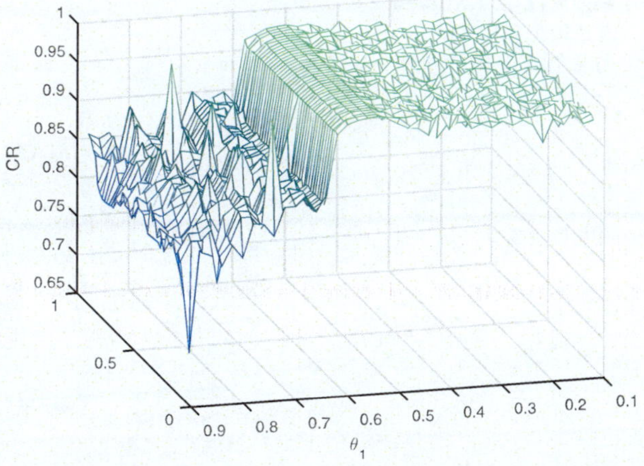

Fig. 8.10 CR for $\lambda \in [0.1, 0.9]$ and $\theta_1 \in [0.1, 0.9]$

$$CR = \frac{\sum_{n=0}^{n=N(t)} f_n}{\sum_{n=0}^{n=N(t)} f_n + \sum_{n=0}^{n=N(t)} l_n},$$

$$ETR = \frac{\sum_{n=0}^{N(t)} k(n) + N(t) + \sum_{n=0}^{\hat{N}(t)} \hat{k}(n) + \hat{N}(t)}{(\sum_{n=0}^{n=N(t)} f_n + \sum_{n=0}^{n=N(t)} g_n)/h}.$$

The CR values distribution displays obviously two area Fig. 8.10. The low value area is about during $\theta_1 \in [0.5, 0.9]$ while the high value area is about during $\theta_1 \in [0.1, 0.4]$. The low value area indicates that the concealing switch law can preserve system stable with less online time. The $1 - \theta_1$ is the weight of convex combination on Σ_2 which indicates the possibility of switching to the redundancy control channel. The CR is positive proportional to the weight $1 - \theta_1$. In other word, the large weight $1 - \theta_1$ increases the online time of the redundancy channel. Thus, it is better to choose the large θ_1 to get better concealing switch law. For given θ_1 in low value area, the CR value has slight wave for different values of λ. For given θ_1 in high value area, the CR value has heavy wave for different values of λ. Even though it is better to choose the large θ_1, the influence of λ on CR can not be ignored. Thus, we suggest that the lower value CR can be obtained by picking suitable value of λ in low area of θ_1.

Figure 8.11 shows the value distribution of the triggering rate under various (θ_1, λ). The low value area is during $\theta_1 \in [0.1, 0.6]$ while the high value area during $\theta_1 \in [0.7, 0.9]$. At the low value area, for given θ_1, the triggering rate has sight wave for various λ. At the high value area, for given θ_1, the triggering rate has serious wave for different λ. Thus, to get better triggering efficiency, it is suggested to choose the low value of θ_1.

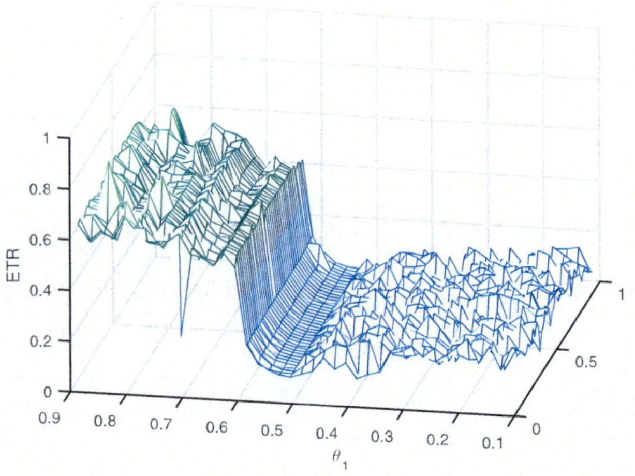

Fig. 8.11 ETR for $\lambda \in [0.1, 0.9]$ and $\theta_1 \in [0.1, 0.9]$

From the above analysis, we find that the conclusion from Figs. 8.10 and 8.11 exist tradeoff for selecting θ_1. Because the large θ_1 is benefit to obtain small CR referring to a better concealing switch law, but will lead to large ETR referring to a worse event-triggering efficiency. Conversely, the small value θ_1 is benefit to obtain small ETR referring to a better event-triggering efficiency, but will lead to large CR referring to a worse concealing switch law. Thus, it is better to first choose a θ_1 at middle area of interval $[0.1, 0.9]$ and then search a suitable value of λ.

8.6 Conclusion

The primary-redundancy control framework under bandwidth allocation has been proposed to design dynamic triggering control for linear continue-time system under DoS attacks. The concealing switch law and correlated delay model have been proposed. A switched delay system with three modes has been established. Sufficient conditions have been derived as the criterion of exponential stability and the design method of dynamic ETM and feedback control gains has been presented. Finally, two simulation examples has been carried out to illustrate the validness of our theory. In future, we will seek to design more flexible concealing switch law to make better use of redundancy control channel.

References

1. S. Hu, Z. Cheng, D. Yue, C. Dou, Y. Xue, Bandwidth allocation-based switched dynamic triggering control against DoS attacks, in *IEEE Transactions on Systems, Man, and Cybernetics: Systems* (Early Access, 2019), pp. 1–12. https://doi.org/10.1109/TSMC.2019.2956945

2. X. Ge, Q. Han, Z. Wang, A dynamic event-triggered transmission scheme for distributed set-membership estimation over wireless sensor networks. IEEE Trans. Cybern. **49**(1), 171–183 (2019)
3. C.D. Persis, P. Tesi, Input-to-state stabilizing control under denial-of-service. IEEE Trans. Autom. Control **60**(11), 2930–2944 (2015)
4. S. Hu, D. Yue, X. Xie, X. Chen, X. Yin, Resilient event-triggered controller synthesis of networked control systems under periodic DoS jamming attacks. IEEE Trans. Cybern. **49**(12), 4271–4281 (2019)
5. Y. Zhu, L. Zhang, W.X. Zheng, Distributed H_∞ filtering for a class of discrete-time markov jump lure systems with redundant channels. IEEE Trans. Indust. Electron. **63**(3), 1876–1885 (2015)
6. H. Yuan, Y. Xia, Resilient strategy design for cyber-physical system under DoS attack over a multi-channel framework. Inf. Sci. **454**, 312–327 (2018)
7. L. Su, D. Ye, A cooperative detection and compensation mechanism against denial-of-service attack for cyber-physical systems. Inf. Sci. **444**, 122–134 (2018)
8. H.S. Foroush, S. Martinez, On triggering control of single-input linear systems under pulse-width modulated DoS signals. Siam J. Control Optim. **54**(6), 3084–3105 (2016)
9. S. Kim, S.A. Campbell, X. Liu, Stability of a class of linear switching systems with time delay. IEEE Trans. Circuits Syst. I: Regul. Pap. **53**(2), 384–393 (2006)
10. X.-M. Sun, W. Wang, G.-P. Liu, J. Zhao, Stability analysis for linear switched systems with time-varying delay. IEEE Trans. Syst., Man, Cybern. Part B (Cybern.) **38**(2), 528–533 (2008)
11. Z. Cheng, D. Yue, S. Hu, C. Huang, C. Dou, L. Chen, Resilient load frequency control design: Dos attacks against additional control loop. Int. J. Electr. Power Energy Syst. **115**, 105496 (2020)

Part IV
Application of Secure Control Methods on Power Systems Under Network Attacks

Chapter 9
Attack-Resilient Event-Triggered Controller Design of DC Microgrids Under Nonperiodic DoS Attacks

This chapter is concerned with the attack-resilient event-triggered controller design problem of a DC microgrid with multiple nonlinear constant power loads and intermittent denial-of-service (DoS) attacks. First, for a resource efficiency purpose, an event-triggering communication scheme is delicately devised in such a way to only invoke the data transmission over the communication line when the DoS attack is inactive. Second, via characterizing the DoS active and inactive time intervals, a new switching piecewise system model for the nonlinear DC microgrid system is presented. Third, a numerically efficient design criterion on the existence of the desired attack-resilient event-triggered controller is established. It is further shown that various performance indices including resource efficiency, attack resilience, robustness against disturbance, sampling performance of the DC microgrid system can be evaluated in a unified framework. Finally, an illustrative example is given to verify the effectiveness of the proposed control design method for the DC microgrid [1].

The chapter is organized as follows. DC microgrids modeling and attack-resilient event-triggered communication are presented in Sect. 9.1. Stability analysis and synthesis are given in Sect. 9.2. A simulation example is illustrated to demonstrate the effectiveness of our theory in Sect. 9.3. Section 9.4 concludes this chapter.

9.1 Problem Formulation

9.1.1 System Modeling of DC Microgrids

The typical structure and circuit diagram of a DC microgrid comprising several CPLs are shown in Figs. 9.1 and 9.2 (see [2] for more details), respectively, where the DC microgrid is decoupled into $Q + 1$ subsystems including Q CPLs and one energy storage system (ESS). The state equation of the jth CPL is given by [2]

$$\dot{x}_j(t) = A_j x_j(t) + d_j h_j\left(x_j(t)\right) + A_{js} x_s(t) \tag{9.1}$$

D. Yue et al., *Secure Control of Networked Control Systems and Its Applications*,
https://doi.org/10.1007/978-981-33-6730-2_9

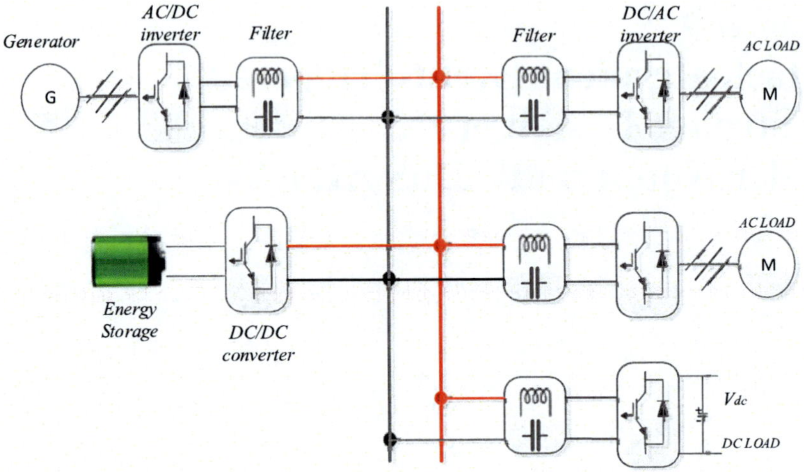

Fig. 9.1 The structure of DC microgrid with Q CPLs

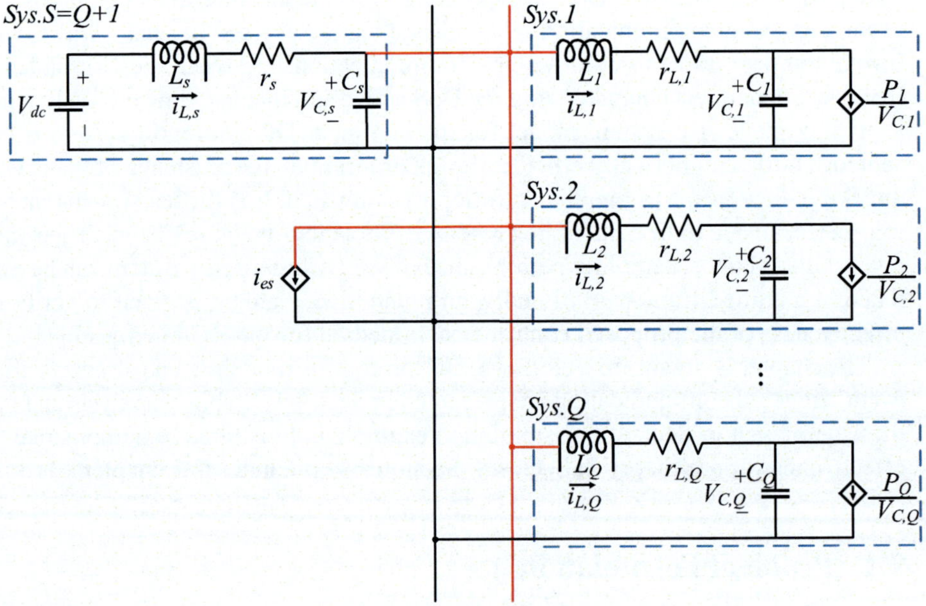

Fig. 9.2 The Circuit diagram of DC microgrid with Q CPLs

where $x_j(t) = [i_{L,j}\ v_{C,j}]^T$ with $i_{L,j}$ and $v_{C,j}$ denoting the current of the inductor and the voltage of the capacitor, respectively, in the jth CPL, and

$$A_j = \begin{bmatrix} -\frac{r_{L,j}}{L_j} & -\frac{1}{L_j} \\ \frac{1}{C_j} & 0 \end{bmatrix}, d_j = \begin{bmatrix} 0 \\ -\frac{1}{C_j} \end{bmatrix}$$

$$A_{js} = \begin{bmatrix} 0 & \frac{1}{L_j} \\ 0 & 0 \end{bmatrix}, h_j\left(x_j(t)\right) = \frac{P_j}{v_{C,j}}$$

$$j = \{1, 2, \ldots, Q\}, s = Q + 1.$$

where $r_{L,j}$ denotes the resistance of the inductor in the jth CPL, L_j denotes the inductance of the jth filter connected to the jth CPL, and P_j is the power of the jth CPL.

It is assumed that the CPLs are ideal and the value of the power is constant. Then, the DC source subsystem can be written as

$$\dot{x}_s(t) = A_s x_s(t) + b_s V_{dc} + b_{es}\tilde{i}_{es}(t) + \sum_{j=1}^{Q} A_{cn} x_j(t) \tag{9.2}$$

where $x_s = [i_{L,s}\ v_{C,s}]^T$ with $i_{L,s}$ and $v_{C,s}$ denoting the current of inductor and the voltage of the capacitor, respectively, in the ESS, and

$$A_s = \begin{bmatrix} -\frac{r_s}{L_s} & -\frac{1}{L_s} \\ \frac{1}{C_s} & 0 \end{bmatrix}, b_s = \begin{bmatrix} \frac{1}{L_s} \\ 0 \end{bmatrix}$$

$$A_{cn} = \begin{bmatrix} 0 & 0 \\ -\frac{1}{C_s} & 0 \end{bmatrix}, b_{es} = \begin{bmatrix} 0 \\ -\frac{1}{C_s} \end{bmatrix}.$$

where r_s denotes the resistance of the filter connected to the DC source.

Augmenting the Q CPLs and the ESS yields the following nonlinear dynamics of the DC microgrid

$$\dot{x}(t) = Ax(t) + DH(x(t)) + B_{es}\tilde{i}_{es}(t) + B_s V_{dc} \tag{9.3}$$

where $x(t) = [x_1^T(t)\quad x_2^T(t)\quad \ldots\quad x_Q^T(t)\quad x_s^T(t)]^T$, $H(x(t)) = [h_1(x_1(t))$ $h_2(x_2(t)) \ldots h_Q\left(x_Q(t)\right)]^T$, and

$$A = \begin{bmatrix} A_1 & 0 & \cdots & 0 & A_{1s} \\ 0 & A_2 & \cdots & 0 & A_{2s} \\ \vdots & \vdots & \ddots & \vdots & \vdots \\ 0 & 0 & \cdots & A_Q & A_{Qs} \\ A_{cn} & A_{cn} & \cdots & A_{cn} & A_s \end{bmatrix}, B_{es} = \begin{bmatrix} 0 \\ \vdots \\ 0 \\ b_{es} \end{bmatrix}, B_s = \begin{bmatrix} 0 \\ \vdots \\ 0 \\ b_s \end{bmatrix}, D = \begin{bmatrix} d_1 & 0 & \cdots & 0 \\ 0 & d_2 & \cdots & 0 \\ \vdots & \vdots & \ddots & \vdots \\ 0 & 0 & \cdots & d_Q \\ 0 & 0 & \cdots & 0 \end{bmatrix}.$$

Applying a coordinate transformation, the equilibrium point of the system (9.3) can be shifted to the origin. Let the energy storage current $\tilde{i}_{es}$ be the control input, the new coordinated DC microgrid is of the following form [3]

$$\dot{\tilde{x}}(t) = A\tilde{x}(t) + DH(\tilde{x}(t)) + B_{es}\tilde{i}_{es}(t) \tag{9.4}$$

where $\tilde{x}(t) = [\tilde{x}_1^T(t) \quad \tilde{x}_2^T(t) \quad \dots \quad \tilde{x}_Q^T(t) \quad \tilde{x}_s^T(t)]^T = x(t) - x_0$, $H(\tilde{x}(t)) = [h_1(\tilde{x}_1(t))\ h_2(\tilde{x}_2(t)) \dots h_Q(\tilde{x}_Q(t))]^T$ and

$$h_j(\tilde{x}_j(t)) = \frac{P_j \tilde{v}_{C,j}}{v_{c0,j}\left(\tilde{v}_{C,j} + v_{c0,j}\right)} \tag{9.5}$$

with x_0 being the equilibrium point of the DC microgrid and $v_{c0,j}$ being the equilibrium point of the $v_{c,j}$.

9.1.2 Controller Structure and Control Objective

Consider the DC microgrid system (9.4), in which the nonlinear vector $H(\tilde{x}(t))$ satisfies

$$\begin{aligned} H^T(\tilde{x}(t))H(\tilde{x}(t)) &= \sum_{j=1}^{j=Q} h_j^T(\tilde{x}_j(t))\, h_j(\tilde{x}_j(t)) \\ &\le \alpha^2 \tilde{x}^T(t) F^T F \tilde{x}(t) \end{aligned} \tag{9.6}$$

with $\alpha > 0$ representing the robustness index and the diagonal matrix F being given by

$$F = \begin{bmatrix} F_1 & \cdots & 0 & 0 \\ \vdots & \ddots & \vdots & \vdots \\ 0 & \cdots & F_Q & 0 \\ 0 & \cdots & 0 & 0 \end{bmatrix}, \ F_i = \begin{bmatrix} 0 & 0 \\ 0 & 1 \end{bmatrix}, \ i \in 1, 2, \dots, Q.$$

In (9.4), $\tilde{i}_{es}(t)$ stands for the injecting current, which stabilizes the DC microgrid and increases the closed-loop system robustness against the nonlinear disturbance and external malicious attacks. To design $\tilde{i}_{es}(t)$, a linear state feedback controller of the following form is adopted:

$$\tilde{i}_{es}(t) = K\tilde{x}(t) \tag{9.7}$$

where K is the gain matrix to be designed later.

The main objective of this chapter is to design K in (9.7) such that the system (9.4) under (9.7) is globally asymptotically stable regardless of the presence of malicious attacks on the communication line.

9.1.3 Nonperiodic DoS Attack Model

In this chapter, we consider a general DoS attack model that employing two typical DoS characteristics in terms of DoS frequency and duration, while requiring no assumptions regarding the underlying attack strategies, which applies to wide classes of DoS attacks. Moreover, it is assumed that the malicious attacker launches the DoS attacks on the communication line to prevent the signal $\tilde{i}_{es}(t)$ from being applied intermittently, as shown in Fig. 9.3. To describe the impact of the DoS attacks quantitatively, let $[h_n, h_n + l_n) \triangleq H_n$ $(h_n \geq 0, n \in \mathbb{N})$ denote the DoS-free interval over which the jamming attack signal is inactive and the communication is allowed, and let the time interval $\left[h_n + l_n, h_{n+1}\right) \triangleq L_n$ $(h_n + l_n < h_{n+1}, n \in \mathbb{N})$ represent the $n + 1$th DoS interval where the jamming attack signal is active and no data can be transmitted during this period. For ease development, we use $\mathcal{S}_{DoS}(t)$ to denote the nonperiodic DoS attack signal as in [4]:

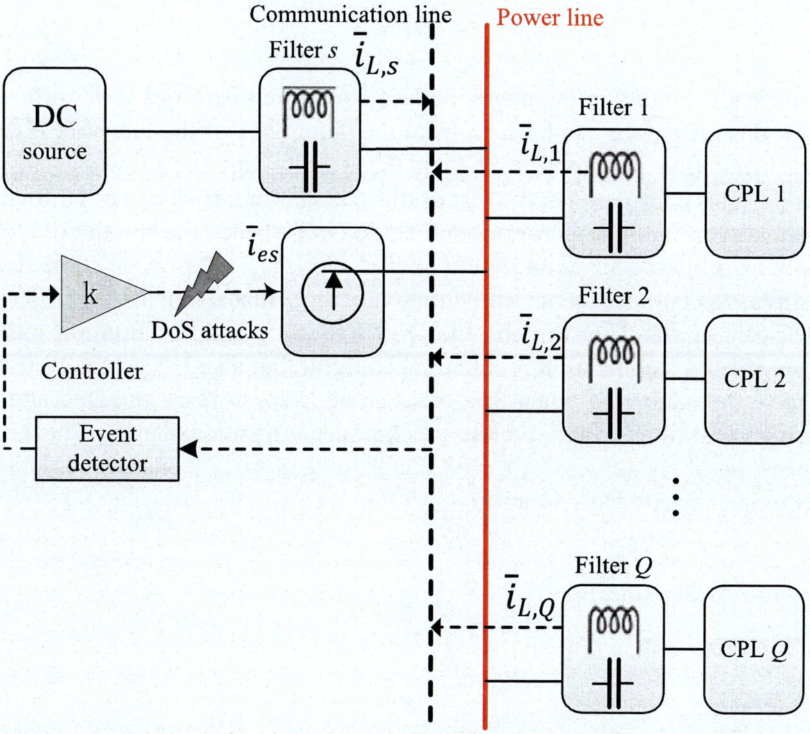

Fig. 9.3 Schematic of event-triggered control of a nonlinear DC MG under nonperiodic DoS attacks

$$\mathcal{S}_{DoS}(t) = \begin{cases} 0, & t \in H_n \\ 1, & t \in L_n. \end{cases} \tag{9.8}$$

For the DoS attack signal $\mathcal{S}_{DoS}(t)$, $\forall t \geq 0$, let $n(t)$ denote the number of DoS off/on transitions happening on the interval $[0, t]$.

Assumption 9.1 (*DoS attack frequency*) There exist two real scalars $\eta \geq 0$ and $\tau_D > 0$ such that for all $t \geq 0$

$$n(t) \leq \eta + \frac{t}{\tau_D}. \tag{9.9}$$

Besides the DoS frequency of the jamming attacker is constrained, the DoS duration of the jamming attacker also needs to be restricted. Consider the sequence $\{h_n + l_n\}$, let

$$\Xi(t) = \left\{\cup_{n=1}^{n(t)-1} L_n\right\} \cup \left[h_{n(t)} + l_{n(t)}, \min\left\{h_{n(t)+1}, t\right\}\right]$$

denote the total DoS intervals up to the current time t, and, for any given a time interval I, its length is denoted by $|I|$.

Assumption 9.2 (*DoS duration*) There exist two real scalars $\kappa \geq 0$ and $\Gamma > 1$ such that for all $t \geq 0$

$$|\Xi(t)| \leq \kappa + \frac{t}{\Gamma}. \tag{9.10}$$

Remark 9.1 Note that the attack pattern considered here is a little different from [5, 6], this difference can be seen from the definitions of the intervals H_n, L_n, and $\Xi(t)$. In addition, Assumptions 9.1 and 9.2 actually limit the allowable class of DoS attack signals considered in the rest of this chapter. Intuitively, in order to guarantee the stability of the DC microgrids subject to DoS attacks, the amount of DoS attack cannot be arbitrary. Inspired by the elegant work [5], we impose some reasonable conditions on both frequency and duration of DoS attack signals $\mathcal{S}_{DoS}(t)$ (see (9.8)) in DC microgrids. Constraining the DoS frequency and duration not only makes the considered stabilization problem meaningful, but also has actual motivation. In practice, there are many measures that can be taken to resist the DoS attacks, see, e.g., high-pass filtering and spreading techniques as mentioned in [5]. These measures reduces the likely success of a DoS attack, and, as such, constraint in practice the DoS frequency and DoS duration.

9.1.4 Design of Attack-Resilient Event-Triggered Scheme

When the communication channel is secure, namely without any DoS attack, whether the data packet should be transmitted or not through the communication line is determined by the following widely used triggering condition [7]:

$$\begin{aligned}&\left[\tilde{x}\left(t_{k_i}h\right)-\tilde{x}\left(t_k h\right)\right]^T \Omega\left[\tilde{x}\left(t_{k_i}h\right)-\tilde{x}\left(t_k h\right)\right]\\&>\sigma\tilde{x}^T\left(t_k h\right)\Omega\tilde{x}\left(t_k h\right)\end{aligned} \tag{9.11}$$

where $\sigma \in (0, 1)$ is a threshold and $\Omega > 0$ is a weighting matrix. In (9.11), $t_k h$ denotes the last event-triggering instant, h is the sampling period, and $t_{k_i}h \triangleq t_k h + ih$ $(k, i \in \mathbb{N})$ denotes the subsequent sampling instant. In the presence of intermittent DoS attack signals, however, some data packet cannot be transmitted to the destination successfully in the DoS interval, which may depress the system performance. In order to compensate for the unavailable data packet stemmed from the sporadic DoS attacks, we next modify the event triggering condition (9.11) as

$$t_{k,n}h \in \left\{t_{k_i}h \text{ satisfying } (9.11) \mid t_{k_i}h \in H_{n-1}\right\} \cup \{h_n\}. \tag{9.12}$$

9.1.5 Modelling of Switched DC Microgrids

Under the resilient triggering scheme (9.12) and intermittent DoS jamming attacks (9.8), the expression for $\tilde{i}_{es}$ in (9.7) can be represented as

$$\tilde{i}_{es}(t) = \begin{cases} K\tilde{x}(t_{k,n}h), & t \in [t_{k,n}h, t_{k+1,n}h) \cap H_n \\ 0, & t \in L_n \end{cases} \tag{9.13}$$

where the set $\{t_{k,n}h\}$ denotes the sequence of successful transmission instants $\left(t_{0,n}h \triangleq h_n\right)$, and $k \in \{0, 1, 2, \ldots, k(n)\} \triangleq K(n)$ with $k(n) = \sup\left\{k \in \mathbb{N} \mid t_{k,n}h < h_n + l_n\right\}$, which implies that $t_{k(n)+1,n}h \geq h_n + l_n$.

Substituting (9.13) into (9.4) yields

$$\dot{\tilde{x}}(t) = \begin{cases} A\tilde{x}(t) + DH(\tilde{x}(t)) + B_{es}K\tilde{x}(t_{k,n}h), \\ \qquad t \in [t_{k,n}h, t_{k+1,n}h) \cap H_n, k \in K(n) \\ A\tilde{x}(t) + DH(\tilde{x}(t)), \quad t \in L_n, n \in \mathbb{N}. \end{cases} \tag{9.14}$$

In what follows, we divide the event intervals $J_{k,n}$ into subintervals as

$$\begin{aligned}J_{k,n} = &\left\{\cup_{m=1}^{\lambda_{k,n}}\left[t_{k,n}h + (m-1)h, t_{k,n}h + mh\right)\right\}\\&\cup\left[t_{k,n}h + \lambda_{k,n}h, t_{k+1,n}h\right)\end{aligned} \tag{9.15}$$

where $k \in K(n)$, $n \in \mathbb{N}$, and

$$\lambda_{k,n} = \sup \left\{ m \in \mathbb{N} \mid t_{k,n}h + mh < t_{k+1,n}h \right\}.$$

Define

$$\begin{cases} F_{k,n}^{m} = \left[t_{k,n}h + (m-1)h, t_{k,n}h + mh \right), \\ \quad m \in \left\{ 1, 2, \cdots, \lambda_{k,n} \right\} \\ F_{k,n}^{\lambda_{k,n}+1} = \left[t_{k,n}h + \lambda_{k,n}h, \;\; t_{k+1,n}h \right). \end{cases} \tag{9.16}$$

It is easy to see that

$$H_n = \cup_{k=0}^{k(n)} \left\{ J_{k,n} \cap H_n \right\} \subseteq \cup_{k=0}^{k(n)} J_{k,n}. \tag{9.17}$$

Combining (9.15), (9.16) and (9.17), the interval H_n can be described as $H_n = \cup_{k=0}^{k(n)} \cup_{m=1}^{\lambda_{k,n}+1} \left\{ F_{k,n}^{m} \cap H_n \right\}$. Let $\varphi_{k,n}^{m} = F_{k,n}^{m} \cap H_n$, then $H_n = \cup_{k=0}^{k(n)} \cup_{m=1}^{\lambda_{k,n}} \varphi_{k,n}^{m}$.

Now, for $k \in K(n)$, $n \in \mathbb{N}$, two piecewise functions are defined as

$$\tau_{k,n}(t) = \begin{cases} t - t_{k,n}h, t \in \varphi_{k,n}^{1} \\ t - t_{k,n}h - h, t \in \varphi_{k,n}^{2} \\ \vdots \\ t - t_{k,n}h - \lambda_{k,n}h, t \in \varphi_{k,n}^{\lambda_{k,n}+1} \end{cases} \tag{9.18}$$

and

$$e_{k,n}(t) = \begin{cases} 0, t \in \varphi_{k,n}^{1} \\ \tilde{x}\left(t_{k,n}h\right) - \tilde{x}\left(t_{k,n}h + h\right), t \in \varphi_{k,n}^{2} \\ \vdots \\ \tilde{x}\left(t_{k,n}h\right) - \tilde{x}\left(t_{k,n}h + \lambda_{k,n}h\right), t \in \varphi_{k,n}^{\lambda_{k,n}+1}. \end{cases} \tag{9.19}$$

Based on the definitions of $\tau_{k,n}(t)$ and $e_{k,n}(t)$, we have $\tau_{k,n}(t) \in [0, h)$. Furthermore, the event-triggered sampled signal $\tilde{x}\left(t_{k,n}h\right)$ can be rewritten as

$$\tilde{x}\left(t_{k,n}h\right) = e_{k,n}(t) + \tilde{x}\left(t - \tau_{k,n}(t)\right), t \in J_{k,n} \cap H_n, \tag{9.20}$$

where the error vector $e_{k,n}(t)$ satisfies

$$e_{k,n}^{T}(t)\,\Omega e_{k,n}(t) \le \sigma \tilde{x}^{T}\left(t_{k,n}h\right) \Omega \tilde{x}\left(t_{k,n}h\right). \tag{9.21}$$

Substituting (9.20) into (9.14), the switched system formulation of the DC microgrid system (9.4) can be expressed as

$$\begin{cases} \dot{\tilde{x}}(t) = \begin{cases} A\tilde{x}(t) + DH(\tilde{x}(t)) + B_{es}K\tilde{x}\left(t - \tau_{k,n}(t)\right) \\ \quad +B_{es}Ke_{k,n}(t), \quad t \in J_{k,n} \cap H_n \\ A\tilde{x}(t) + DH(\tilde{x}(t)), t \in L_n \end{cases} \\ \tilde{x}(t) = \phi(t), \quad t \in [-h, 0]. \end{cases} \tag{9.22}$$

9.2 Stability Analysis

We first present the following technical lemma which estimates the upper bound of the candidate Lyapunov–Krasovskii function in the DoS-free and DoS intervals, respectively, and serves as the basis of deriving the main results of this chapter.

Lemma 9.1 *Consider an attack-resilient event-triggering scheme (9.12), along with DoS attacks satisfying Assumptions 9.1 and 9.2. For given parameters* $\eta \geq 0$, $\kappa \geq 0$, $\tau_D > 0$, $\Gamma > 1$, $\alpha > 0$, $a_i \in (0, +\infty)$, $\sigma \in (0, 1)$, $h \in (0, +\infty)$, *if there exist symmetric positive define matrices* $P_i \in \mathbb{R}^{n\times n}$, $Q_{1i} \in \mathbb{R}^{n\times n}$, $Q_{2i} \in \mathbb{R}^{n\times n}$, $R_i \in \mathbb{R}^{n\times n}$, $T_{\omega i} \in \mathbb{R}^{n\times n}$, $\Omega \in \mathbb{R}^{n\times n}$, *matrices* M_i *and* N_i $(i = 1, 2)$ *such that*

$$\Psi_1 = \begin{bmatrix} \Psi_{11}^1 & * & * & * \\ \sqrt{h}M_1^T & \Psi_{22}^1 & * & * \\ \sqrt{h}N_1^T & 0 & \Psi_{33}^1 & * \\ \sqrt{h}R_1\mathcal{F}_1 & 0 & 0 & -R_1 \end{bmatrix} < 0 \tag{9.23}$$

$$\Psi_2 = \begin{bmatrix} \Psi_{11}^2 & * & * & * \\ \sqrt{h}M_2^T & -R_2 & * & * \\ \sqrt{h}N_2^T & 0 & -R_2 & * \\ \sqrt{h}R_2\mathcal{F}_2 & 0 & 0 & -R_2 \end{bmatrix} < 0 \tag{9.24}$$

where $\Psi_{11}^1 = \Upsilon_1 + \Lambda_1 + \Lambda_1^T$, $\Psi_{11}^2 = \Upsilon_2 + \Lambda_2 + \Lambda_2^T$, $\Psi_{22}^1 = \Psi_{33}^1 = -e^{-a_1 h}R_1$, $\varpi_{21} = e^{-a_1 h+1}$, $\varpi_{22} = e^{a_2 h+1}$ *with*

$$\Upsilon_1 = \begin{bmatrix} \Upsilon_{11}^1 & * & ** & * \\ D^T P_1 & -I & ** & * \\ K^T B_{es}^T P_1 & 0 & \sigma\Omega - \varpi_{21}T_{\omega 1}* & * \\ 0 & 0 & \varpi_{21}T_{\omega 1} - \varpi_{21}\left(Q_{21} + T_{\omega 1}\right) & * \\ K^T B_{es}^T P_1 & 0 & \sigma\Omega 0 & (\sigma - 1)\,\Omega \end{bmatrix}$$

$$\Upsilon_2 = \begin{bmatrix} \Upsilon_{11}^2 & * & * & * \\ D^T P_2 & -I & * & * \\ 0 & 0 & -\varpi_{22}T_{\omega 2} & * \\ 0 & 0 & \varpi_{22}T_{\omega 2} & -\varpi_{22}\left(Q_{22} + T_{\omega 2}\right) \end{bmatrix}$$

$$\begin{aligned}
\Upsilon_{11}^{1} &= a_1 P_1 + P_1 A + A^T P_1 + Q_{11} + \alpha^2 F^T F \\
\Upsilon_{11}^{2} &= -a_2 P_2 + P_2 A + A^T P_2 + Q_{12} + \alpha^2 F^T F \\
\Lambda_1 &= \left[M_1 \; 0 \; -M_1 + N_1 \; -N_1 \; 0 \right] \\
\Lambda_2 &= \left[M_2 \; 0 \; -M_2 + N_2 \; -N_2 \right] \\
\mathcal{F}_1 &= \left[A \; D \; B_{es} K \; 0 \; B_{es} K \right] \\
\mathcal{F}_2 &= \left[A \; D \; 0 \; 0 \right]
\end{aligned}$$

Then we can get

$$V(t) \leq \begin{cases} e^{-a_1(t-h_n)} V_1(h_n), & t \in H_n \\ e^{a_2(t-h_n-l_n)} V_2(h_n + l_n), & t \in L_n. \end{cases} \tag{9.25}$$

where $V(t)$ is a piecewise Lyapunov–Krasovskii functional (PLKF) for the system (9.22) defined as

$$V(t) = \begin{cases} V_1(t), t \in H_n \cap J_{k,n} \\ V_2(t), t \in L_n \end{cases}$$

with

$$\begin{aligned}
V_i(t) &= \tilde{x}^T(t) P_i \tilde{x}(t) + \int_{t-\tau_{k,n}(t)}^{t} \tilde{x}^T(s) \varpi_{1i} Q_{1i} \tilde{x}(s) ds \\
&\quad + \int_{t-h}^{t-\tau_{k,n}(t)} \tilde{x}^T(s) \varpi_{2i} Q_{2i} \tilde{x}(s) ds + \int_{-h}^{0} \int_{t+\theta}^{t} \dot{\tilde{x}}^T(s) \varpi_{1i} R_i \dot{\tilde{x}}(s) ds d\theta \\
&\quad + \int_{t-h}^{t-\tau_{k,n}(t)} \varpi_{2i} \left[\tilde{x}(s) - \tilde{x}(t - \tau_{k,n}(t)) \right]^T \\
&\quad T_{\omega_i} \left[\tilde{x}(s) - \tilde{x}(t - \tau_{k,n}(t)) \right] ds
\end{aligned} \tag{9.26}$$

where $\varpi_{1i} \triangleq e^{(-1)^i a_i (t-s)}$, $\varpi_{2i} \triangleq e \cdot \varpi_{1i}$, with $a_i > 0$, $P_i > 0$, $Q_{1i} > 0$, $Q_{2i} > 0$, $R_i > 0$.

Proof Taking derivation of (9.26) with $i = 1$ and by using free weighting matrix technique [8], one has

$$\begin{aligned}
\dot{V}_1(t) \leq& -a_1 V_1(t) + a_1 \tilde{x}^T(t) P_1 \tilde{x}(t) + 2\tilde{x}^T(t) P_1 \dot{\tilde{x}}(t) + \tilde{x}^T(t) Q_{11} \tilde{x}(t) \\
&- \tilde{x}^T(t-h) e^{-a_1 h + 1} Q_{21} \tilde{x}(t-h) + h \dot{\tilde{x}}^T(t) R_1 \dot{\tilde{x}}(t) \\
&- \int_{t-h}^{t} \dot{\tilde{x}}^T(s) e^{-a_1 h} R_1 \dot{\tilde{x}}(s) ds - e^{-a_1 h + 1} \tilde{x}^T(t-h) T_{\omega 1} \tilde{x}(t-h) \\
&+ 2 e^{-a_1 h + 1} \tilde{x}^T(t-h) T_{\omega 1} \tilde{x}(t - \tau_{k,n}(t)) \\
&- e^{-a_1 h + 1} \tilde{x}^T(t - \tau_{k,n}(t)) T_{\omega 1} \tilde{x}(t - \tau_{k,n}(t)) + 2\xi_1^T(t)(M_1 \vartheta_1 + N_1 \vartheta_2).
\end{aligned}$$

where $\vartheta_1 \triangleq \tilde{x}(t) - \tilde{x}(t - \tau_{k,n}(t)) - \int_{t-\tau_{k,n}(t)}^{t} \dot{\tilde{x}}(s) ds$, $\vartheta_2 \triangleq \tilde{x}(t - \tau_{k,n}(t)) - \tilde{x}(t-h) - \int_{t-h}^{t-\tau_{k,n}(t)} \dot{\tilde{x}}(s) ds$, $\xi_1(t) = \text{col}\left\{ \tilde{x}(t), H(\tilde{x}(t)), \tilde{x}(t - \tau_{k,n}(t)), \tilde{x}(t-h), e_{k,n}(t) \right\}$.

Applying the element inequality $-2a^T b \leq a^T X^{-1} a + b^T X b$ ($\forall a, b \in \mathbb{R}^n$, $0 < X \in \mathbb{R}^{n\times n}$, $0 < X \in \mathbb{R}^{n\times n}$) to estimate the integral terms in the last term of (9.27) and inserting the event-triggering condition (9.21) into (9.27), we can obtain

$$\dot{V}_1(t) \leq -a_1 V_1(t) + \xi_1^T(t)[\Psi_{11}^1 + hM_1 e^{a_1 h} R_1^{-1} M_1^T \\ +hN_1 e^{a_1 h} R_1^{-1} N_1^T + h\mathcal{F}_1^T R_1 \mathcal{F}_1]\xi_1(t).$$

Taking the Schur complement of the matrix $\Psi_1 < 0$ in (9.23), we get $\Psi_{11}^1 + hM_1 e^{a_1 h} R_1^{-1} M_1^T + hN_1 e^{a_1 h} R_1^{-1} N_1^T + h\mathcal{F}_1^T R_1 \mathcal{F}_1 < 0$, which implies that

$$\dot{V}_1(t) \leq -a_1 V_1(t)\ , t \in J_{k,n} \cap H_n. \tag{9.27}$$

Considering the arbitrariness of k, we can deduce that for $t \in H_n$

$$V_1(t) \leq e^{-a_1(t-h_n)} V_1(h_n). \tag{9.28}$$

Similar to the above analysis, differentiating $V_2(t)$ with respect to $t \in L_n$ along the trajectory of the system (9.26) with $i = 2$ yields

$$\dot{V}_2(t) \leq a_2 V_2(t) + \xi_2^T(t)[\Psi_{11}^2 + hM_2 R_2^{-1} M_2^T \\ +hN_2 R_2^{-1} N_2^T + h\mathcal{F}_2^T R_2 \mathcal{F}_2]\xi_2(t).$$

where $\xi_2(t) = \text{col}\left\{\tilde{x}(t), \tilde{x}(t - \tau_{k,n}(t)), H(\tilde{x}(t)), \tilde{x}(t-h)\right\}$.

According to the condition $\Psi_2 < 0$ in (9.24), it holds that

$$\dot{V}_2(t) \leq a_2 V_2(t), t \in L_n,$$

which implies that for $t \in L_n$

$$V_2(t) \leq e^{a_2(t-h_n-l_n)} V_2(h_n + l_n). \tag{9.29}$$

This completes the proof.

Based on Lemma 9.1, we are in a position to present the following theorem through which the closed-loop stability of system (9.22) under the proposed attack-resilient event-triggering controller can be preserved.

Theorem 9.1 *Consider the switched system (9.22) with known K. For given parameters $\eta \geq 0$, $\kappa \geq 0$, $\tau_D > h$, $\Gamma > 1$, $\alpha > 0$, $a_i \in (0, +\infty)$, $\upsilon_i \in (0, +\infty)$, $\upsilon_1\upsilon_2 \geq 1$, $\sigma \in (0, 1)$, and $h \in (0, l_n)$, there exist symmetric positive define matrices $P_i \in \mathbb{R}^{n\times n}$, $Q_{1i} \in \mathbb{R}^{n\times n}$, $Q_{2i} \in \mathbb{R}^{n\times n}$, $R_i \in \mathbb{R}^{n\times n}$, $T_{\omega i} \in \mathbb{R}^{n\times n}$, $\Omega \in \mathbb{R}^{n\times n}$, matrices M_i and N_i ($i = 1, 2$) of appropriate dimensions such that the linear matrix inequalities (LMIs) (9.23), (9.24) and the following LMIs are satisfied:*

$$P_1 \leq \upsilon_2 P_2 \tag{9.30}$$
$$P_2 \leq \upsilon_1 e^{(a_1+a_2)h} P_1 \tag{9.31}$$
$$Q_{1i} \leq \upsilon_{3-i} Q_{1(3-i)} \tag{9.32}$$
$$Q_{2i} \leq \upsilon_{3-i} Q_{2(3-i)} \tag{9.33}$$
$$R_i \leq \upsilon_{3-i} R_{3-i} \tag{9.34}$$
$$T_{\omega_i} \leq \upsilon_{3-i} T_{\omega_{3-i}} \tag{9.35}$$
$$\frac{\frac{\ln(\upsilon_1\upsilon_2)}{a_1+a_2}+h}{\tau_D} + \frac{1}{\Gamma} < \frac{a_1}{a_1+a_2}, \tag{9.36}$$

then system (9.22) is globally asymptotically stable for any intermittent DoS jamming attack (9.8) satisfying Assumptions 9.1 and 9.2.

Proof Construct a PLKF candidate as in Lemma 9.1: $V(t) = V_i(t)$, $i = 1, 2$. According to Lemma 9.1 and (9.30)–(9.35), it follows that

$$\begin{cases} V_1(h_n) \leq \mu_2 V_2(h_n^-) \\ V_2(h_n + l_n) \leq \mu_1 e^{(a_1+a_2)h} V_1\left((h_n + l_n)^-\right). \end{cases}$$

When $t \in H_n$, applying the preceding inequalities recursively, one has

$$\begin{aligned} & V_1(t) \\ \leq\, & e^{-a_1(t-h_n)} V_1(h_n) \\ \leq\, & \upsilon_2 e^{-a_1(t-h_n)} V_2((h_n)^-) \\ & \vdots \\ \leq\, & e^{n(t)[(a_1+a_2)h+\ln(\upsilon_1\upsilon_2)]+(a_1+a_2)|\Xi(t)|-a_1 t} V_1(0) \end{aligned} \tag{9.37}$$

Similarly, when $t \in L_n$, we can obtain

$$\begin{aligned} & V_2(t) \\ \leq\, & e^{a_2(t-h_n-l_n)} V_2(h_n + l_n) \\ \leq\, & \upsilon_1 e^{(a_1+a_2)h+a_2(t-h_n-l_n)} V_1((h_n + l_n)^-) \\ \leq\, & \frac{1}{\upsilon_2} e^{n(t)[(a_1+a_2)h+\ln(\upsilon_1\upsilon_2)]+(a_1+a_2)|\Xi(t)|-a_1 t} V_1(0) \end{aligned} \tag{9.38}$$

Let $\upsilon = \max\left\{1, \frac{1}{\upsilon_2}\right\}$, it follows from (9.37)–(9.38) that

$$V(t) \leq \upsilon e^{n(t)[(a_1+a_2)h+\ln(\upsilon_1\upsilon_2)]+(a_1+a_2)|\Xi(t)|-a_1 t} V_1(0). \tag{9.39}$$

Substituting (9.9) and (9.10) into (9.39), we have

$$V(t) \leq \upsilon e^{\Delta_1} e^{\Delta_2(t)} V_1(0)$$

where $\Delta_1 = [\ln(v_1 v_2) + (a_1 + a_2)h]\eta + (a_1 + a_2)\kappa$,
$\Delta_2(t) = [\ln(v_1 v_2) + (a_1 + a_2)h]\frac{t}{\tau_D} + (a_1 + a_2)\frac{t}{\Gamma} - a_1 t$.

As a result, $\lim_{t\to\infty} V(t) = 0$ if and only if $\frac{\ln(v_1 v_2)+(a_1+a_2)h}{\tau_D} + \frac{a_1+a_2}{\Gamma} < a_1$. Thus, the system (9.22) is globally asymptotically stable. The proof is completed.

Remark 9.2 From (9.36), it can be seen that $\frac{1}{\Gamma} < \frac{a_1}{a_1+a_2} < 1$, which implies that $\frac{\frac{\ln(v_1 v_2)}{a_1+a_2}+h}{\tau_D} < 1$, and thus $h < \tau_D - \frac{\ln(v_1 v_2)}{a_1+a_2} \le \tau_D$. Notice that τ_D can be deemed as the average dwell-time between two consecutive DoS off/on transitions. Then it follows from (9.36) that $\tau_D > \frac{\frac{\ln(v_1 v_2)}{a_1+a_2}+h}{\frac{a_1}{a_1+a_2}-\frac{1}{\Gamma}} \triangleq \bar{\tau}_D$, where $\bar{\tau}_D$ is the lower bound of τ_D. Since $0 < \frac{a_1}{a_1+a_2} - \frac{1}{\Gamma} < 1$, one has $\bar{\tau}_D = \frac{\frac{\ln(v_1 v_2)}{a_1+a_2}+h}{\frac{a_1}{a_1+a_2}-\frac{1}{\Gamma}} > \frac{\frac{\ln(v_1 v_2)}{a_1+a_2}+h}{1} > \frac{\ln(v_1 v_2)}{a_1+a_2}$. Furthermore, denote the jamming attack rate by γ. Then the stability condition (9.36) can be replaced by $\gamma < \bar{\gamma}$, where $\bar{\gamma} = \frac{a_1}{a_1+a_2} - \frac{\frac{\ln(v_1 v_2)}{a_1+a_2}+h}{\tau_D}$. According to this relation, it can be seen that the upper bound $\bar{\gamma}$ of the jamming attack rate γ is explicitly dependent on the convergence rate a_1, the divergence rate a_2, the sampling period h, the average DoS duration τ_D, and the tuning parameters v_1 and v_2. More specifically, if a_1, a_2, h, v_1, and v_2 are fixed, a larger τ_D results in a larger $\bar{\gamma}$, and vice versa. Similarly, when a_1, a_2, τ_D, v_1, and v_2 are fixed, $\bar{\gamma}$ is inverse relation with h. Therefore, in order to tolerate more DoS attacks (i.e., tolerate bigger $\bar{\gamma}$), we can reduce the value of sampling period h (other parameter are fixed). A smaller h may result in better performance of the DC microgrid system under consideration while increasing the communication burden. Considering the event-triggered communication mechanism can effectively reduce the network loads, one should choose a smaller h and bigger σ (for balance) carefully to design a feasible event-triggered controllers such that the resulting DC microgrid system can achieve the desired performance. Notice that for a larger h and σ, it is more likely that the following Theorem 9.2 is not feasible.

9.3 Attack-Resilient Event-Triggered Controller Design

Theorem 9.2 *Consider the switched system (9.22). For given parameters $\eta \ge 0$, $\kappa \ge 0$, $\tau_D > h$, $\Gamma > 1$, $a_i \in (0, +\infty)$, $v_i \in (0, +\infty)$, $v_1 v_2 \ge 1$, $\sigma \in (0, 1)$, $h \in (0, l_n)$, α, ϵ_i, ϕ_i, φ_i, $\theta_i \in (0, +\infty)$, if there exist symmetric positive define matrices $P_i \in \mathbb{R}^{n\times n}$, $\bar{Q}_{1i} \in \mathbb{R}^{n\times n}$, $\bar{Q}_{2i} \in \mathbb{R}^{n\times n}$, $\bar{R}_i \in \mathbb{R}^{n\times n}$, $\bar{T}_{\omega i} \in \mathbb{R}^{n\times n}$, $\bar{\Omega} \in \mathbb{R}^{n\times n}$, matrices $\bar{M}_i$ and $\bar{N}_i$ $(i = 1, 2)$ of appropriate dimensions such that (9.36) and the following LMIs hold*

$$\bar{\Psi}_1 = \begin{bmatrix} \bar{\Psi}_{11}^1 & * & * & * & * \\ \sqrt{h}\bar{M}_1^T & \bar{\Psi}_{22}^1 & * & * & * \\ \sqrt{h}\bar{N}_1^T & 0 & \bar{\Psi}_{33}^1 & * & * \\ \sqrt{h}\bar{\mathcal{F}}_1 & 0 & 0 & \check{\varepsilon}_1 & * \\ FW_1 & 0 & 0 & 0 & -\alpha^{-2}I \end{bmatrix} < 0 \tag{9.40}$$

$$\bar{\Psi}_2 = \begin{bmatrix} \bar{\Psi}_{11}^2 & * & * & * & * \\ \sqrt{h}\bar{M}_2^T & -\bar{R}_2 & * & * & * \\ \sqrt{h}\bar{N}_2^T & 0 & -\bar{R}_2 & * & * \\ \sqrt{h}\bar{\mathcal{F}}_2 & 0 & 0 & \breve{\varepsilon}_2 & * \\ FW_2 & 0 & 0 & 0 & -\alpha^{-2}I \end{bmatrix} < 0 \tag{9.41}$$

$$\begin{bmatrix} -\upsilon_2 W_2 & * \\ W_2 & -W_1 \end{bmatrix} \leq 0 \tag{9.42}$$

$$\begin{bmatrix} -\upsilon_1 e^{(a_1+a_2)h} W_1 & * \\ W_1 & -W_2 \end{bmatrix} \leq 0 \tag{9.43}$$

$$\begin{bmatrix} -\upsilon_{3-i}\bar{Q}_{1(3-i)} & * \\ W_{3-i} & \phi_i^2\bar{Q}_{1i} - 2\phi_i W_i \end{bmatrix} \leq 0 \tag{9.44}$$

$$\begin{bmatrix} -\upsilon_{3-i}\bar{Q}_{2(3-i)} & * \\ W_{3-i} & \theta_i^2\bar{Q}_{2i} - 2\theta_i W_i \end{bmatrix} \leq 0 \tag{9.45}$$

$$\begin{bmatrix} -\upsilon_{3-i}\bar{T}_{\omega_{3-i}} & * \\ W_{3-i} & \varphi_i^2\bar{T}_{\omega_i} - 2\varphi_i W_i \end{bmatrix} \leq 0 \tag{9.46}$$

$$\begin{bmatrix} -\upsilon_{3-i}\bar{R}_{3-i} & * \\ W_{3-i} & \epsilon_i^2\bar{R}_i - 2\epsilon_i W_i \end{bmatrix} \leq 0 \tag{9.47}$$

where $\bar{\Psi}_{11}^1 = \bar{\Upsilon}_1 + \bar{\Lambda}_1 + \bar{\Lambda}_1^T$, $\bar{\Psi}_{11}^2 = \bar{\Upsilon}_2 + \bar{\Lambda}_2 + \bar{\Lambda}_2^T$, $\bar{\Psi}_{22}^1 = \bar{\Psi}_{33}^1 = -e^{-a_1 h}\bar{R}_1$, $\breve{\varepsilon}_1 = \epsilon_1^2\bar{R}_1 - 2\epsilon_1 W_1$, $\breve{\varepsilon}_2 = \epsilon_2^2\bar{R}_2 - 2\epsilon_2 W_2$, $\varpi_{21} = e^{-a_1 h+1}$, $\varpi_{22} = e^{a_2 h+1}$ *with*

$$\bar{\Upsilon}_1 = \begin{bmatrix} \bar{\Upsilon}_{11}^1 & * & * & * & * \\ D^T & -I & * & * & * \\ S^T B_{es}^T & 0 & \sigma\bar{\Omega} - \varpi_{21}\bar{T}_{\omega 1} & * & * \\ 0 & 0 & \varpi_{21}\bar{T}_{\omega 1} & -\varpi_{21}\bar{Q}_{21} - \varpi_{21}\bar{T}_{\omega 1} & * \\ S^T B_{es}^T & 0 & \sigma\bar{\Omega} & 0 & (\sigma-1)\bar{\Omega} \end{bmatrix}$$

$$\bar{\Upsilon}_2 = \begin{bmatrix} \bar{\Upsilon}_{11}^2 & * & * & * \\ D^T & -I & * & * \\ 0 & 0 & -\varpi_{22}\bar{T}_{\omega 2} & * \\ 0 & 0 & \varpi_{22}\bar{T}_{\omega 2} & -\varpi_{22}\bar{Q}_{22} - \varpi_{22}\bar{T}_{\omega 2} \end{bmatrix}$$

$$\begin{aligned}
\bar{\Upsilon}_{11}^{1} &= a_1 W_1 + W_1 A^T + A W_1 + \bar{Q}_{11} \\
\bar{\Upsilon}_{11}^{1} &= -a_2 W_2 + A W_2 + W_2 A^T + \bar{Q}_{12} \\
\bar{\Lambda}_1 &= \begin{bmatrix} \bar{M}_1 \; 0 \; -\bar{M}_1 + \bar{N}_1 \; -\bar{N}_1 \; 0 \end{bmatrix} \\
\bar{\Lambda}_2 &= \begin{bmatrix} \bar{M}_2 \; 0 \; -\bar{M}_2 + \bar{N}_2 \; -\bar{N}_2 \end{bmatrix} \\
\bar{\mathcal{F}}_1 &= \begin{bmatrix} A W_1 \; D \; B_{es} S \; 0 \; B_{es} S \end{bmatrix} \\
\bar{\mathcal{F}}_2 &= \begin{bmatrix} A W_2 \; D \; 0 \; 0 \end{bmatrix},
\end{aligned}$$

then system (9.22) with $K = SW_1^{-1}$ is globally asymptotically stable in the presence of the intermittent DoS jamming attacks (9.8) satisfying Assumptions 9.1 and 9.2.

Proof The desired result can be obtained along the standard procedure that is widely used in the LMI-based controller synthesis, and thus the detailed proof is omitted for concise.

9.4 Simulation Results

In this section, we verify the presented control design method using a well-known benchmark example in the DC microgrid literature [2, 3, 9]. The circuit parameters of the considered DC microgrid with two CPLs are listed in Table 9.1. Based on these values, the nominal robustness index is chosen as $\alpha = 0.0055$.

First, we will design the event-triggering parameters Ω in (9.21) and control gain K in (9.13) simultaneously, so that the system (9.3) is asymptotically stable in the presence of the intermittent DoS jamming attacks. Toward this end, set the DoS parameters $\tau_D = 0.05$, $\Gamma = 1.5$, $\upsilon_1 = \upsilon_2 = 1.05$, $a_1 = 0.65$, $a_2 = 0.2$, $h = 0.01$ satisfying (9.36). In this simulation process, it can be inferred that the upper bound of attack frequency is $\frac{1}{\tau_D} = 20$Hz and the upper bound of attack duration is $\frac{1}{\tau_D} = 0.6667s$. From Fig. 9.1, in practice, we count the attack duration $|\Xi(t)| = 0.5411s$ and attack frequency $n(t) = 13$, which is smaller than the theoretical values. For given $\sigma = 0.08$, $\epsilon_i = \theta_i = \varphi_i = \phi_i = 250$ $(i = 1, 2)$, by solving Theorem 9.2, the event-triggering parameter Ω and a low control gain K are found and given as

Table 9.1 Circuit parameters

r_1	1.1 Ω	r_2	0.5 Ω	r_s	1 Ω
L_1	39.5 mH	L_2	19.5 mH	L_s	17 mH
C_1	500 μF	C_2	550 μF	C_s	550 μF
P_1	300 W	P_2	400 W	V_{dc}	200 V

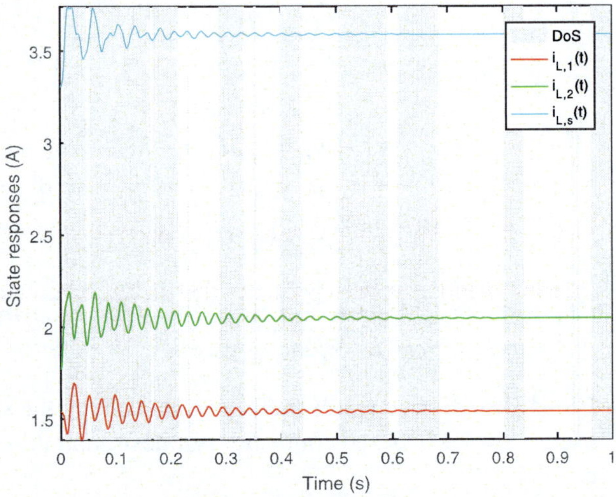

Fig. 9.4 Current (A) responses of the DC microgrid

$$\Omega = 10^{-3} \times \begin{bmatrix} 0.2146 & 0.0022 & -0.0072 & 0.0013 & -0.0009 & 0.0010 \\ 0.0022 & 0.0030 & -0.0003 & 0.0000 & 0.0003 & 0.0001 \\ -0.0072 & -0.0003 & 0.0982 & 0.0012 & 0.0004 & 0.0007 \\ 0.0013 & 0.0000 & 0.0012 & 0.0031 & 0.0009 & 0.0001 \\ -0.0009 & 0.0003 & 0.0004 & 0.0009 & 0.0828 & 0.0018 \\ 0.0010 & 0.0001 & 0.0007 & 0.0001 & 0.0018 & 0.0030 \end{bmatrix}$$

$$K = 10^{-3} \times \begin{bmatrix} 0.1321 & 0.0075 & 0.0194 & 0.0235 & 0.4017 & 0.2140 \end{bmatrix}.$$

The initial condition of the DC microgrid system is taken as $x_0 = \begin{bmatrix} -0.003 & 0.283 & -0.277 & 0.418 & -0.28 & 0.28 \end{bmatrix}^T$ and the simulation time is assumed to be $50s$. The triggering instants and intervals release, the injecting current, the current and voltage responses of the system (9.3) under intermittent DoS jamming signal are plotted in Figs. 9.4, 9.5, 9.6 and 9.7 (note that we only display the figures in $[0, 1s]$), respectively. According to these simulation results, it can be seen that (i) the nonlinear DC microgrid system is still stable despite the presence of the intermittent DoS jamming attacks; (ii) the proposed attack-resilient event-triggering strategy can greatly reduce the frequency of control updates and save the calculating resources in the DC microgrid system while offsetting the impact of nonlinear disturbance and intermittent DoS jamming attacks as well.

Next, we will demonstrate the relationship between robustness index α and various performance indices of the system. Let $h_{\max}$ denote the maximum allowable sampling period, and the effect of robustness index α on $h_{\max}$ is shown in Fig. 9.8. For this purpose, we choose different values of the index α and fix $\sigma = 0.08$, the parameters τ_D, Γ, v_1, v_2, a_1, a_2, ϵ_i, ϕ_i, φ_i, θ_i $(i = 1, 2)$ and simulation time are chosen as same as before. From Fig. 9.8, it is seen that the sampling period is gradually decreasing with the increasing of the robustness index α. When the robustness index α is beyond

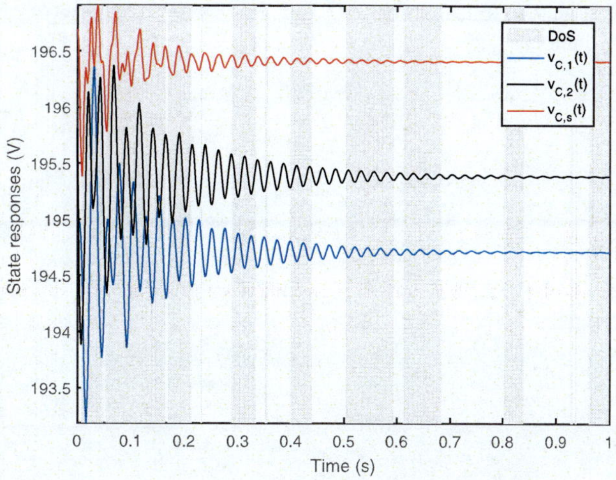

Fig. 9.5 Voltage (V) responses of the DC microgrid

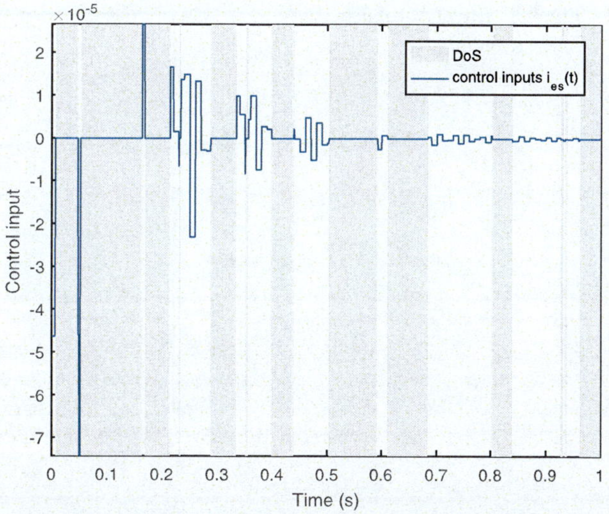

Fig. 9.6 Injecting current

0.01, the sampling period is decreasing sharply and fast. It means that high frequency feedback control is necessary to stabilize the system with large nonlinear disturbance. Also, it should be pointed out that although the maximum value of robustness index $\alpha = 0.045$ obtained in [9] is larger than that obtained here ($\alpha \approx 0.014$), the feedback gain K obtained in [9] is $K = \begin{bmatrix} 1.406 & -0.116 & 0.47 & 0.354 & 0.653 & 0.121 \end{bmatrix}$, which is clearly higher than that obtained here. This means that the energy consumption of the proposed event-triggered control method is less than that in [9]. It is noteworthy that a direct comparison between the related control approaches is in general difficult. Besides, the robustness index should not be considered as the only relevant measure of system performance. Some methods result in better results in the transient-state than in the steady-state, while the proposed resilient control method incorporates several performance measures, such as the amount of data transmis-

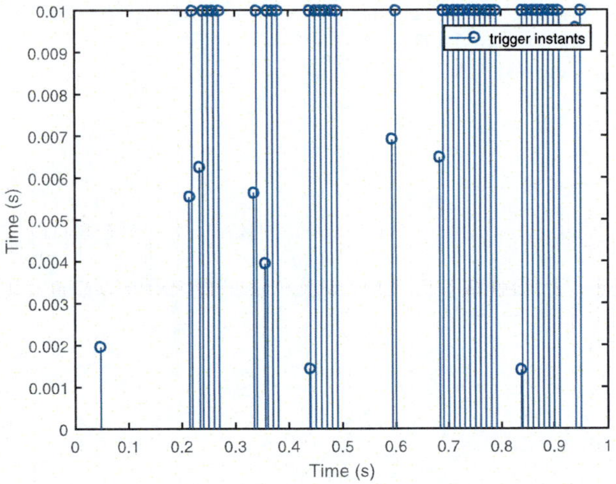

Fig. 9.7 Release time intervals

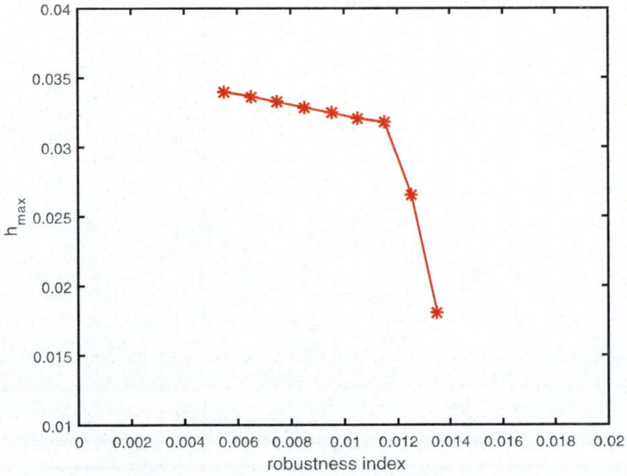

Fig. 9.8 Effect of robustness index α on $h_{\max}$

sion in communication network and the tolerance of the control system against DoS attack. Finally, the effect of robustness index α on a_1 is plotted in Fig. 9.9. In this case, we choose different values of the index α, and fix $\sigma = 0.08$, other parameters $\tau_D, \Gamma, \upsilon_1, \upsilon_2, a_2, \sigma, h, \epsilon_i, \phi_i, \varphi_i, \theta_i$ $(i = 1, 2)$ and simulation time are chosen as same as before. It is observed from Fig. 9.9 that the robustness index α is in inverse ratio to the value of a_1, which implies that the robustness can be improved by decreasing the convergence rate. The simulation results above indicate that there is a tradeoff between the nonlinearity robustness index and the degree of attack tolerance, as claimed in Remark 9.2.

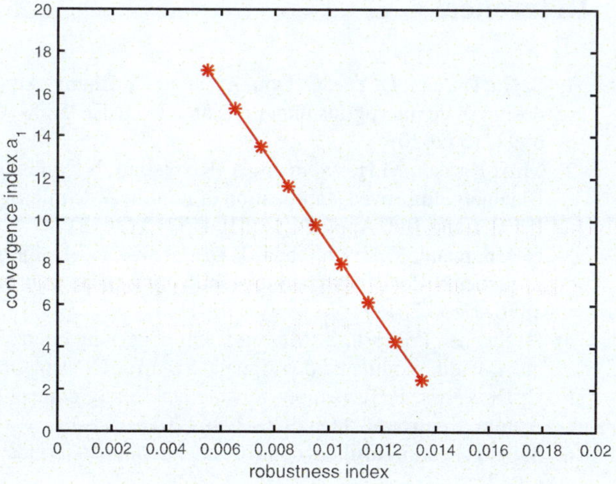

Fig. 9.9 Effect of robustness index α on a_1

9.5 Conclusion

We have proposed a novel attack-resilient event-triggered control synthesis approach for a networked nonlinear DC microgrid system under intermittent DoS jamming attacks. To address the controller design problem, a new kind of resilient event-triggered scheme has been presented to reduce the computation load while compensating the impact of the DoS attacks. An event-triggered switched system model of the nonlinear DC microgrid system has been established. By employing average dwell-time method and piecewise Lyapunov functional method, the asymptotic stability criterion of the resultant DC microgrid system has been derived. LMIs-based design method has been presented to obtain explicitly the controller gain matrix and event-triggering parameters. Finally, simulation results have been presented to verify that the nonlinear DC microgrid with multiple CPLs can be stabilized with the developed event-triggered control strategy even in the presence of intermittent DoS attacks. Our future work will focus on the observer-based secure control design following the ideas of [10, 11]. Moreover, the current study on event-triggering scheme achieved improvement in energy-consuming by reducing unnecessary communication. The lifetime of the wireless sensors, however, can be further prolonged by taking the income of energy into account as well, e.g., by using the energy harvesting technology proposed in [12]. Based on [12], how to investigate the event-triggered stabilization problem of nonlinear DC microgrid with an energy harvesting sensor is also an interesting topic.

References

1. S. Hu, P. Yuan, D. Yue, C. Dou, Z. Cheng, Y. Zhang, Attack-resilient event-triggered controller design of dc microgrids under dos attacks. IEEE Trans. Circuits Syst. I: Regular Papers **67**(2), 699–710 (2020)
2. M.A. Kardan, M.H. Asemani, A. Khayatian, N. Vafamand, M.H. Khooban, T. Dragičević, F. Blaabjerg, Improved stabilization of nonlinear dc microgrids: cubature kalman filter approach. IEEE Trans. Ind. Appl. **54**(5), 5104–5112 (2018)
3. N. Vafamand, M.H. Khooban, T. Dragičević, F. Blaabjerg, Networked fuzzy predictive control of power buffers for dynamic stabilization of dc microgrids. IEEE Trans. Ind. Electr. **66**(2), 1356–1362 (2019)
4. H. Shisheh Foroush, S. Martinez, On triggering control of single-input linear systems under pulse-width modulated dos signals. SIAM J. Control Opt. **54**(6), 3084–3105 (2016)
5. C. De Persis, P. Tesi, Input-to-state stabilizing control under denial-of-service. IEEE Trans. Autom. Control **60**(11), 2930–2944 (2015)
6. S. Feng, P. Tesi, Resilient control under denial-of-service: Robust design. Automatica **79**, 42–51 (2017)
7. D. Yue, E. Tian, Q.-L. Han, A delay system method for designing event-triggered controllers of networked control systems. IEEE Trans. Autom. Control **58**(2), 475–481 (2013)
8. Y. He, Q.G. Wang, L. Xie, C. Lin, Further improvement of free-weighting matrices technique for systems with time-varying delay. IEEE Trans. Autom. Control **52**(2), 293–299 (2007)
9. L. Herrera, W. Zhang, J. Wang, Stability analysis and controller design of dc microgrids with constant power loads. IEEE Trans. Smart Grid **8**(2), 881–888 (2017)
10. Y. Wan, J. Cao, G. Chen, W. Huang, Distributed observer-based cyber-security control of complex dynamical networks. IEEE Trans. Circuits Syst. I: Regular Papers **64**(11), 2966–2975 (2017)
11. Y. Yang, H. Xu, D. Yue, Observer-based distributed secure consensus control of a class of linear multi-agent systems subject to random attacks. IEEE Trans. Circuits Syst. I: Regular Papers **66**(8), 3089–3099 (2019)
12. B. Shen, Z. Wang, D. Wang, J. Luo, H. Pu, Y. Peng, Finite-horizon filtering for a class of nonlinear time-delayed systems with an energy harvesting sensor. Automatica **100**, 144–152 (2019)

Chapter 10
Resilient Dynamic Event-Triggered Control for Multi-area Power Systems with Renewable Energy Penetration Under Nonperiodic DoS Attacks

Resilient dynamic event-triggered control design problem is concerned for frequency stabilization of multi-area power systems with renewable energy penetration under transmission delay and denial-of-service (DoS) attacks. The main works are presented as followings: (1) A novel state-space model is established to describe multi-area power systems with solar and wind generation disturbance under load frequency control and virtual inertia control (LFC-VIC). (2) The influence of DoS attacks is considered and first modeled by model based average dwell time (MDADT). A criterion of preserving desired exponential H_∞ performance is obtained about the required lower bound of average dwell time (ADT) of sleep intervals, namely $\tau_F > ln\lambda_0/\alpha_1$ and the tolerable upper bound of ADT of attack intervals, namely $\tau_D < -(ln\lambda_1 + (\alpha_0 + \alpha_1)d_M)/\alpha_0$. (3) To reduce communication load, a dynamic event-triggered mechanism is proposed and designed jointly with resilient LFC-VIC scheme by using piecewise Lyapunov–krasovskii functional method and LMIs technique. (4) To defense DoS attacks actively, based on the resulting MDADT criterion, a dynamic prediction strategy is proposed to compensate the packet drops with unknown bound subjected to attack strategy. Finally, a numerical example for a two-area power system is carried out to verify the designed resilient LFC-VIC scheme and the dynamic event-triggered mechanism [1].

The chapter is organized as follows. The power system with RESs under attack and DETM are presented in Sect. 10.1. Section 10.2 proofs the DETM based LFC-VIC scheme to guarantee the exponential stability with H_∞ performance. Active secure scheme is presented in Sect. 10.3. A two-area power system is simulated to demonstrate the effectiveness of our theory in Sect. 10.4. Finally, the conclusion is stated in Sect. 10.5.

D. Yue et al., *Secure Control of Networked Control Systems and Its Applications*,
https://doi.org/10.1007/978-981-33-6730-2_10

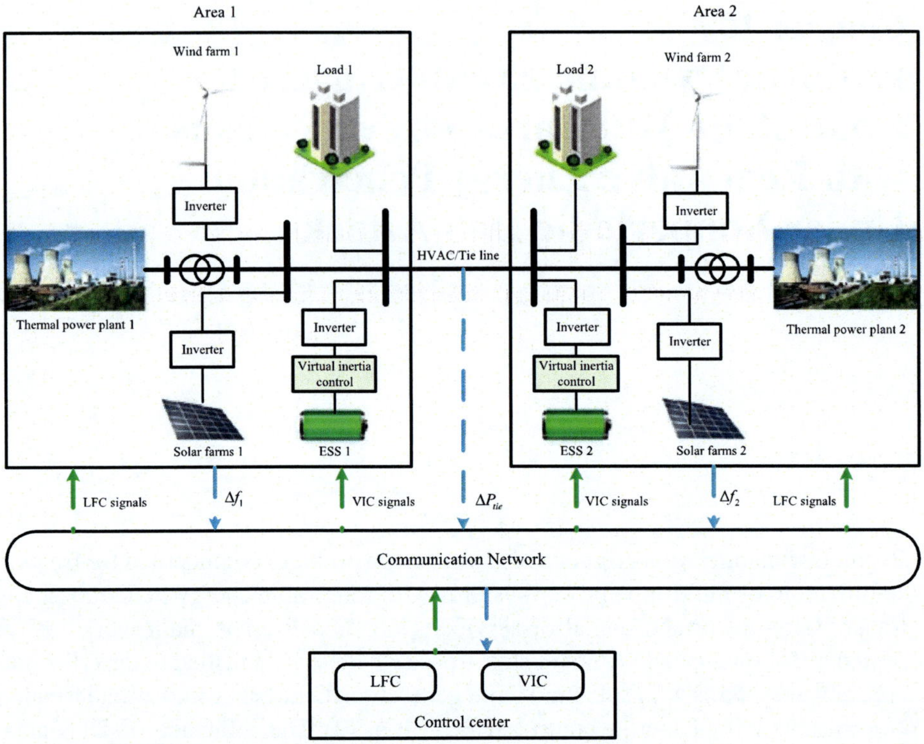

Fig. 10.1 Diagram of modern interconnected power system with RESs under LFC and virtual inertial control (VIC)

10.1 Problem Formulation

10.1.1 Modeling of Multi-area Power Systems with Penetrations of Solar and Wind Energy

To show the interconnected power system with penetration of RESs, Fig. 10.1 is depicted as an example about a two-area power system with the LFC scheme and the VIC scheme. More detailed, the structure of the ith area power system is illustrated in Fig. 10.2.

The dynamic of the ith area power system is represented by

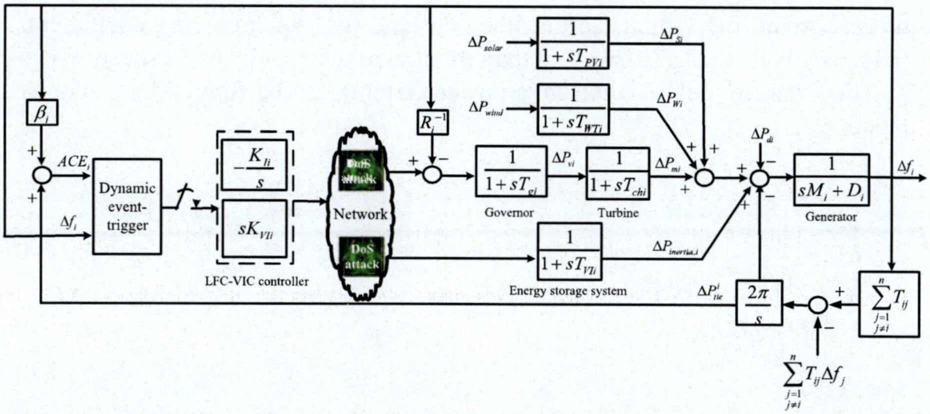

Fig. 10.2 The ith area power system with dynamic event-triggered mechanism based LFC-VIC scheme under DoS attacks and time delay

$$
\begin{cases}
\Delta \dot{f}_i = \dfrac{1}{M_i}\big(-D_i\Delta f_i + \Delta P_{mi} + \Delta P_{Wi} + \Delta P_{Si} \\
\qquad\qquad + \Delta P_{inertia,i} - \Delta P_{tie,i} - \Delta P_{di}\big), \\
\Delta \dot{P}_{mi} = \dfrac{1}{T_{chi}}\big(-\Delta P_{mi} + \Delta P_{vi}\big), \\
\Delta \dot{P}_{vi} = \dfrac{1}{T_{gi}}\Big(-\dfrac{1}{R_i}\Delta f_i - \Delta P_{vi} - K_{Ii}\displaystyle\int ACE_i(t-d_i)\Big), \\
\Delta \dot{P}_{Wi} = \dfrac{1}{T_{WTi}}\big(-\Delta P_{Wi} + \Delta P_{wind,i}\big), \\
\Delta \dot{P}_{Si} = \dfrac{1}{T_{PVi}}\big(-\Delta P_{Si} + \Delta P_{solar,i}\big), \\
\Delta \dot{P}_{inertia,i} = \dfrac{1}{T_{VIi}}\big(-\Delta P_{inertia,i} + K_{VIi}\Delta \dot{f}_i(t-d_i)\big), \\
\Delta \dot{P}_{tie,i} = \displaystyle\sum_{j=1, j\neq i}^{n} 2\pi T_{ij}\big(\Delta f_i - \Delta f_j\big), \\
ACE_i = \beta_i\Delta f_i + \Delta P_{tie,i},
\end{cases}
\tag{10.1}
$$

where Δf_i, ΔP_{mi}, ΔP_{vi}, ΔP_{Wi}, ΔP_{Si}, $\Delta P_{inertai,i}$, $\Delta P_{tie,i}$, ΔP_{di}, $\Delta P_{wind,i}$, $\Delta P_{solar,i}$, ACE_i are the deviation of frequency, the generator mechanical output, the governor valve position, the wind farm power, the solar farm power, the virtual inertia power of energy storage system (ESS), the tie-line power, the load, the wind farm disturbance, the solar farm disturbance and the area control error, respectively as seen in Fig. 10.1; M_i, D_i, T_{chi}, T_{gi}, R_i, T_{WTi}, T_{PVi}, T_{VIi}, T_{ij}, β_i, K_{Ii} and K_{VIi} are the generator inertia constant, the generator damping coefficient, the turbine time constant, the governor time constant, the drop constant, the wind turbine time constant, the solar system time constant, the virtual inertia time constant, the synchronizing coefficient, the frequency bias, the LFC gain and the virtual inertia control gain, respectively.

Then, the ith area of multi-area power systems can be formed by a state-space model

$$\begin{cases} \dot{x}_i(t) = A_{ii}x_i(t) + \sum\limits_{j=1, j\neq i}^{n} A_{ij}x_j(t) + B_i u_i(t-d_i) + F_i w_i(t), \\ y_i(t) = C_i x_i(t) + D_i w_i(t), \\ z_i(t) = E_i x_i(t), \end{cases} \tag{10.2}$$

where

$$x_i(t) = col\Big[\Delta f_i \quad \Delta P_{mi} \quad \Delta P_{vi} \quad \Delta P_{Wi} \quad \Delta P_{Si} \quad \Delta P_{inertia,i} \quad \Delta P_{tie,i} \quad \int ACE_i\Big],$$

$$y_i(t) = col\left[\int ACE_i \ \Delta \dot{f}_i\right],$$
$$z_i(t) = col\left[ACE_i \ \int ACE_i\right],$$
$$w_i(t) = col\left[\Delta P_{di} \ \Delta P_{wind,i} \ \Delta P_{solar,i}\right],$$

$$A_{ii} = \begin{bmatrix} \frac{-D_i}{M_i} & \frac{1}{M_i} & 0 & \frac{1}{M_i} & \frac{1}{M_i} & \frac{1}{M_i} & \frac{-1}{M_i} & 0 \\ 0 & \frac{-1}{T_{chi}} & \frac{1}{T_{chi}} & 0 & 0 & 0 & 0 & 0 \\ \frac{-1}{R_i T_{gi}} & 0 & -\frac{1}{T_{gi}} & 0 & 0 & 0 & 0 & 0 \\ 0 & 0 & 0 & \frac{-1}{T_{WTi}} & 0 & 0 & 0 & 0 \\ 0 & 0 & 0 & 0 & \frac{-1}{T_{PVi}} & 0 & 0 & 0 \\ 0 & 0 & 0 & 0 & 0 & \frac{-1}{T_{VIi}} & 0 & 0 \\ A_{ii}(7,1) & 0 & 0 & 0 & 0 & 0 & 0 & 0 \\ \beta_i & 0 & 0 & 0 & 0 & 0 & 1 & 0 \end{bmatrix},$$

$$A_{ii}(7,1) = \sum_{j=1, j\neq i}^{n} 2\pi T_{ij}, \ A_{ij} = \mathbf{0}_{8\times 8}, \ A_{ij}(7,1) = -2\pi T_{ij},$$

$$B_i = \begin{bmatrix} 0 & 0 & \frac{-1}{T_{gi}} & 0 & 0 & 0 & 0 & 0 \\ 0 & 0 & 0 & 0 & 0 & \frac{1}{T_{VIi}} & 0 & 0 \end{bmatrix}^T,$$

$$F_i = \begin{bmatrix} \frac{-1}{M_i} & 0 & 0 & 0 & 0 & 0 & 0 & 0 \\ 0 & 0 & 0 & \frac{1}{T_{WTi}} & 0 & 0 & 0 & 0 \\ 0 & 0 & 0 & 0 & \frac{1}{T_{PVi}} & 0 & 0 & 0 \end{bmatrix}^T,$$

$$C_i = \begin{bmatrix} 0 & 0 & 0 & 0 & 0 & 0 & 0 & 1 \\ \frac{-D_i}{M_i} & \frac{1}{M_i} & 0 & \frac{1}{M_i} & \frac{1}{M_i} & \frac{1}{M_i} & \frac{-1}{M_i} & 0 \end{bmatrix},$$

$$D_i = \begin{bmatrix} 0 & 0 & 0 \\ \frac{-1}{M_i} & 0 & 0 \end{bmatrix}, \ E_i = \begin{bmatrix} \beta_i & 0 & 0 & 0 & 0 & 0 & 1 & 0 \\ 0 & 0 & 0 & 0 & 0 & 0 & 0 & 1 \end{bmatrix}.$$

Combining the PI based load frequency control (LFC) scheme and the derivative control based virtual inertia control (VIC) scheme, a static output feedback control law is adopted in ith area power system

$$u_i(t) = K_i y_i(t),$$

where $K_i = \begin{bmatrix} K_{Ii} & 0 \\ 0 & K_{VIi} \end{bmatrix}$. Thus, a multi-area power systems interconnected with n subsystems can be written by an augmented form

$$\begin{cases} \dot{x}(t) = Ax(t) + Bu(t - d(t)) + Fw(t), \\ y(t) = Cx(t) + Dw(t), \\ z(t) = Ex(t). \end{cases} \tag{10.3}$$

where $x(t) = col\left[x_1(t)\ x_2(t)\ \dots\ x_n(t)\right]$, $y(t) = col\left[y_1(t)\ y_2(t)\ \dots\ y_n(t)\right]$, $z(t) = col\left[z_1(t)\ z_2(t)\ \dots\ z_n(t)\right]$, $u(t) = col\left[u_1(t)\ u_2(t)\ \dots\ u_n(t)\right]$, $w(t) = col\left[w_1(t)\ w_2(t)\ \dots\ w_n(t)\right]$ and

$$\begin{aligned} A &= \begin{bmatrix} A_{11} & A_{12} & \dots & A_{1n} \\ A_{21} & A_{22} & \dots & A_{2n} \\ \vdots & \vdots & \ddots & \vdots \\ A_{n1} & A_{n2} & \dots & A_{nn} \end{bmatrix}, \\ B &= diag\left[B_1\ B_2\ \cdots\ B_n\right], \\ C &= diag\left[C_1\ C_2\ \cdots\ C_n\right], \\ D &= diag\left[D_1\ D_2\ \cdots\ D_n\right], \\ E &= diag\left[E_1\ E_2\ \cdots\ E_n\right], \\ F &= diag\left[F_1\ F_2\ \cdots\ F_n\right], \end{aligned}$$

With sample periodic h, the control inputs of multi-area power systems (10.3) are implemented by actuators through zero-order-hold (ZOH) mechanism

$$u(t) = Ky(kh), t \in [kh, (k+1)h) \tag{10.4}$$

where $\{kh\}, k \in N$ is the set of sampled sequence and $K = diag\{K_1, K_2, \dots, K_n\}$ is the control gain to be designed.

10.1.2 DoS Attack Model

In this study, the influence of DoS attacks on S-C channel and C-A channel is characterized as the missing of3control input signal $u(kh)$. Then, the nonperiodic attack strategy can be represented by a switching signals (10.5) taking value of 1 and 0, where 1 represents the normal communication and 0 the launching of DoS attacks.

$$s(t) = \begin{cases} 1, & t \in [g_n, g_n + b_n), \\ 0, & t \in [g_n + b_n, g_{n+1}), \end{cases} \tag{10.5}$$

where $g_n \in R_{\geq 0}, n \in N$ denotes the beginning time of the nth sleep interval $\mathcal{I}_n = [g_n, g_n + b_n)$ with $g_0 = 0$;b_n the length of the nth sleep interval; $g_n + b_n \in R_{>0}, n \in N$ the beginning time of the nth attack interval $\mathcal{S}_n = [g_n + b_n, g_{n+1})$.

Further, from the viewpoint of switched system including unstable subsystem, it is interesting to explore the upper bound of the average dwell time of attack interval and the lower bound of the average dwell time of sleep interval. Thus, model-based average dwell time (MDADT) concept is first introduced to characterize DoS attacks. The attack model is represented by the following definitions.

Definition 10.1 ([2]) Denote $N_F(t, t_0)$ the numbers of sleep intervals and $\Xi_F(t, t_0)$ the sum of sleep intervals length during $[t_0, t]$. There exist scalars $\zeta_F \in R_{\geq 0}, \tau_F \in R_{\geq h}$ satisfying

$$N_F(t, t_0) \leq \zeta_F + \frac{\Xi_F(t, t_0)}{\tau_F} \tag{10.6}$$

Definition 10.2 ([2]) Denote $N_D(t, t_0)$ the numbers of attack intervals and $\Xi_D(t, t_0)$ the sum of attack intervals length during $[t_0, t]$. There exist scalars $\zeta_D \in R_{\geq 0}, \tau_D \in R_{\geq h}$ satisfying

$$N_D(t, t_0) \geq \zeta_D + \frac{\Xi_D(t, t_0)}{\tau_D} \tag{10.7}$$

Remark 10.1 The proposed DoS attack model (10.6), (10.7) can not only represent periodic DoS attacks [3–5] by $\tau_F = T_{off}$ and $\tau_D = T - T_{off}$ but also represent aperiodic DoS attacks [6] by $\tau_F = b_{min}$ and $\tau_D = l_{max}$. Compared with ADT model in [7, 8], the lower bound of average dwell time $\frac{t-\tau}{n(\tau,t)-\eta} \geq \tau_D$ is defined for the mixed interval composed of one attack interval and its adjacent sleep interval. Such definition represents the stable subsystem operating in the mixed interval and treats the influence of attack as the beginning transient dynamic which will be eliminated with a sufficient large τ_D. In Definition 10.1, however, the average dwell time of sleep interval purely indicating the operation of stable subsystem is denoted as τ_F. On the other hand, the duty cycle of attack duration should be less than its upper bound $\frac{\Xi(\tau,t)-\kappa}{t-\tau} \leq 1/T$. In Definition 10.2, however, the average dwell time of attack interval is denoted as τ_D which is more flexible than the concept of $1/T$. The proposed MDADT model is more general and flexible than the previous models.

According to (10.5), the availability of control signals $u(t)$ generated by load frequency control and virtual inertia control (LFC-VIC) scheme is subjected to the switching signal $s(t)$, which means that the control signal $u(t)$ is available during sleep intervals $\mathcal{I}_n$ while the zero inputs strategy is adopted during attack interval $\mathcal{S}_n$.

$$u(t) = \begin{cases} Ky(kh), & t \in \mathcal{I}_n, \\ 0, & t \in \mathcal{S}_n. \end{cases} \tag{10.8}$$

10.1.3 Design of Dynamic Event-Triggered Mechanism and Problem Statement

To improve the communication efficiency during the sleep interval $\mathcal{I}_n$, a dynamic event-triggered mechanism (DETM) is introduced to determine whether to transmit the measurement output signals $y(kh)$.

Denote $\{t_{k,n}h\}$ as the set of triggering instants which satisfy

$$t_{k,n}h = \{t_{k_j,n}h | t_{k_j,n}h \ satisfying \ (10.9) \ and \ t_{k_j,n}h \in \mathcal{I}_n\} \cup \{h_n\}$$

where $n, j, k_j \in N$, $\{t_{k,n}h\} \subseteq \{kh\}$. By denoting $t_{k_j,n}h \triangleq t_{k,n}h + jh$, the triggering condition is given by

$$\begin{aligned} &[y(t_{k_j,n}h) - y(t_{k,n}h)]^T \Omega_c [y(t_{k_j,n}h) - y(t_{k,n}h)] \\ &> \varepsilon y^T(t_{k,n}h)\Omega_c y(t_{k,n}h) + \frac{1}{\varphi}\eta(t_{k_j,n}h) \end{aligned} \tag{10.9}$$

where positive triggering parameters ε, φ and Ω_c. Moreover, the definition of scalar function $\eta(t)$ is given by

$$\begin{cases} \dfrac{d\eta(t)}{dt} = -\mu\eta(t) + \varepsilon y^T(t_{k,n}h)\Omega_c y(t_{k,n}h) \\ - [y(t_{k_j,n}h) - y(t_{k,n}h)]^T \Omega_c [y(t_{k_j,n}h) - y(t_{k,n}h)], \\ \mu > 0, \eta(0) = \eta_0 > 0, \eta(g_n) = 0, n > 1. \end{cases} \tag{10.10}$$

For simplification and convenience, the dynamic event-triggered condition (10.9) with (10.10) will be represented by the form of state vector.

$$\begin{cases} [x(t_{k_j,n}h) - x(t_{k,n}h)]^T \Omega [x(t_{k_j,n}h) - x(t_{k,n}h)] \\ > \varepsilon x^T(t_{k,n}h)\Omega x(t_{k,n}h) + \dfrac{1}{\varphi}\eta(t_{k_j,n}h), \\ \dfrac{d\eta(t)}{dt} = -\mu\eta(t) + \varepsilon x^T(t_{k,n}h)\Omega x(t_{k,n}h) \\ - [x(t_{k_j,n}h) - x(t_{k,n}h)]^T \Omega [x(t_{k_j,n}h) - x(t_{k,n}h)], \end{cases} \tag{10.11}$$

where $\Omega = C^T \Omega_c C$.

Remark 10.2 Note that $\eta(t) \geq 0$ if $ln(1+\mu\varphi) > \mu h$. The detail proof can be seen in [9] and omitted here. Due to the existence of positive function of $\eta(t) \geq 0$, the considered DETM would be more efficient than the traditional ETM [3, 10]. Because the value of the right side of triggering condition is larger than that of traditional ETM.

According to the triggered sequence, the sleep interval can be divided by the triggering intervals

$$\mathcal{I}_n = \cup_{k=0}^{k(n)} \mathcal{I}_{k,n},$$

with

$$\mathcal{I}_{k,n} = \begin{cases} [h_n,\ t_{1,n}h + d(t_{1,n}h)),\ k = 0, \\ [t_{k,n}h + d(t_{k,n}h), t_{k+1,n}h + d(t_{k+1,n}h)), k = 1, \ldots, k(n)-1, \\ [t_{k(n),n}h + d(t_{k(n),n}h),\ h_n + g_n),\ k = k(n), \end{cases}$$

where $k(n) = sup\{k \in N | t_{k,n}h + d(t_{k,n}h) < h_n + g_n\}$ and $d(t_{k,n}h) \in [0, \bar{d}]$.

Within the triggering interval $\mathcal{I}_{k,n}$, one can further have

$$\mathcal{I}_{k,n} = \cup_{j=0}^{j(k)} \mathcal{I}_{k,n}^j,$$

with

$$\mathcal{I}_{k,n}^j = \begin{cases} [t_{k,n}h + d(t_{k,n}h),\ t_{k,n}h + \bar{d} + h),\ j = 0, \\ [t_{k,n}h + \bar{d} + jh,\ t_{k,n}h + \bar{d} + (j+1)h), j = 1, \ldots, j(k)-1, \\ [t_{k,n}h + \bar{d} + j(k)h,\ t_{k+1,n}h + d(t_{k+1,n}h)), j = j(k), \end{cases}$$

where $j(k) = sup\{j \in N | t_{k,n}h + \bar{d} + jh < t_{k+1,n}h + d(t_{k+1,n}h)\}$.

For $t \in \mathcal{I}_{k,n}^j$, $j = 0, \ldots, j(k)$, a virtual input delay is introduced

$$d(t) = \begin{cases} t - t_{k,n}h, & t \in \mathcal{I}_{k,n}^0, \\ t - t_{k,n}h - jh, & t \in \mathcal{I}_{k,n}^j,\ j = 1, \ldots, j(k)-1 \\ t - t_{k,n}h - j(k)h, & t \in \mathcal{I}_{k,n}^{j(k)}. \end{cases} \tag{10.12}$$

which indicates

$$d(t_{k,n}h) \leq \bar{d} \leq d(t) \leq d_M,\ t \in \mathcal{I}_{k,n},$$

where $d_M = \bar{d} + h$.

Furthermore, a piecewise function to represent the triggered error is defined by

$$e(t) = \begin{cases} x(t_{k,n}h) - x(t_{k,n}h), & t \in \mathcal{I}_{k,n}^0, \\ x(t_{k,n}h) - x(t_{k,n}h + jh), & t \in \mathcal{I}_{k,n}^j,\ j = 1, \ldots, j(k)-1, \\ x(t_{k,n}h) - x(t_{k,n}h + j(k)h), & t \in \mathcal{I}_{k,n}^{j(k)}. \end{cases} \tag{10.13}$$

Then, the final control inputs implemented by actuators can be rewritten by

$$u(t) = \begin{cases} KC(x(t-d(t)) + e(t)), & t \in \mathcal{I}_{k,n}^{j} \cap \mathcal{I}_{k,n} \cap \mathcal{I}_n, \\ 0, & t \in \mathcal{S}_n. \end{cases} \tag{10.14}$$

Based on the resilient control scheme (10.14), the multi-area power systems (10.3) under DoS attacks is described by a switched time delay system $\Sigma_{s(t)}, s(t) \in \{0, 1\}$

$$\begin{cases} \Sigma_1 \begin{cases} \dot{x}(t) = Ax(t) + BKC(x(t-d(t)) + e(t)) + Fw(t), \\ z(t) = Ex(t), t \in \mathcal{I}_{k,n}^{j} \cap \mathcal{I}_{k,n} \cap \mathcal{I}_n, \end{cases} \\ \Sigma_0 \begin{cases} \dot{x}(t) = Ax(t) + Fw(t), \\ z(t) = Ex(t), t \in \mathcal{S}_n, \end{cases} \end{cases} \tag{10.15}$$

with input delay $d(t) \in [0, d_M]$, $\dot{d}(t) = 1$, initial conditions $x(t_0)=x_0$, $t_0 \in [-d_M, 0]$ and triggered error satisfying

$$\begin{cases} e^T(t)\Omega e(t) \le \varepsilon[x(t-d(t)) + e(t)]^T \Omega [x(t-d(t)) + e(t)] + \dfrac{1}{\varphi}\eta(t_{k_j,n}h), \\ \dfrac{d\eta(t)}{dt} = -\mu\eta(t) + \varepsilon[x(t-d(t)) + e(t)]^T \Omega [x(t-d(t)) + e(t)] - e(t)^T \Omega e(t). \end{cases} \tag{10.16}$$

For switched time delay system (10.15), the objective of this study is to design output control gain K and triggering matrix Ω to preserve the exponential H_∞ performance.

(1) The closed-loop system (10.15) with $w(t) = 0$ is exponentially stable;
(2) Under zero initial state condition, for any nonzero $w(t) \in \mathcal{L}_2[0, \infty)$, it holds $\|z(t)\| \le \gamma \|w(t)\|$ for prescribed level $\gamma > 0$.

10.2 Resilience Analysis

In order to facilitate the theory development, we first establish a technical result on decay rate estimation of piecewise Lyapunov–Krasovskii functional.

Theorem 10.1 *Given positive scalars $d_M > 0$, $\alpha_0 > 0$, $\alpha_1 > 0$, $\varepsilon > 0$, $\varphi > 0$, $\mu > 0$ along each subsystem (10.15), respectively, it follows*

$$\dot{V}_1(t) \le -\alpha_1 V_1(t) - z^T(t)z(t) + \gamma^2 w^T(t)w(t), \tag{10.17}$$

$$\dot{V}_0(t) \le \alpha_0 V_0(t) - z^T(t)z(t) + \gamma^2 w^T(t)w(t), \tag{10.18}$$

if there exist positive definite matrices $P_i > 0$, $Q_i > 0$, $R_i > 0$, $Z_i > 0$, $i = 0, 1$, $\Omega > 0$, and appropriate dimension matrices K, X_0, X_1 satisfying

$$\Phi \le 0, \quad \Gamma \le 0, \tag{10.19}$$

$$\begin{bmatrix} R_1 & X_1 \\ X_1^T & R_1 \end{bmatrix} \ge 0, \begin{bmatrix} R_0 & X_0 \\ X_0^T & R_0 \end{bmatrix} \ge 0. \tag{10.20}$$

where

$$\Phi = \begin{bmatrix} \Phi_0 & * & * & * \\ d_M \Phi_1 & -R_1^{-1} & * & * \\ d_M \Phi_1 & 0 & -Z_1^{-1} & * \\ \Phi_2 & 0 & 0 & -I \end{bmatrix},$$

$$\Phi_0 = \begin{bmatrix} \Phi_0(1,1) & \Phi_0(1,2) & -\rho_1 X_1 & P_1 BKC & P_1 F \\ * & \Phi_0(2,2) & \Phi_0(2,3) & \varepsilon\psi\Omega & 0 \\ * & * & \Phi_0(3,3) & 0 & 0 \\ * & * & * & \Phi_0(4,4) & 0 \\ * & * & * & * & -\gamma^2 I \end{bmatrix},$$

$$\Phi_0(1,1) = He(P_1 A) + \alpha_1 P_1 + Q_1 - \frac{\pi^2}{4} Z_1 + \rho_1 R_1,$$

$$\Phi_0(1,2) = P_1 BKC - \rho_1 R_1 + \rho_1 X_1 + \frac{\pi^2}{4} Z_1,$$

$$\Phi_0(2,2) = \varepsilon\psi\Omega + \rho_1(2R_1 - X_1 - X_1^T) - \frac{\pi^2}{4} Z_1,$$

$$\Phi_0(2,3) = \rho_1(X_1 - R_1), \Phi_0(3,3) = -e^{-\alpha_1 d_M} Q_1 + \rho_1 R_1,$$

$$\Phi_0(4,4) = (\varepsilon - 1)\psi\Omega, \psi = \varphi(\mu - \alpha_1) + 1,$$

$$\Phi_1 = [A, BKC, 0, BKC, F], \ \Phi_2 = [E, 0, 0, 0, 0],$$

$$\Gamma = \begin{bmatrix} \Gamma_0 & * & * & * \\ d_M \Gamma_1 & -R_0^{-1} & * & * \\ d_M \Gamma_1 & 0 & -Z_0^{-1} & * \\ \Gamma_2 & 0 & 0 & -I \end{bmatrix},$$

$$\Gamma_0 = \begin{bmatrix} \Gamma_0(1,1) & \Gamma_0(1,2) & -\rho_0 X_0 & P_0 F \\ * & \Gamma_0(2,2) & \rho_0(X_0 - R_0) & 0 \\ * & * & -e^{\alpha_0 d_M} Q_0 + \rho_0 R_0 & 0 \\ * & * & * & -\gamma^2 I \end{bmatrix},$$

$$\Gamma_0(1,1) = He(P_0 A) - \alpha_0 P_0 + Q_0 + \rho_0 R_0 - \frac{\pi^2}{4} Z_0,$$

$$\Gamma_0(1,2) = -\rho_0 R_0 + \rho_0 X_0 + \frac{\pi^2}{4} Z_0,$$

$$\Gamma_0(2,2) = \rho_0(2R_0 - X_0 - X_0^T) - \frac{\pi^2}{4} Z_0,$$

$$\Gamma_1 = [A, 0, 0, F], \ \Gamma_2 = [E, 0, 0, 0].$$

Proof Choose a piecewise Lyapunov–Krasovskii functional

$$V(t)=\begin{cases}V_1(t), & t\in\mathcal{I}_{k,n}^j\cap\mathcal{I}_{k,n}\cap\mathcal{I}_n,\\ V_0(t), & t\in\mathcal{S}_n,\end{cases}\tag{10.21}$$

where

$$\begin{aligned}V_i(t)=\;& i*\eta(t)|_{t=(t_{k_j,n}h)^+}+x^T(t)P_ix(t)\\&+\int_{t-d_M}^{t}x^T(s)exp(\diamond)Q_ix(s)ds\\&+d_M\int_{-d_M}^{0}\int_{t+\theta}^{t}\dot{x}^T(s)exp(\diamond)R_i\dot{x}(s)dsd\theta\\&+d_M^2\int_{t-d(t)}^{t}\dot{x}^T(s)exp(\diamond)Z_i\dot{x}(s)ds\\&-\frac{\pi^2}{4}\int_{t-d(t)}^{t}\Delta x(s)^Texp(\diamond)Z_i\Delta x(s)ds,\end{aligned}$$

where $exp(\diamond)=e^{(-1)^i\alpha_i(t-s)}$ and $\Delta x(s)=x(s)-x(t-d(t))$.

Case I: for $t\in\mathcal{I}_{k,n}^j$, taking derivation of $V_1(t)$ (10.21) along subsystem Σ_1 with triggering constraint (10.16), one has

$$\begin{aligned}\dot{V}_1(t)\le&-\alpha_1V_1(t)+(\alpha_1-\mu)\eta(t_{k_j,n}h)+\alpha_1x^T(t)P_1x(t)\\&+\varepsilon[x(t-d(t))+e(t)]^T\Omega[x(t-d(t))+e(t)]\\&-e^T(t)\Omega e(t)+2x^T(t)P_1\dot{x}(t)\\&+x^T(t)Q_1x(t)-e^{-\alpha_1d_M}x^T(t-d_M)Q_1x(t-d_M)\\&+d_M^2\dot{x}^T(t)(R_1+Z_1)\dot{x}(t)\\&-d_M\int_{t-d_M}^{t}\dot{x}^T(s)e^{-\alpha_1(t-s)}R_1\dot{x}(s)ds\\&-\frac{\pi^2}{4}[x(t)-x(t-d(t))]^TZ_1[x(t)-x(t-d(t))]\\&+z^T(t)z(t)-\gamma^2w^T(t)w(t)\\&-z^T(t)z(t)+\gamma^2w^T(t)w(t).\end{aligned}$$

According to Lemmas 1 and 2 in [6], we have

$$\begin{aligned}&-d_M\int_{t-d_M}^{t}\dot{x}^T(s)e^{-\alpha_1(t-s)}R_1\dot{x}(s)ds\\&\le\rho_1\begin{bmatrix}\Delta x(t)\\-\Delta x(t-d_M)\end{bmatrix}^T\begin{bmatrix}R_1 & X_1\\X_1^T & R_1\end{bmatrix}\begin{bmatrix}\Delta x(t)\\-\Delta x(t-d_M)\end{bmatrix}\end{aligned}$$

where $\rho_1 = \alpha_1 d_M/(1 - e^{\alpha_1 d_M})$.

Let $\xi_1(t) = col[x(t), x(t - d(t)), x(t - d_M), e(t), w(t)]$ and $\mu \geq \alpha_1$. By using Schur complement lemma, it is further obtained

$$\dot{V}_1(t) \leq -\alpha_1 V_1(t) + \xi_1^T(t)\Phi\xi_1(t) - z^T(t)z(t) + \gamma^2 w^T(t)w(t).$$

According to $\Phi \leq 0$ (10.19), the decay estimation of $V_1(t)$ (10.17) can be obtained.

Case II: for $t \in \mathcal{S}_n$, taking derivation of $V_0(t)$ along subsystem Σ_0, it can be obtained that

$$\begin{aligned}\dot{V}_0(t) \leq& \alpha_0 V_0(t) - \alpha_0 x^T(t)P_0x(t)\\ &+ 2x^T(t)P_0[Ax(t) + Fw(t)]\\ &+ x^T(t)Q_0x(t) - e^{\alpha_0 d_M}x^T(t - d_M)Q_0x(t - d_M)\\ &+ d_M^2\dot{x}^T(t)(R_0 + Z_0)\dot{x}(t)\\ &- d_M\int_{t-d_M}^{t} e^{\alpha_0(t-s)}\dot{x}^T(s)R_0\dot{x}(s)ds\\ &- \frac{\pi^2}{4}[x(t) - x(t - d(t))]^T Z_0[x(t) - x(t - d(t))]\\ &+ z^T(t)z(t) - \gamma^2 w^T(t)w(t)\\ &- z^T(t)z(t) + \gamma^2 w^T(t)w(t),\end{aligned}$$

According to Lemmas 1 and 2 in [6], we have

$$\begin{aligned}&- d_M\int_{t-d_M}^{t} e^{\alpha_0(t-s)}\dot{x}^T(s)R_0\dot{x}(s)ds\\ &\leq \rho_0\begin{bmatrix}\Delta x(t)\\ -\Delta x(t - d_M)\end{bmatrix}^T\begin{bmatrix}R_0 & X_0\\ X_0^T & R_0\end{bmatrix}\begin{bmatrix}\Delta x(t)\\ -\Delta x(t - d_M)\end{bmatrix}\end{aligned}$$

where $\rho_0 = \alpha_0 d_M/(e^{-\alpha_0 d_M} - 1)$.

By denoting $\xi_0(t) = col[x(t), x(t - d(t)), x(t - d_M), w(t)]$ and using Schur complement lemma, one has

$$\dot{V}_0(t) \leq \alpha_0 V_0(t) + \xi_0^T(t)\Gamma\xi_0(t) - z^T(t)z(t) + \gamma^2 w^T(t)w(t).$$

According to $\Gamma \leq 0$ (10.19), the decay estimation of $V_0(t)$ (10.18) can be obtained. This completes the proof.

Based on Theorem 10.1, the resilience analysis, exponentially stable with H_∞ performance, of switched delay system (10.15) will be given by using switched system approach.

Theorem 10.2 *Given positive scalars* $\gamma > 0, 1 > \lambda_1 > 0, \ \lambda_0 > 1,$ *under DoS attacks (10.5) with* τ_D, τ_F*, switched time delay system (10.15) is exponentially stable with* H_∞ *performance level* γ *if Theorem 6.1 holds and there exist positive definite matrices* $P_i > 0, Q_i > 0, R_i > 0, Z_i > 0, i = 0, 1,$ *satisfying*

$$\begin{aligned} &P_1 \le \lambda_0 P_0, \ P_0 \le \lambda_1 e^{(\alpha_1+\alpha_0)d_M} P_1, \ Q_1 \le \lambda_0 Q_0, \ Q_0 \le \lambda_1 Q_1, \\ &R_1 \le \lambda_0 R_0, \ R_0 \le \lambda_1 R_1, \ Z_1 \le \lambda_0 Z_0, \ Z_0 \le \lambda_1 Z_1, \end{aligned} \tag{10.22}$$

$$\tau_F > \frac{ln\lambda_0}{\alpha_1}, \tau_D < \frac{-(ln\lambda_1 + (\alpha_0 + \alpha_1)d_M)}{\alpha_0}. \tag{10.23}$$

Proof According to (10.22), the inequality relationship of Lyapunov–Krasovskii functional (10.21) in switching points are derived

$$\begin{aligned} &V_1(g_n^+) \le \lambda_0 V_0(g_n^-), \\ &V_0((g_n + b_n)^+) \le \lambda_1 e^{(\alpha_1+\alpha_0)d_M} V_1((g_n + b_n)^-). \end{aligned} \tag{10.24}$$

First, the exponential stability of the switched delay system (10.15) will be analyzed under the assumption of $w(t) = 0$. Without loss of generality, we develop the decay process of piecewise LKF (10.21) from $t \in [g_n, g_n + b_n)$. Based on (10.17), (10.18) and (10.24), one has

$$\begin{aligned} V(t) \le& e^{-\alpha_1(t-g_n)} V_1(g_n^+) \\ \le& e^{-\alpha_1(t-g_n)+\alpha_0(g_n-g_{n-1}-b_{n-1})-\alpha_1 b_{n-1}} \\ & \lambda_0\lambda_1 e^{(\alpha_1+\alpha_0)d_M} \lambda_0 V_0(g_{n-1}^-) \\ \le& (\lambda_1 e^{(\alpha_0+\alpha_1)d_M})^{N_D(t,0)} \lambda_0^{N_F(t,0)} e^{-\alpha_1 \Xi_F(t,0)} \\ & e^{\alpha_0 \Xi_D(t,0)} V(0) \end{aligned}$$

According to (10.6) and (10.7), by denoting $\mathbf{exp}\{\diamond\} = e^\diamond$, we further obtain

$$\begin{aligned} V(t) \le& \mathbf{exp}\{(ln\lambda_1 + (\alpha_0 + \alpha_1)d_M)\zeta_D + \zeta_F ln\lambda_0\} \\ & \mathbf{exp}\left\{\left(\frac{ln\lambda_1 + (\alpha_0 + \alpha_1)d_M}{\tau_D} + \alpha_0\right) \Xi_D(t,0)\right\} \\ & \mathbf{exp}\left\{\left(\frac{ln\lambda_0}{\tau_F} - \alpha_1\right) \Xi_F(t,0)\right\} V(0) \\ \le& \phi \mathbf{exp}\left\{max\left(\frac{ln\lambda_1 + (\alpha_0 + \alpha_1)d_M}{\tau_D} + \alpha_0, \frac{ln\lambda_0}{\tau_F} - \alpha_1\right)\right. \\ & \left.(\Xi_F(t,0) + \Xi_D(t,0))\right\} V(0) \\ \le& \phi e^{-\beta t} V(0) \end{aligned}$$

where $\phi = \mathbf{exp}\{(ln\lambda_1 + (\alpha_0 + \alpha_1)d_M)\zeta_D + \zeta_F ln\lambda_0\}$, $\beta = max\{-\frac{ln\lambda_1+(\alpha_0+\alpha_1)d_M}{\tau_D} - \alpha_0, \alpha_1 - \frac{ln\lambda_0}{\tau_F}\}$. According to (10.23), it has $\beta > 0$. Thus, when $t \to \infty$, the switched delay system (10.15) is exponentially stable.

Furthermore, we analyze the H_∞ performance of system (10.15) with $x(t_0) = 0$. According to (10.17), (10.18) and (10.24), for $t \in [g_n, g_n + b_n)$, we have the following recursion steps

$$
\begin{aligned}
V(t) \leq & e^{-\alpha_1(t-g_n)}\lambda_0 V_0(g_n^-) + \int_{g_n}^{t} e^{-\alpha_1(t-s)}\Delta(s)ds \\
\leq & e^{-\alpha_1(t-g_n)+\alpha_0(g_n-g_{n-1}-b_{n-1})-\alpha_1 b_{n-1}} \\
& \lambda_0\lambda_1 e^{(\alpha_1+\alpha_0)d_M}\lambda_0 V_0(g_{n-1}^-) + \int_{g_n}^{t} e^{-\alpha_1(t-s)}\Delta(s)ds \\
& + \int_{g_{n-1}+b_{n-1}}^{g_n} e^{-\alpha_1(t-g_n)+\alpha_0(g_n-s)}\lambda_0\Delta(s)ds \\
& + \int_{g_{n-1}}^{g_{n-1}+b_{n-1}} e^{-\alpha_1(t-g_n)+\alpha_0(g_n-g_{n-1}-b_{n-1})} \\
& e^{-\alpha_1(g_{n-1}+b_{n-1}-s)}\lambda_0\lambda_1 e^{(\alpha_1+\alpha_0)d_M}\Delta(s)ds \\
\leq & (\lambda_1 e^{(\alpha_0+\alpha_1)d_M})^{N_D(t,g_{n-1})}\lambda_0^{N_F(t,g_{n-1})}e^{-\alpha_1\Xi_F(t,g_{n-1})} \\
& e^{\alpha_0\Xi_D(t,g_{n-1})}V(g_{n-1}^-) + \int_{g_{n-1}}^{t}(\lambda_1 e^{(\alpha_0+\alpha_1)d_M})^{N_D(t,s)} \\
& \lambda_0^{N_F(t,s)}e^{-\alpha_1\Xi_F(t,s)+\alpha_0\Xi_D(t,s)}\Delta(s)ds \\
\leq & (\lambda_1 e^{(\alpha_0+\alpha_1)d_M})^{N_D(t,0)}\lambda_0^{N_F(t,0)}e^{-\alpha_1\Xi_F(t,0)} \\
& e^{\alpha_0\Xi_D(t,0)}V(0) + \int_{0}^{t}(\lambda_1 e^{(\alpha_0+\alpha_1)d_M})^{N_D(t,s)} \\
& \lambda_0^{N_F(t,s)}e^{-\alpha_1\Xi_F(t,s)+\alpha_0\Xi_D(t,s)}\Delta(s)ds
\end{aligned}
\tag{10.25}
$$

where $\Delta(s) = -z^T(s)z(s) + \gamma^2 w^T(s)w(s)$.

For convenience, define a propagation operator

$$
\begin{aligned}
\mathcal{P}(t,\tau) = & \mathbf{exp}\{(ln\lambda_1 + (\alpha_0+\alpha_1)d_M)N_D(t,\tau) \\
& + ln\lambda_0 N_F(t,\tau) - \alpha_1\Xi_F(t,\tau) + \alpha_0\Xi_D(t,\tau)\}, \forall t \geq \tau.
\end{aligned}
\tag{10.26}
$$

Then, we rewrite (10.25) as

$$
V(t) \leq \mathcal{P}(t,0)V(0) + \int_0^t \mathcal{P}(t,s)\Delta(s)ds. \tag{10.27}
$$

From (10.23), it has $\mathcal{P}(t, \tau) \leq \phi e^{-\beta(t-\tau)}$. Under zero initial condition, one has

$$0 \leq \int_0^t \phi e^{-\beta(t-s)} \Delta(s) ds. \tag{10.28}$$

Integrating both sides of (10.28) from $t = 0$ to ∞, one has

$$\begin{aligned} & 0 \leq \int_0^\infty \int_0^t \phi e^{-\beta(t-s)} \Delta(s) ds dt. \\ \Leftrightarrow\ & 0 \leq \int_0^\infty \int_s^\infty \phi e^{-\beta(t-s)} \Delta(s) dt ds. \\ \Leftrightarrow\ & 0 \leq \int_0^\infty \frac{\phi}{\beta} \Delta(s) ds. \end{aligned}$$

It finally has $\int_0^\infty z^T(t) z(t) dt \leq \gamma^2 \int_0^\infty w^T(t) w(t) dt$, namely $\|z(t)\| \leq \gamma \|w(t)\|$. The proof is completed that system (10.15) is exponentially stable with H_∞ disturbance attenuant level γ.

Remark 10.3 The satisfaction of the resulting conditions (10.22) guarantees the decay at the switching points of attack intervals and sleep intervals. The condition about τ_F exhibits the lower bound of the ADT of sleep intervals to guarantee frequency stability of power system. The lower bound is in inverse proportion to the exponential decay rate α_1 during sleep interval. It means that if power system has a fast exponential convergence with a large value of α_1 during sleep interval a short ADT τ_F would be required. It indicates the excellent performance during sleep intervals store much resilience for power system against DoS attacks. The condition about τ_D exhibits the upper bound of the ADT of attack intervals. The upper bound is in inverse proportion to the exponential decay rate α_0 during attack interval. It means that if power system has a large exponential divergence with a large value of α_0 during attack interval a short ADT τ_D would be tolerant. It indicates the poor performance during attack intervals would lead to power system only tolerating short DoS attacks. On the other hand, the exponential decay rate β depends not only on the adjusting parameters $\alpha_i, \lambda_i (i = 0, 1)$ but also on delay bound d_M and model based average dwell time (MDADT) parameters τ_D, τ_F. And the criterion (10.23) about the tolerable ADT of DoS attacks and the required ADT of normal transmission is our main contribution. The MDADT criterion (10.23) is more flexible and practical than that in [6, 7, 11] from the aspects of constraining attack interval and sleep interval independently, respectively.

10.3 Resilient Controller Design Based LFC-VIC Scheme

According to Theorems 10.1 and 10.2, the co-design method of resilient load frequency control and virtual inertia control (LFC-VIC) gain and DETM parameter is presented by using LMI technique.

Theorem 10.3 *Given positive scalars* $d_M > 0, \gamma > 0, \alpha_0 > 0, \alpha_1 > 0, 1 > \lambda_1 > 0, \lambda_0 > 1, \delta > 0, \varepsilon > 0, \varphi > 0, \mu > 0$, *under DoS attacks (10.5) with* τ_D, τ_F, *switched time delay system (10.15) is exponentially stable with* H_∞ *performance level* γ *if (10.23) holds and there exist positive definite matrices* $\tilde{P}_i, \tilde{Q}_i, \tilde{R}_i, \tilde{Z}_i, (i = 0, 1)$, $\tilde{\Omega}$ *and appropriate dimension matrices* $\tilde{K}, \tilde{X}_0, \tilde{X}_1, \Upsilon$ *satisfying*

$$\tilde{\Phi} \le 0, \quad \tilde{\Gamma} \le 0, \tag{10.29}$$

$$\begin{bmatrix} \tilde{R}_1 & \tilde{X}_1 \\ \tilde{X}_1^T & \tilde{R}_1 \end{bmatrix} \ge 0, \begin{bmatrix} \tilde{R}_0 & \tilde{X}_0 \\ \tilde{X}_0^T & \tilde{R}_0 \end{bmatrix} \ge 0. \tag{10.30}$$

$$\begin{cases} \begin{bmatrix} -\epsilon I & * \\ C\tilde{P}_1 - \Upsilon C & -I \end{bmatrix} < 0, \\ \epsilon \to 0 \end{cases} \tag{10.31}$$

$$\begin{aligned} &\begin{bmatrix} -\lambda_0 \tilde{P}_0 & * \\ \tilde{P}_0 & -\tilde{P}_1 \end{bmatrix} < 0, \begin{bmatrix} -\lambda_1 e^{(\alpha_1+\alpha_0)d_M} \tilde{P}_1 & * \\ \tilde{P}_1 & -\tilde{P}_0 \end{bmatrix} < 0, \\ &\begin{bmatrix} -\lambda_0 \tilde{Q}_0 & * \\ \tilde{P}_0 & \hat{Q}_1 \end{bmatrix} < 0, \begin{bmatrix} -\lambda_1 \tilde{Q}_1 & * \\ \tilde{P}_1 & \hat{Q}_0 \end{bmatrix} < 0, \end{aligned} \tag{10.32}$$

$$\begin{bmatrix} -\lambda_0 \tilde{R}_0 & * \\ \tilde{P}_0 & \hat{R}_1 \end{bmatrix} < 0, \begin{bmatrix} -\lambda_1 \tilde{R}_1 & * \\ \tilde{P}_1 & \hat{R}_0 \end{bmatrix} < 0,$$

$$\begin{bmatrix} -\lambda_0 \tilde{Z}_0 & * \\ \tilde{P}_0 & \hat{Z}_1 \end{bmatrix} < 0, \begin{bmatrix} -\lambda_1 \tilde{Z}_1 & * \\ \tilde{P}_1 & \hat{Z}_0 \end{bmatrix} < 0.$$

where

$$\tilde{\Phi} = \begin{bmatrix} \tilde{\Phi}_0 & * & * & * \\ d_M \tilde{\Phi}_1 & \hat{R}_1 & * & * \\ d_M \tilde{\Phi}_1 & 0 & \hat{Z}_1 & * \\ \tilde{\Phi}_2 & 0 & 0 & -I \end{bmatrix},$$

$$\tilde{\Phi}_0 = \begin{bmatrix} \tilde{\Phi}_0(1,1) & \tilde{\Phi}_0(1,2) & -\rho_1 \tilde{X}_1 & B\tilde{K}C & F \\ * & \tilde{\Phi}_0(2,2) & \tilde{\Phi}_0(2,3) & \varepsilon\psi\tilde{\Omega} & 0 \\ * & * & \tilde{\Phi}_0(3,3) & 0 & 0 \\ * & * & * & \tilde{\Phi}_0(4,4) & 0 \\ * & * & * & * & -\gamma^2 I \end{bmatrix},$$

$$\tilde{\Phi}_0(1,1) = He(A\tilde{P}_1) + \alpha_1\tilde{P}_1 + \tilde{Q}_1 - \frac{\pi^2}{4}\tilde{Z}_1 + \rho_1\tilde{R}_1,$$
$$\tilde{\Phi}_0(1,2) = B\tilde{K}C - \rho_1\tilde{R}_1 + \rho_1\tilde{X}_1 + \frac{\pi^2}{4}\tilde{Z}_1,$$
$$\tilde{\Phi}_0(2,2) = \varepsilon\psi\tilde{\Omega} + \rho_1(2\tilde{R}_1 - \tilde{X}_1 - \tilde{X}_1^T) - \frac{\pi^2}{4}\tilde{Z}_1,$$
$$\tilde{\Phi}_0(2,3) = \rho_1(\tilde{X}_1 - \tilde{R}_1), \tilde{\Phi}_0(3,3) = -e^{-\alpha_1 d_M}\tilde{Q}_1 + \rho_1\tilde{R}_1,$$
$$\tilde{\Phi}_0(4,4) = (\varepsilon - 1)\psi\tilde{\Omega}, \psi = \varphi(\mu - \alpha_1) + 1,$$
$$\tilde{\Phi}_1 = [A\tilde{P}_1, B\tilde{K}C, 0, B\tilde{K}C, F], \ \tilde{\Phi}_2 = [E\tilde{P}_1, 0, 0, 0, 0],$$
$$\tilde{\Gamma} = \begin{bmatrix} \tilde{\Gamma}_0 & * & * & * \\ d_M\tilde{\Gamma}_1 & \hat{R}_0 & * & * \\ d_M\tilde{\Gamma}_1 & 0 & \hat{Z}_0 & * \\ \tilde{\Gamma}_2 & 0 & 0 & -I \end{bmatrix},$$
$$\tilde{\Gamma}_0 = \begin{bmatrix} \tilde{\Gamma}_0(1,1) & \tilde{\Gamma}_0(1,2) & -\rho_0\tilde{X}_0 & F \\ * & \tilde{\Gamma}_0(2,2) & \rho_0(\tilde{X}_0 - \tilde{R}_0) & 0 \\ * & * & -e^{\alpha_0 d_M}\tilde{Q}_0 + \rho_0\tilde{R}_0 & 0 \\ * & * & * & -\gamma^2 I \end{bmatrix},$$

$$\tilde{\Gamma}_0(1,1) = He(A\tilde{P}_0) - \alpha_0\tilde{P}_0 + \tilde{Q}_0 + \rho_0\tilde{R}_0 - \frac{\pi^2}{4}\tilde{Z}_0,$$
$$\tilde{\Gamma}_0(1,2) = -\rho_0\tilde{R}_0 + \rho_0\tilde{X}_0 + \frac{\pi^2}{4}\tilde{Z}_0,$$
$$\tilde{\Gamma}_0(2,2) = \rho_0(2\tilde{R}_0 - \tilde{X}_0 - \tilde{X}_0^T) - \frac{\pi^2}{4}\tilde{Z}_0,$$
$$\tilde{\Gamma}_1 = [A\tilde{P}_0, 0, 0, F], \ \tilde{\Gamma}_2 = [E\tilde{P}_0, 0, 0, 0].$$

The LFC-VIC control gain and DETM parameter can be obtained by

$$K = \tilde{K}\Upsilon^{-1}, \ \Omega = \tilde{P}_1^{-1}\tilde{\Omega}\tilde{P}_1^{-1}.$$

Proof See the Appendix 10.6.3.

Proof By denoting $\tilde{P}_1 = P_1^{-1}$, $\tilde{Q}_1 = \tilde{P}_1 Q_1 \tilde{P}_1$, $\tilde{R}_1 = \tilde{P}_1 R_1 \tilde{P}_1$, $\tilde{Z}_1 = \tilde{P}_1 Z_1 \tilde{P}_1$, $\tilde{X}_1 = \tilde{P}_1 X_1 \tilde{P}_1$, $\tilde{\Omega} = \tilde{P}_1 \Omega \tilde{P}_1$, $\hat{Q}_1 = \delta^2\tilde{Q}_1 - 2\delta\tilde{P}_1$, $\hat{R}_1 = \delta^2\tilde{R}_1 - 2\delta\tilde{P}_1$ and $\hat{Z}_1 = \delta^2\tilde{Z}_1 - 2\delta\tilde{P}_1$ pre-and pro-multiply $diag\{\tilde{P}_1, \tilde{P}_1, \tilde{P}_1, \tilde{P}_1, I, I, I, I\}$ and its transpose on both sides of $\Phi < 0$. To deal with the nonlinear term $BKC\tilde{P}_1$, (10.31) is introduced by denoting $BKC\tilde{P}_1 = BK\Upsilon C = B\tilde{K}C$. By using inequality technique $-R_1^{-1} < \delta^2\tilde{R}_1 - 2\delta\tilde{P}_1$, LMI $\tilde{\Phi} < 0$ (10.29) is derived.

Similarly, denoting $\tilde{P}_0 = P_0^{-1}$, $\tilde{Q}_0 = \tilde{P}_0 Q_0 \tilde{P}_0$, $\tilde{R}_0 = \tilde{P}_0 R_0 \tilde{P}_0$, $\tilde{Z}_0 = \tilde{P}_0 Z_0 \tilde{P}_0$, $\hat{Q}_0 = \delta^2\tilde{Q}_0 - 2\delta\tilde{P}_0$, $\hat{R}_0 = \delta^2\tilde{R}_0 - 2\delta\tilde{P}_0$ and $\hat{Z}_0 = \delta^2\tilde{Z}_0 - 2\delta\tilde{P}_0$. Pre-and pro multiply $diag\{\tilde{P}_0, \tilde{P}_0, \tilde{P}_0, I, I, I, I\}$ and its transpose on both sides of $\Gamma < 0$, one has LMI $\tilde{\Gamma} < 0$ in (10.29). Moreover, pre-and pro multiply $diag\{\tilde{P}_1, \tilde{P}_1\}$, $diag\{\tilde{P}_0, \tilde{P}_0\}$

and its transpose on both sides of (10.20), respectively, one has LMI in (10.30). By using Schur complement lemma, (10.22) are converted to LMIs (10.32) This completes the proof.

From Remark 10.7, it can be found that the design of resilient LFC-VIC gain and DETM parameters relates to many parameters. Thus, it is necessary to provide an algorithm for the concrete implementation of Theorem 10.3.

Algorithm 3.1 Solving LFC-VIC gain K and DETM parameter Ω

Initialization:
Set the prescribed d_M,τ_F,τ_D, $\epsilon \to 10^{-6}$, $k = 0$, $k_{end} >> 0$;
Set larger initial values $\gamma \in (0, \infty)$, $\delta \in (0, \infty)$, $\lambda_1 \in (0, 1)$, $\varphi \in (0, \infty)$, $\mu \in (0, \infty)$, $\alpha_0 \in (0, \infty)$ and smaller initial values $\alpha_1 \in (0, \infty)$, $\lambda_0 \in (1, \infty)$, $\varepsilon \in (0, \infty)$;
Iteration:
1: Choose nonnegative deviation variables $\Delta\gamma(k)$, $\Delta\delta(k)$, $\Delta\lambda_1(k)$, $\Delta\varphi(k)$, $\Delta\mu(k)$, $\Delta\alpha_1(k)$, $\Delta\alpha_0(k)$, $\Delta\lambda_0(k)$, $\Delta\varepsilon(k)$;
2: **if** the condition (10.23) holds **then** Go to Step 4
3: **else** $\Delta\gamma(k) = 0$, $\Delta\delta(k) = 0$, $\Delta\varepsilon(k) = 0$, $\Delta\varphi(k) = 0$, $\Delta\mu(k) = 0$ and go to Step 6;
4: **end if**
5: **if** the LMIs (10.29)-(10.32) hold **then** Go to Step 7
6: **else** Go to Step 6;
7: **end if**
8: $\gamma^* = \gamma + \Delta\gamma(k)$, $\delta^* = \delta + \Delta\delta(k)$; $\alpha_1^* = \alpha_1 - \Delta\alpha_1(k)$, $\alpha_0^* = \alpha_0 + \Delta\alpha_0(k)$, $\varepsilon^* = \varepsilon - \Delta\varepsilon(k)$, $\varphi^* = \varphi + \Delta\varphi(k)$, $\mu^* = \mu + \Delta\mu(k)$, $\gamma = \gamma^*$, $\delta = \delta^*$, $\alpha_1 = \alpha_1^*$, $\alpha_0 = \alpha_0^*$, $\varepsilon = \varepsilon^*$, $\varphi = \varphi^*$, $\mu = \mu^*$, $k = k + 1$ and return to Step 1;
9: **if** $k < k_{end}$ **then** $\gamma^* = \gamma - \Delta\gamma(k)$, $\alpha_1^* = \alpha_1 + \Delta\alpha_1(k)$, $\alpha_0^* = \alpha_0 - \Delta\alpha_0(k)$, $\varepsilon^* = \varepsilon + \Delta\varepsilon(k)$, $\varphi^* = \varphi - \Delta\varphi(k)$, $\mu^* = \mu - \Delta\mu(k)$, $\gamma = \gamma^*$, $\alpha_1 = \alpha_1^*$, $\alpha_0 = \alpha_0^*$, $\varepsilon = \varepsilon^*$, $\varphi = \varphi^*$, $\mu = \mu^*$, $k = k + 1$ and return to Step 1
10: **else**
11: Go to Step 10;
12: **end if**
13: **return** $K = \tilde{K}\Upsilon^{-1}$, $\Omega = \tilde{P}_1^{-1}\tilde{\Omega}\tilde{P}_1^{-1}$.

10.4 Active Secure Scheme

Until now, the analysis and synthesis of resilient DETM based LFC-VIC of multi-area power system under DoS attacks is completed. One desired exponential H_∞ performance will be guaranteed under the novel criterion (10.23) and the designed resilient controller. However, the criterion (10.23) can not be satisfied automatically because it is subjected to not only system secure scheme (τ_F) but also attack strategy (τ_D). Thus, to realize the criterion (10.23), an active secure scheme is proposed on the basis of model based predictive control against the unknown but observable attack strategy.

This secure scheme exhibits a dynamic prediction horizon. The idea of dynamic horizon is original from the consideration for uncontrollable and changeable attack strategy and limited computation times and resources. This is different from the conventional MPC with uniform prediction horizon referring to upper bound of packet drop and time delay. First, there is no idea about the upper bound of packet drop caused by DoS attacks . Second, uniform prediction horizon is hard to be realized under the limited computation resources and computation time.

In this study, the dynamic prediction horizon $N_P(t_{k,n}h)$ is scheduled according to the observed value $\tilde{\tau}_D(t_{k,n}h)$, the theoretical value $\tilde{\tau}_F(t_{k,n}h)$ and the triggering interval length $\Delta T(t_{k,n}h) = t_{k,n}h - t_{k-1,n}h$. Define $\Delta\tilde{T}(t_{k,n}h)$ as the mean of $\Delta T(t_{k,n}h)$. Based on the secure criterion (10.23), the scheduling scheme of dynamic prediction horizon is proposed as follows.

$$\begin{aligned} \tilde{\tau}_F(t_{k,n}h) &> \frac{d_M ln\lambda_0}{-ln\lambda_1 - \alpha_0(d_M + \tilde{\tau}_D(t_{k,n}h))}, \\ N_P(t_{k,n}h) &= \frac{\Delta T(t_{k,n}h)}{\Delta\tilde{T}(t_{k,n}h)}\tilde{\tau}_F(t_{k,n}h). \end{aligned} \tag{10.33}$$

Under the scheduling scheme, the prediction controls are generated and transmitted to actuators.

$$U(t_{k,n}h) = \{u(t_{k,n}h), \hat{u}(t_{k_1,n}h), \ldots, \hat{u}(t_{k_{N_P},n}h)\}.$$

Remark 10.4 The $\frac{\Delta T(t_{k,n}h)}{\Delta\tilde{T}(t_{k,n}h)}$ represents the time efficiency of the triggering signal at $t_{k,n}h$. The small value $\frac{\Delta T(t_{k,n}h)}{\Delta\tilde{T}(t_{k,n}h)}$ indicates that the feedback signal at $t_{k,n}h$ is the required one with high real-time for system. Thus, it is better to have a short prediction horizon to save the computation time. Compared with traditional predictive control (TPC) supported by the knowledge of upper bound of data dropout [12, 13], the proposed dynamic prediction horizon (DPH) deals with DoS attacks with priori unknown of bound of data loss. DPH strategy regulates variable prediction horizon around average dwell time of sleep intervals while TPC generates uniform prediction horizon every time. Both TPC and DPH are kinds of compensation strategy by increasing the redundancy of control inputs. Considered that the redundant control commands will consume additional computation and communication which would influences the real-time of control, DPH can enhance the real-time of control by automatically regulating prediction horizon according to the information of the current triggered interval length $\frac{\Delta T(t_{k,n}h)}{\Delta\tilde{T}(t_{k,n}h)}$.

Remark 10.5 An overall block diagram of the proposed approach is given as shown in Fig. 10.3. The proposed approach consists of the resilient conditions obtained offline and resilient result based dynamic prediction horizon as active secure scheme of power system in online operation.

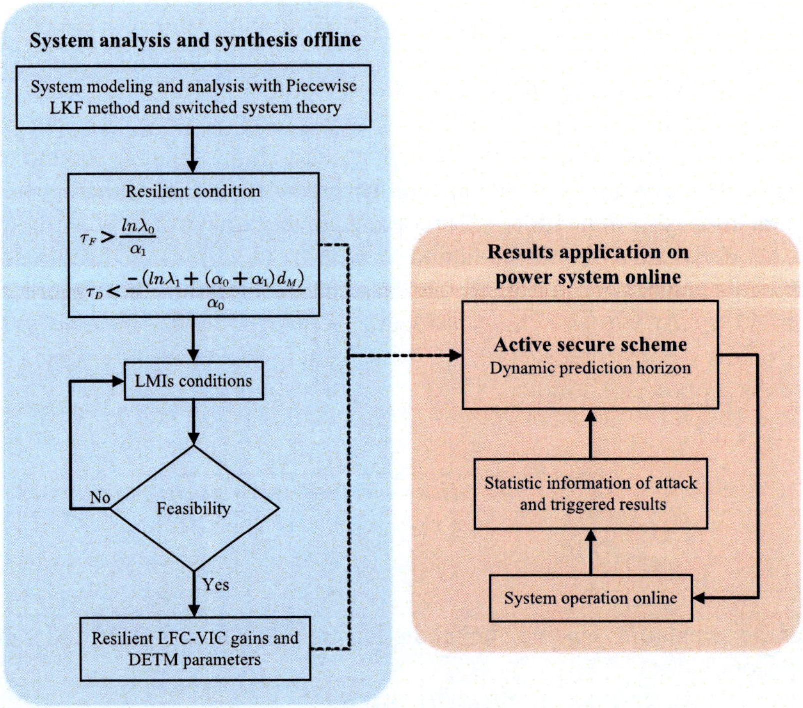

Fig. 10.3 An overall block diagram of the proposed approach

10.5 Simulation Example

This section presents a simulation example to verify our resilient design method for model based average dwell time (DETM) based load frequency control and virtual inertia control (LFC-VIC) scheme. A two-area power system under DoS attacks is considered in following simulation with parameters from [14] Table 10.1.

10.5.1 Simulation of a Two-Area System Under DoS Attacks and Time Delay

Let $\gamma = 320, d_M = 0.5, \delta = 10, \alpha_0 = \alpha_1 = 0.1, \lambda_0 = 1.8, \lambda_1 = 0.6, \varepsilon = 0.13, \varphi = \mu = 10$ and $\epsilon = 10^{-6}$. In following, set the simulation step $ts = 0.001s$ and the sample periodic $h = 0.1s$. The parameters of RESs are the solar system time constant $T_{PV1} = T_{PV2} = 1.3$, the wind turbine time constant $T_{WT1} = T_{WT2} = 1.5$ and the virtual inertia time constant $T_{VI1} = T_{VI2} = 10$.

According to the Theorem 10.3, the results are obtained in following Tables. From Table 10.2, the allowable delay bound d_M is decreased with the increasing

Table 10.1 Parameters of two-area power system

Parm	M	D	R	$T_g(s)$	$T_{ch}(s)$	β	T_{ij}
Area 1	0.166	0.015	3	0.08	0.4	0.3493	0.08
Area 2	0.202	0.016	2.73	0.06	0.44	0.3827	0.08

Table 10.2 The trigger parameter ε for various delay upper bound d_M

ε	0.13	0.18	0.2	0.22	0.23
d_M	0.5	0.47	0.4	0.29	0.1

Table 10.3 The average dwell time of attack intervals τ_D for various delay upper bound d_M

d_M	0.1	0.2	0.3	0.4	0.5
τ_D	4.9083	4.7083	4.5083	4.3083	4.1083

Table 10.4 The MDADT index τ_F, τ_D for various α_1

α_1	0.2	0.4	0.6	0.8	1.2
τ_F	29.3893	14.6947	9.7964	7.3473	5.8779
τ_D	4.5083	4.4083	4.3083	4.2083	4.1083

of the triggering threshold ε. The reduction of transmission by tuning to a large threshold lower the system tolerable ability in delay. The delay stability is the primary performance of system. Thus, a suitable tradeoff should be concerned for system designer between transmission efficiency and system performance.

Table 10.3 shows that for the increased delay bound d_M there is a tendency toward decrease in allowable attack ADT. The simultaneous existence of delay and DoS attacks, the coupling influence from network, bring double deterioration on system performance. Therefore, power system with the determinate stability margin will has less resilience to DoS attacks under the large time delay.

According to the resilient condition (10.23), a relationship bridged by α_1 between sleep ADT and attack ADT is given in Table 10.4, which shows the negative correlation between α_1 and τ_F or τ_D. However, the negative correlation level in (α_1, τ_F) is larger than (α_1, τ_D). It means that for the large α_1, sleep subsystem is allowed to be with less duration while the tolerable attack ADT still keep the same level. Thus, it is better to choice a relative large α_1.

From Table 10.5, as the triggering threshold ε increases, the three indices T_n, T_a, T_p decrease for the communication load can be reduced by choosing a large triggering threshold ε.

Table 10.6 shows the average triggering periodic T_a obtained from experiment statistic under various weight value φ of dynamic function in DETM condition. With the increasing of φ, the average triggering periodic T_a is decreased to transmit more

Table 10.5 The triggered times T_n, the average release periods T_a, and the transmission percentage T_p for various ε with $\varphi = 10$ and $\mu = 10$

ε	0	0.01	0.03	0.06	0.09	0.12
T_n	389	206	147	102	85	73
T_a	0.0946	0.1786	0.2502	0.3606	0.4327	0.5039
T_p (%)	97.98	51.89	37.03	25.69	21.41	18.39

Table 10.6 The average release periods T_a for various φ with $\varepsilon = 0.13$, $\mu = 10$

φ	0.001	0.01	0.1	1	10
T_a	0.6130	0.6130	0.5490	0.5109	0.4777

Table 10.7 The average release periods T_a for various μ with $\varepsilon = 0.13$, $\varphi = 10$

μ	0.001	0.01	0.1	1	10
T_a	0.5331	0.5109	0.4904	0.4840	0.4777

Table 10.8 Comparison among various different event-triggered schemes with $\varepsilon = 0.1$, $h = 0.1$ and $d_M = 0.1$

Schemes	Trigger numbers	Average periodic
ST	397	0.100
ET	105	0.3503
DET (this study)	83	0.4432

packets, which proves the negative correlation between the triggering periodic and φ in triggering condition (10.11). Though the smaller φ can save more communication load, there is a dead zone $\varphi \in (0, 0.01]$.

Table 10.7 shows the negative correlation between average triggering periodic T_a and μ. Because the tuning function $\eta(t)$ with large value μ is decaying so fast that the triggering periodic is reduced.

The communication numbers are shown in Table 10.8 under sampled transmission (ST), event-triggered transmission (ET) and dynamic event-triggered transmission (DET). It illustrates DET is better than ET and ST on transmission efficiency.

On the other hand, based on the given parameters, the resilient control gain K can be obtained by using MATLAB Toolbox YALMIP with solver MOSEK.

$$K = \begin{bmatrix} 0.0231 & 0.0079 & 0.0293 & -0.0020 \\ -0.0033 & -0.0197 & 0.0025 & 0.0203 \\ 0.0364 & -0.0009 & 0.0357 & 0.0094 \\ 0.0040 & 0.0240 & -0.0031 & -0.0247 \end{bmatrix}$$

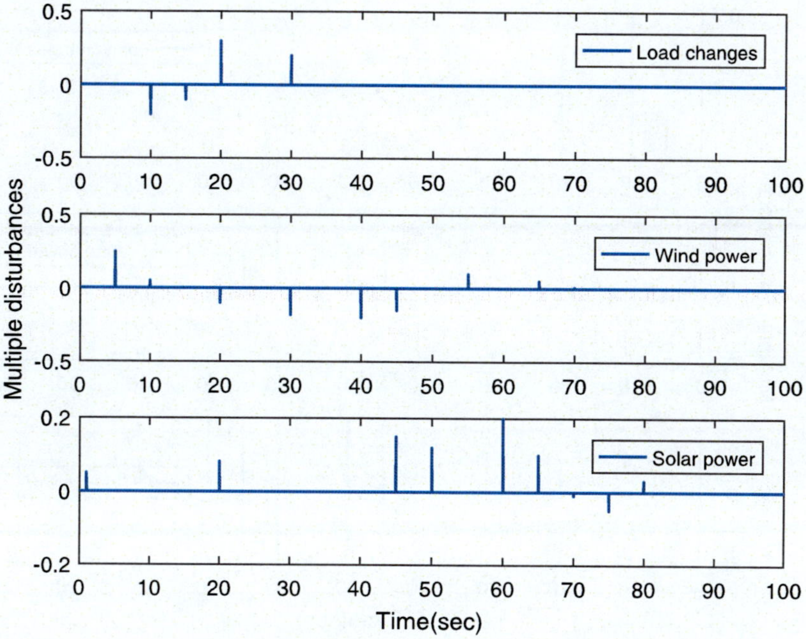

Fig. 10.4 Multiple disturbances: load changes, wind power and solar power in area 1

By calculating (10.23), the theoretical values of model based average dwell time (MDADT) are obtained $\tau_F^* > 5.8779, \quad \tau_D^* < 4.1083$. The solved MDADT index illustrates that to guarantee the desired H_∞ performance sleep intervals should be larger than $5.8779s$ and attack intervals should be less than $4.1083s$ in mean sense.

In this process, the multiple disturbances are depicted in Figs. 10.4 and 10.5. According to the eigenvalue analysis (EA) method in [14], a stabilizing control gain is chosen as $K = diag[0.3, 1.5, 0.2, 1.7]$. From Figs. 10.6 and 10.7, under the network circumstance of transmission delay and DoS attacks, the dynamic of frequency derivations and tie-lie power derivations in the two-area power system are finally approaching zeros under our proposed method while the dynamic responses under the control gain derived by the eigenvalue analysis method in [14] are serious oscillation and divergence. With the designed control inputs depicted in Fig. 10.8, the H_∞ disturbance attenuant level can be obtained by $\gamma* = \|z\|/\|w\| = 0.0443/0.6049 = 0.0733$, which is far smaller the desired value $\gamma = 320$.

The exhibition of release intervals is depicted in Fig. 10.9 to show the efficiency transmission of dynamic ETM. First, DETM do not trigger transmission during DoS attacks to ignore wasting energy. Second, high frequency triggering communication happens to attenuate the multiple disturbances distributing in $[0, 80s]$. Low frequency triggering communication service for steady state interval $[80s, 100s]$ to reduce transmission numbers. The average triggering interval is $0.5255s$ in statistics. With $\varphi \longrightarrow \infty$, the dynamic event-triggered scheme is drawback to traditional event-triggered scheme and the average triggering interval is 0.4840 in statistics.

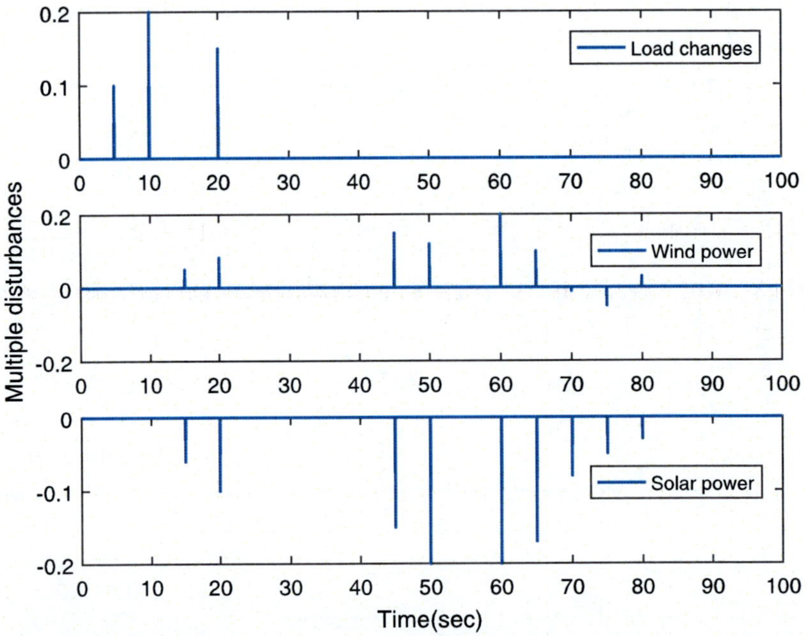

Fig. 10.5 Multiple disturbances: load changes, wind power and solar power in area 2

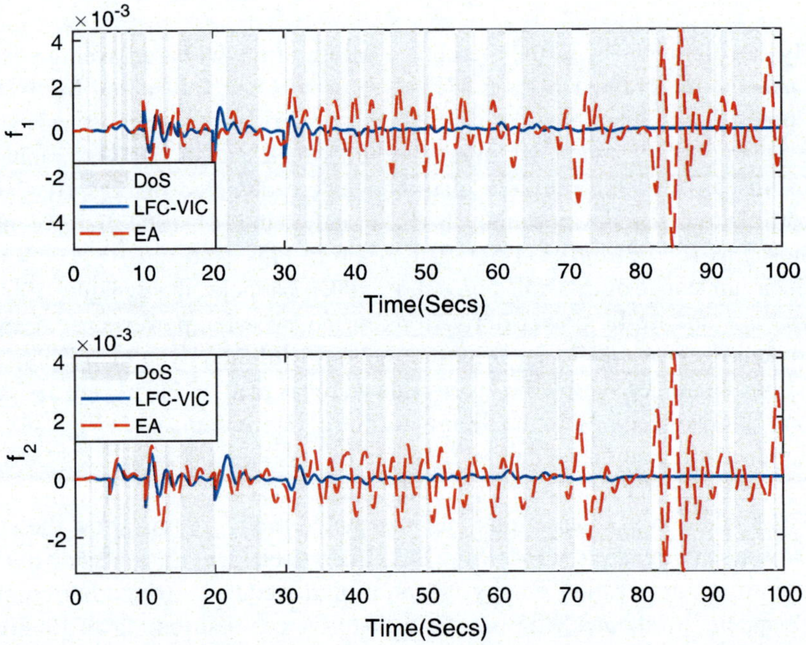

Fig. 10.6 Frequency derivations of the two-area power system under DoS attacks and time delay $d_M = 0.5$ (LFC-VIC designed by this study, EA provided by eigenvalue analysis (EA) method from [14])

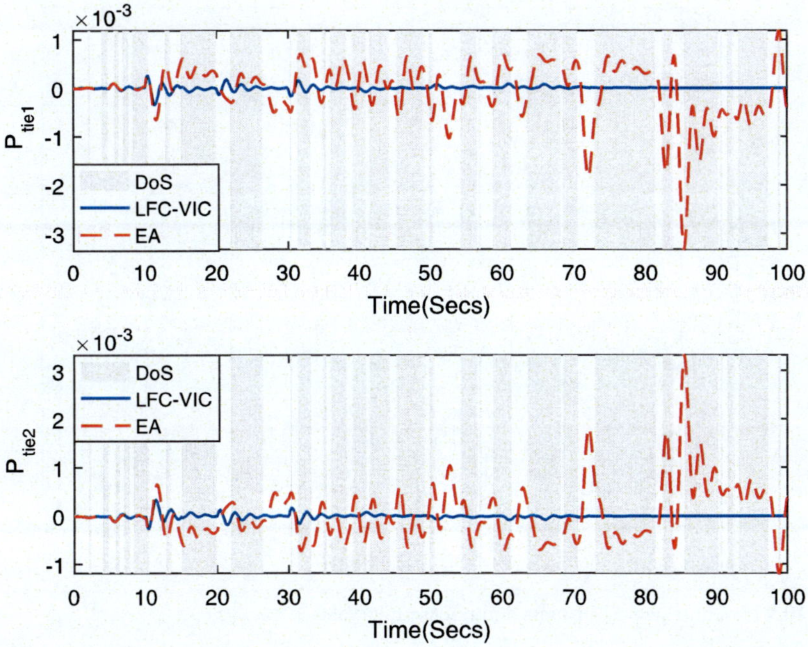

Fig. 10.7 Tie-line power derivations of the two-area power system under DoS attacks and time delay $d_M = 0.5$ (LFC-VIC designed by this study, EA provided by eigenvalue analysis (EA) method from [14])

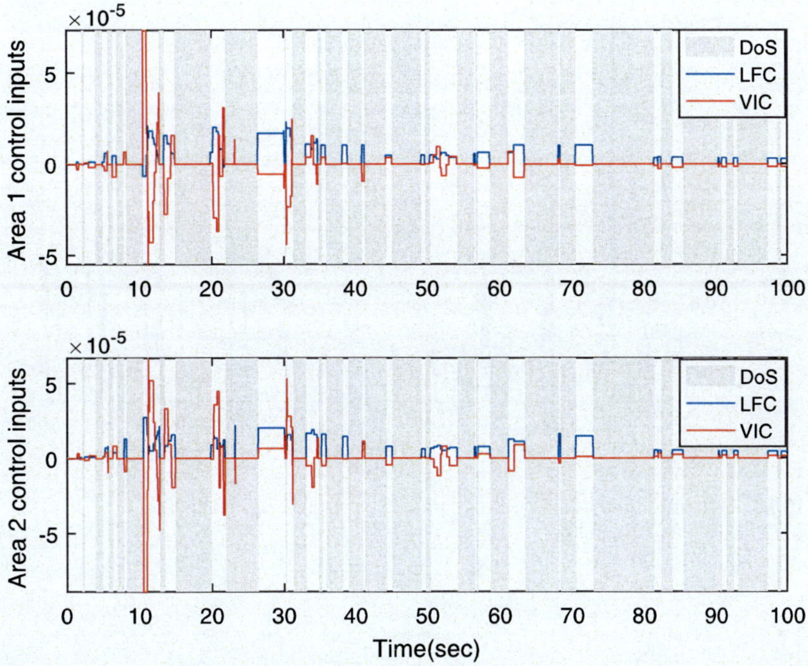

Fig. 10.8 Control inputs of the two-area power system

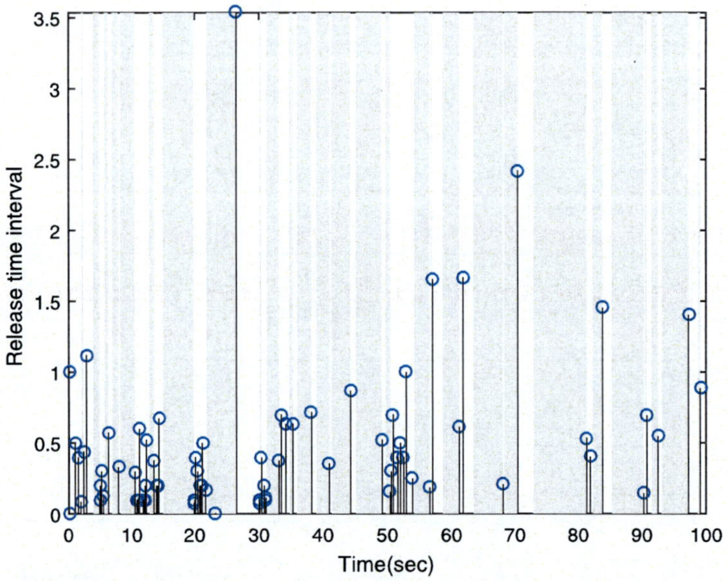

Fig. 10.9 Event triggered instants and release intervals

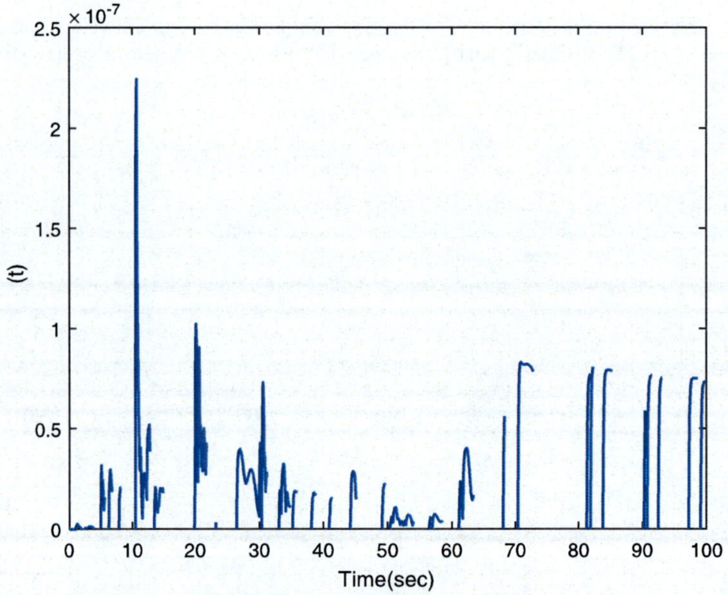

Fig. 10.10 Dynamic function $\eta(t)$ of dynamic event-triggered mechanism

Figure 10.10 shows the trajectory of dynamic function $\eta(t)$. As can be seen, the function is piecewise positive, which plays the key role to increase the triggered interval compared with traditional event-triggered mechanism. It also verifies the conclusion in Remark 10.2. Further, the trajectory of $\eta(t)$ exhibits the tendency of decay and oscillation as said in (10.10). This decay tendency guarantees the necessary triggering of feedback signals to preserve system performance.

10.5.2 Simulation of a Two-Area Power System with Inertia Reduction

Let $d_M = 0.23$ and reduce 30% of the system inertia M_i. Other parameters are same with before. On the one hand, the frequency derivations and the tie-line power derivations (the blue lines in Figs. 10.11 and 10.12) happen to more serious oscillation than that in Figs. 10.6 and 10.7. Because the reduction of system inertia influences the frequency stability. However, under the designed LFC-VIC scheme, the frequency derivations and tie-line power derivations finally approach to zeros, which demonstrates the validness of our method. On the other hand, compared to the blue lines, the red lines generated by EA method have serious oscillation without approaching to zero steady states. Thus, it varies the advantage of our proposed LMI based design method.

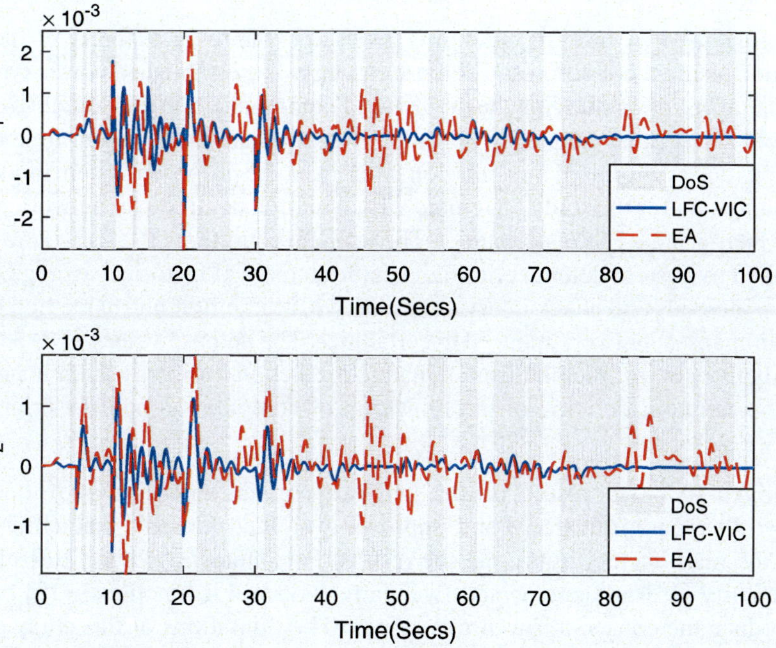

Fig. 10.11 Frequency derivations of the two-area power system with inertia reduction of 30% under DoS attacks and time delay $d_M = 0.23$ (LFC-VIC designed by this study, EA provided by eigenvalue analysis (EA) method from [14])

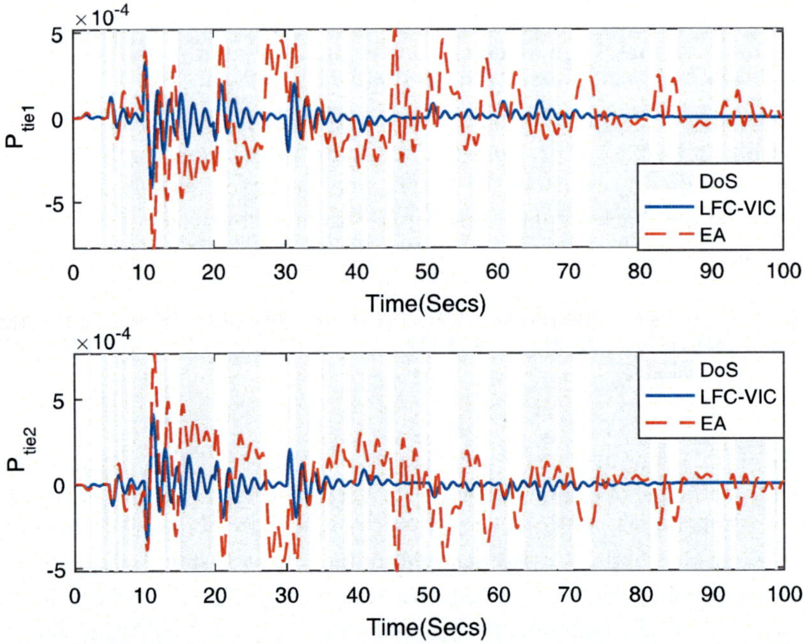

Fig. 10.12 Tie-line power derivations of the two-area power system with inertia reduction of 30% under DoS attacks and time delay $d_M = 0.23$ (LFC-VIC designed by this study, EA provided by eigenvalue analysis (EA) method from [14])

Remark 10.6 A brief discussion is given to explain the usability of the proposed model, resilient condition and defense strategy. The established power system model is mainly used for studying the influence of communication uncertainties on physical power system that frequency deviation dynamics are subjected to time delay, DoS attacks and event-based transmission. Moreover, the proposed system model provides a foundation for the study on remote virtual inertial control over network to deal with low inertia and RESs disturbances. The resilient condition (10.23) is convenient to be applied by system defender through simple statistic of historical average duration of sleep intervals and attack intervals. As known, predictive control method is powerful to compensate network induced uncertainties. The defense strategy (10.33) improves predictive control method from dynamic prediction horizon, which is practical due to the full consideration of the limitation of computation and the priori unknown DoS attacks.

Remark 10.7 The results of this study are derived on the basis of the Wirtinger-based Lyapunov functional and improved weighted Jensen's inequality [6], which are the main factors impacting the results. For interested reader, Bessel–Legendre inequality [15] and augmented Lyapunov–Krasovskii functional [16] can be used to reduce the conservatism of this results. The limitations of this study are that the proposed method is not adequate to deal with the nonlinearities of power system like GDB, GRC. In future, model predictive control with the proposed defense strategy will be employed to actively counter power system nonlinearities and DoS attacks.

10.6 Conclusion

Resilient design problem of interconnected power system with renewable energy sources penetration has been studied under DoS attacks and transmission delay. A switched delay system model has been established to represent the dynamic of interconnected power system under dynamic event-triggered mechanism based load frequency control (LFC) and virtual inertia control (VIC) scheme. By using piecewise LKF method, a criterion about tolerable model based average dwell time (MDADT) ($\tau_F > ln\lambda_0/\alpha_1$, $\tau_D < -(ln\lambda_1 + (\alpha_0 + \alpha_1)d_M)/\alpha_0$) has been derived to preserve exponential H_∞ performance. A co-design method for DETM and resilient LFC-VIC has been proposed with the LMIs technique. Further, according to the criterion, a dynamic prediction compensation strategy has been scheduled as the active secure strategy against DoS attacks. In future, active secure strategy against hybrid network attacks will be studied by using model-based predictive control method with a dynamic prediction scheme.

References

1. Z. Cheng, D. Yue, S. Hu, C. Huang, C. Dou, X. Ding, Resilient dynamic event-triggered control for multi-area power systems with renewable energy penetration under dos attacks. IET Control Theory & Appl. (2020). https://doi.org/10.1049/iet-cta.2019.1478
2. X. Zhao, Y. Kao, B. Niu, T. Wu, *Control Synthesis of Switched Systems* (Springer, Cham, 2017)
3. S. Hu, D. Yue, X. Xie, X. Chen, X. Yin, Resilient event-triggered controller synthesis of networked control systems under periodic DoS jamming attacks. IEEE Trans. Cybern. **49**(12), 4271–4281 (2019)
4. J. Liu, Y. Gu, L. Zha, Y. Liu, J. Cao, Event-triggered h_∞ load frequency control for multiarea power systems under hybrid cyber attacks. IEEE Trans. Syst. Man Cybern.: Syst. **49**(8), 1665–1678 (2019)
5. H. Chen, S. Hu, D. Yue, X. Chen, Resilient event-triggered h_∞ load frequency control of isolated hybird power system under periodic DoS jamming attacks, in *IEEE 28th International Symposium on Industrial Electronics (ISIE)* (IEEE, 2019), pp. 2271–2276
6. Z. Cheng, D. Yue, S. Hu, C. Huang, C. Dou, L. Chen, Resilient load frequency control design: Dos attacks against additional control loop. Int. J. Electr. Power & Energy Syst. **115** (2020). https://doi.org/10.1016/j.ijepes.2019.105496
7. C.D. Persis, P. Tesi, Input-to-state stabilizing control under denial-of-service. IEEE Trans. Autom. Control **60**(11), 2930–2944 (2015)
8. Z. Cheng, D. Yue, S. Hu, X. Xie, C. Huang, Detection-based weighted H_∞ lfc for multi-area power systems under dos attacks. IET Control Theory & Appl. **13**(12), 1909–1919 (2019)
9. S. Hu, Z. Cheng, D. Yue, C. Dou, Y. Xue, Bandwidth allocation-based switched dynamic triggering control against DoS attacks. IEEE Trans. Syst. Man Cybern.: Syst. 1–12 (2019), Early Access. https://doi.org/10.1109/TSMC.2019.2956945
10. S. Wen, X. Yu, Z. Zeng, J. Wang, Event-triggering load frequency control for multiarea power systems with communication delays. IEEE Trans. Ind. Electr. **63**(2), 1308–1317 (2016)
11. S. Hu, D. Yue, X. Chen, Z. Cheng, X. Xie, Resilient H_∞ filtering for event-triggered networked systems under nonperiodic DoS jamming attacks. IEEE Trans. Syst. Man Cybern.: Syst. (2019). https://doi.org/10.1109/TSMC.2019.2896249
12. G.P. Liu, Predictive controller design of networked systems with communication delays and data loss. IEEE Trans. Circuits Syst. II-Express Briefs **57**(6), 481–485 (2010)

13. X. Yin, D. Yue, S. Hu, C. Peng, Y. Xue, MODEL-Based event-triggered predictive control for networked systems with data dropout. SIAM J. Control Optim. **54**(2), 567–586 (2016)
14. T. Kerdphol, F. Rahman, Y. Mitani, Virtual inertia control application to enhance frequency stability of interconnected power systems with high renewable energy penetration. Energies **11**(4), 1–16 (2018)
15. A. Seuret, F. Gouaisbaut, Stability of linear systems with time-varying delays using Bessel-Legendre inequalities. IEEE Trans. Autom. Control **63**(1), 225–232 (2018)
16. X.-M. Zhang, Q.-L. Han, A. Seuret, F. Gouaisbaut, An improved reciprocally convex inequality and an augmented lyapunov-krasovskii functional for stability of linear systems with time-varying delay. Automatica **84**, 221–226 (2017)

Chapter 11
Resilient Load Frequency Control Design: Nonperiodic DoS Attacks Against Additional Control Loop

This paper is concerned with resilient load frequency control (LFC) design problem of a multi-area power system with uncertainty and physical constraint under nonperiodic denial-of-service (DoS) attacks and transmission delay. LFC scheme consists of a local PI control loop and additional control loop to be designed. PI control loop is subjected to constant transmission delay while additional control loop is subjected to DoS attacks. We characterize DoS attacks by attack frequency, uniform lower bound of sleep interval and uniform upper bound of attack interval. Considering DoS attacks and time delay, we describe LFC system dynamic by a switched delay system model. Further, a stability criterion is derived by using delay system method and switched system method. Under the criterion, a design method is presented to solve state feedback control gain. Finally, numerical simulations are given to verify the validness of our proposed design method [1].

The chapter is organized as follows. The power system model and attack scenario are presented in Sect. 11.1. Section 11.2 analyzes the resilient performance of power grid under DoS attacks and presents a design method of LFC gain. Section 11.3 illustrates the simulation results to demonstrate the effectiveness of our theory. Finally, the conclusion is stated in Sect. 11.4.

11.1 Problem Formulation

11.1.1 Modeling of Multi-area Power System with Uncertainty, GRC, and Delay

In this section, we describe a multi-area load frequency control (LFC) system with parameters uncertainty under the influence of time delay and DoS attacks. The structure of the ith area power system is illustrated in Fig. 11.1. The system dynamic of the ith area power system is represented by [2].

D. Yue et al., *Secure Control of Networked Control Systems and Its Applications*,
https://doi.org/10.1007/978-981-33-6730-2_11

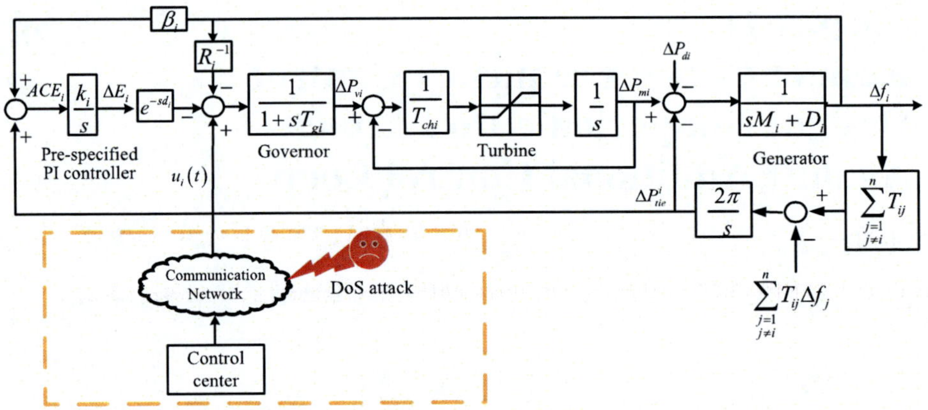

Fig. 11.1 The ith area power system under DoS attacks

$$\begin{cases} \Delta \dot{f}_i = \dfrac{1}{M_i}\left(-D_i \Delta f_i + \Delta P_{mi} - \Delta P^i_{tie} - \Delta P_{di}\right), \\ \Delta \dot{P}^i_{tie} = \displaystyle\sum_{j=1, j\neq i}^{n} 2\pi T_{ij}\left(\Delta f_i - \Delta f_j\right), \\ \Delta \dot{P}_{mi} = \dfrac{1}{T_{chi}}\left(-\Delta P_{mi} + \Delta P_{vi}\right), \\ \Delta \dot{P}_{vi} = \dfrac{1}{T_{gi}}\left(-\dfrac{1}{R_i}\Delta f_i - \Delta P_{vi} - k_i \displaystyle\int ACE_i(t - d_i) + u_i\right), \\ ACE_i = \beta_i \Delta f_i + \Delta P^i_{tie}, \end{cases} \tag{11.1}$$

where ACE_i, Δf_i, ΔP_{vi}, ΔP_{mi}, ΔP_{tie-i} are the area control error, deviation of area frequency, governor valve position, mechanical power output of the alternator, tie-line power, respectively and $\Delta E_i = k_i \int ACE_i(t)$ as seen in Fig. 11.1; M_i, D_i, T_{gi}, R_i, T_{chi}, T_{ij} and β_i are area equivalent inertia, area load frequency characteristic, governor time constant, droop characteristic, turbine time constant, synchronizing coefficient between area i and j and frequency bias, respectively.

Generator rate constraint (GRC) [3]: The physical constraint is posed on $\Delta \dot{P}_{mi}$ which is limited by $\underline{v} \leq \Delta \dot{P}_{mi} \leq \bar{v}$. To achieve LFC goal as well as deal with the physical constraint on generator units (GRC), we adopt the fictitious output vector method [4]:

$$z_i(t) = col\left[\xi^i_1 \Delta f_i \; \xi^i_2 \Delta P^i_{tie} \; \xi^i_3 \int ACE_i \; \xi^i_4 u_i\right]. \tag{11.2}$$

Speed droop coefficient uncertainty [5]: In this study, we consider the uncertainty of power system referring to the speed droop coefficient R_i. The uncertainty of speed droop coefficient is described by $(1+\epsilon_i)R_i$, where $0 \leq \epsilon_i \leq 1$.

Time varying delay of transmission channel: Considering the ACE_i signals with transmission delay, a multi-area power system is a multi-delay system. To fully represent input delay d_i, $i = 1, 2 \ldots n$ of each area in multi-area power systems [6], we introduce a single time varying delay $d(t)$ with satisfying $0 \le d(t) \le d_m$ and $\dot{d}(t) = 1$.

Then, multi-area power systems model can be standardized by

$$\begin{cases} \dot{x}(t) = (A + \Delta A)x(t) + A_d x(t - d(t)) + Bu(t) + Fw(t), \\ z(t) = Dx(t) + Eu(t), \end{cases} \tag{11.3}$$

where $x(t) = \left[x_1(t)\ x_2(t)\ \ldots\ x_n(t)\right]^T$, $w(t) = \left[\Delta P_{d1}(t)\ \Delta P_{d2}(t)\ \ldots\ \Delta P_{dn}(t)\right]^T$, $u(t) = \left[u_1(t)\ u_2(t)\ \ldots\ u_n(t)\right]^T$, $x_i(t) = \left[\Delta f_i\ \Delta P^i_{tie}\ \Delta P_{mi}\ \Delta P_{vi}\ \int ACE_i\right]^T$.

$$\begin{aligned}
A &= \begin{bmatrix} A_{11} & A_{12} & \ldots & A_{1n} \\ A_{21} & A_{22} & \ldots & A_{2n} \\ \vdots & \vdots & \ddots & \vdots \\ A_{n1} & A_{n2} & \ldots & A_{nn} \end{bmatrix}, \\
\Delta A &= diag\left[\Delta A_{11}\ \Delta A_{22} \cdots \Delta A_{nn}\right], \\
A_d &= diag\left[A_{d1}\ A_{d2} \cdots A_{dn}\right], \\
B &= diag\left[B_1\ B_2 \cdots B_n\right], \\
F &= diag\left[F_1\ F_2 \cdots F_n\right], \\
D &= diag\left[D_1\ D_2 \cdots D_n\right], \\
E &= diag\left[E_1\ E_2 \cdots E_n\right], \\
A_{ii} &= \begin{bmatrix} -\frac{D_i}{M_i} & -\frac{1}{M_i} & \frac{1}{M_i} & 0 & 0 \\ \sum\limits_{j=1, j\neq i}^{n} 2\pi T_{ij} & 0 & 0 & 0 & 0 \\ 0 & 0 & -1/T_{chi} & 1/T_{chi} & 0 \\ -\frac{1}{R_i T_{gi}} & 0 & 0 & -\frac{1}{T_{gi}} & 0 \\ \beta_i & 1 & 0 & 0 & 0 \end{bmatrix}, \\
A_{ij} &= \mathbf{0}_{5\times 5},\ A_{ij}(2,1) = -2\pi T_{ij}, \\
\Delta A_{ii} &= \mathbf{0}_{5\times 5},\ \Delta A_{ii}(4,1) = \frac{\epsilon_i}{(1+\epsilon_i) R_i T_{gi}}, \\
A_{di} &= \mathbf{0}_{5\times 5},\ A_{di}(4,5) = -k_i / T_{gi}, \\
B_i &= col\left[0\ 0\ 0\ \tfrac{1}{T_{gi}}\ 0\right],\ F_i = col\left[-\tfrac{1}{M_i}\ 0\ 0\ 0\ 0\right], \\
D_i &= \begin{bmatrix} \xi^i_1 & 0 & 0 & 0 & 0 \\ 0 & \xi^i_2 & 0 & 0 & 0 \\ 0 & 0 & 0 & 0 & \xi^i_3 \\ 0 & 0 & 0 & 0 & 0 \end{bmatrix}, \\
E_i &= col[0, 0, 0, \xi^i_4].
\end{aligned}$$

Then, state feedback control law is adopted in multi-area power systems (11.3).

$$u(t) = Kx(t). \tag{11.4}$$

where $K = diag\{K_1, K_2, \ldots, K_n\}$ are control gain to be designed.

11.1.2 Resilience LFC System Formulation Under DoS Attacks

Due to deep integration of network transmission technique in physical power system, the influence of cyber attacks on LFC system is a noticeable problem. In this study, we consider the influence of nonperiodic DoS attacks which invalidate the transmission channel serving for additional control signals $u(t)$. First, we denote the attack intervals as follows.

$$s(t) = \begin{cases} 1, & t \in [g_n, g_n + b_n), \\ 0, & t \in [g_n + b_n, g_{n+1}), \end{cases} \tag{11.5}$$

where $g_n \in R_{>0}, n \in N$ denotes the nth ending time of DoS attacks with $g_0 = 0$, $g_n + b_n \in R_{>0}, n \in N$ denotes the nth beginning time of DoS attacks. Then, the entire time interval can be divided by the sleep interval $I_{1,n} = [g_n, g_n + b_n)$ and the attack interval $I_{2,n} = [g_n + b_n, g_{n+1})$ (Fig. 11.2).

Because of attacks strategy priori unknown, DoS attacks with limited energy can be constrained by upper bound of attack frequency and uniform upper (lower) bound of attack (sleep) duration. Considering the attack scenario that control center is tempered by DoS attacks (11.5), we model DoS attacks by the following Assumptions.

Assumption 11.1 Denote $N(t, t_0)$ as the numbers of DoS attacks during $[t_0, t]$. There exist scalars $\kappa \in R_{\geq 0}$, $\tau_D \in R_{>h}$ satisfying

$$N(t, t_0) < \kappa + \frac{t - t_0}{\tau_D} \tag{11.6}$$

where h is sample time.

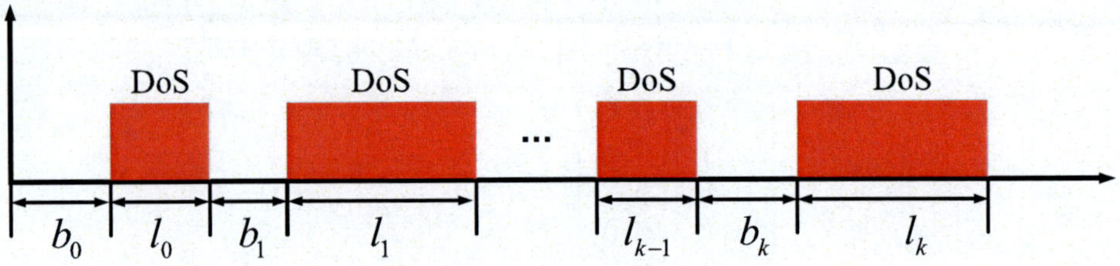

Fig. 11.2 Sleep intervals and attack intervals distribution diagram

Assumption 11.2 For sleep intervals $I_{1,n}, n \in \{0, 1, \ldots, N\}$, there exists a positive scalar $b_{min} = inf_{n \in N}\{b_n\}$. For attacks intervals $I_{2,n}, n \in \{0, 1, \ldots, N\}$, there exists a positive scalar $l_{max} = sup_{n \in N}\{l_n\}$ where $l_n = g_{n+1} - g_n - b_n$.

As shown in Fig. 11.1, additional control signals transmitted from control center to actuator would be lost in presence of DoS attacks. Here, we assume that control center operate in time-driven mode and actuator in event-driven mode. Then, control inputs are finally adopted in actuator as follows.

$$u(t) = \begin{cases} Kx(t), & t \in I_{1,n}, \\ 0, & t \in I_{2,n}, \end{cases} \tag{11.7}$$

Based on the resilient control scheme (11.7), the multi-area power systems (11.3) under DoS attacks is described by a switched time delay system $\Sigma_{s(t)}, s(t) \in \{0, 1\}$

$$\begin{cases} \Sigma_1 : \begin{cases} \dot{x}(t) = (\bar{A}_1 + \Delta A)x(t) + A_d x(t - d(t)) + Fw(t), \\ z(t) = Dx(t) + Eu(t), t \in I_{1,n}, \end{cases} \\ \Sigma_0 : \begin{cases} \dot{x}(t) = (\bar{A}_0 + \Delta A)x(t) + A_d x(t - d(t)) + Fw(t), \\ z(t) = Dx(t), t \in I_{2,n}, \end{cases} \end{cases} \tag{11.8}$$

where $\bar{A}_1 = A + BK$, $\bar{A}_0 = A$ with initial conditions $x(t_0) = x_0$, $t_0 \in [-d_m, 0]$.

Remark 11.1 In practice, control signals $u(t)$ transmitted over network should be sampled-time based signals $u(th)$. For simplification, we view $u(t)$ as continue signals because the sampled time h is far smaller than d_m.

In this study, our aim is to design additional control gain K to stabilize switched time delay system (11.8) and preserve H_∞ disturbance attenuation level γ.

(1) The closed-loop system (11.8) is exponential stable when $w(t) = 0$.
(2) Under zero initial condition, the closed-loop system satisfies

$$\int_0^\infty z^T(t)z(t)dt \le \gamma^2 \int_0^\infty w^T(t)w(t)dt \tag{11.9}$$

where $\gamma > 0$.

11.2 Resilience Analysis of LFC System

Similar to the previous chapter, we first establish a technical result on decay rate estimation of piecewise Lyapunov–Krasovskii functional.

Lemma 11.1 ([7]) *For any given matrix $R \in \mathbb{S}^n_+$, assume that there exists a matrix $X \in \mathbb{R}^{n\times n}$ such that $\begin{bmatrix} R & X \\ X^T & R \end{bmatrix} \geq 0$. Then, the following inequality holds*

$$\begin{bmatrix} \frac{1}{\alpha}R & 0 \\ 0 & \frac{1}{1-\alpha}R \end{bmatrix} \geq \begin{bmatrix} R & X \\ X^T & R \end{bmatrix}, \forall \alpha \in (0, 1).$$

According to [8], we have the following Lemma 11.2. The proof is similar with [8] and is omitted here.

Lemma 11.2 *For a given matrix $R \in \mathbb{S}^n_+$, scalars $\beta, b > a$ and a function $\varphi \in C([a, b], \mathbb{R}^n)$, the following inequality holds*

$$\int_a^b e^{\beta(s-t)}\varphi^T(s)R\varphi(s)ds$$
$$\geq \frac{\beta}{e^{\beta(t-a)} - e^{\beta(t-b)}}\left(\int_a^b \varphi(s)ds\right)^T R\left(\int_a^b \varphi(s)ds\right).$$

Theorem 11.1 *Given positive scalars d_m>0, $\gamma > 0$, α_0>0, $\alpha_1 > 0$, ξ_i>0, $(i = 1, 2, 3, 4)$, along each subsystem (11.8), respectively, it follows*

$$\dot{V}_1(t) \leq -\alpha_1 V_1(t) - z^T(t)z(t) + \gamma^2 w^T(t)w(t), \tag{11.10}$$

$$\dot{V}_0(t) \leq \alpha_0 V_0(t) - z^T(t)z(t) + \gamma^2 w^T(t)w(t), \tag{11.11}$$

if there exist positive definite matrices $P_i > 0$, $Q_i > 0$, $R_i > 0$, $Z_i > 0$, $i = 0, 1$ and appropriate dimension matrices K, X_0, X_1 satisfying

$$\Xi_i \leq 0, i = 1, 0. \tag{11.12}$$

$$\begin{bmatrix} R_1 & X_1 \\ X_1^T & R_1 \end{bmatrix} \geq 0, \begin{bmatrix} R_0 & X_0 \\ X_0^T & R_0 \end{bmatrix} \geq 0. \tag{11.13}$$

where

$$\Xi_1 = \begin{bmatrix} \Phi_0 & * & * & * \\ d_m\Phi_1 & -R_1^{-1} & * & * \\ d_m\Phi_1 & 0 & -Z_1^{-1} & * \\ \Phi_2 & 0 & 0 & -I \end{bmatrix},$$

$$\Phi_0 = \begin{bmatrix} \Phi_0(1,1) & \Phi_0(1,2) & -\rho_1 X_1 & P_1 F \\ * & \Phi_0(2,2) & \rho_1(X_1 - R_1) & 0 \\ * & * & -e^{-\alpha_1 d_m} Q_1 + \rho_1 R_1 & 0 \\ * & * & * & -\gamma^2 I \end{bmatrix},$$

$$\begin{aligned} \Phi_0(1,1) =& He(P_1(\bar{A}_1 + \Delta A)) + \alpha_1 P_1 + Q_1 \\ &+ \rho_1 R_1 - \frac{\pi^2}{4} Z_1, \end{aligned}$$

$$\Phi_0(1,2) = P_1 A_d - \rho_1 R_1 + \rho_1 X_1 + \frac{\pi^2}{4} Z_1,$$

$$\Phi_0(2,2) = \rho_1(2R_1 - X_1 - X_1^T) - \frac{\pi^2}{4} Z_1,$$

$$\Phi_1 = [\bar{A}_1 + \Delta A, A_d, 0, F], \ \Phi_2 = [D + EK, 0, 0, 0],$$

$$\Xi_0 = \begin{bmatrix} \Gamma_0 & * & * & * \\ d_m \Gamma_1 & -R_0^{-1} & * & * \\ d_m \Gamma_1 & 0 & -Z_0^{-1} & * \\ \Gamma_2 & 0 & 0 & -I \end{bmatrix},$$

$$\Gamma_0 = \begin{bmatrix} \Gamma_0(1,1) & \Gamma_0(1,2) & -\rho_0 X_0 & P_0 F \\ * & \Gamma_0(2,2) & \rho_0(X_0 - R_0) & 0 \\ * & * & -e^{\alpha_0 d_m} Q_0 + \rho_0 R_0 & 0 \\ * & * & * & -\gamma^2 I \end{bmatrix},$$

$$\begin{aligned} \Gamma_0(1,1) =& He(P_0(\bar{A}_0 + \Delta A)) - \alpha_0 P_0 + Q_0 \\ &+ \rho_0 R_0 - \frac{\pi^2}{4} Z_0, \end{aligned}$$

$$\Gamma_0(1,2) = P_0 A_d - \rho_0 R_0 + \rho_0 X_0 + \frac{\pi^2}{4} Z_0,$$

$$\Gamma_0(2,2) = \rho_0(2R_0 - X_0 - X_0^T) - \frac{\pi^2}{4} Z_0,$$

$$\Gamma_1 = [\bar{A}_0 + \Delta A, A_d, 0, F], \ \Gamma_2 = [D, 0, 0, 0].$$

Proof See the Appendix A.

In the following, switched system method is utilized to derive sufficient conditions under which system (11.8) is exponential stable with H_∞ performance.

Theorem 11.2 *Given positive scalars* $d_m > 0, \gamma > 0, \alpha_0 > 0, \alpha_1 > 0, \lambda_1 \geq 1, \lambda_0 \geq 1, \xi_i > 0, (i = 1, 2, 3, 4)$, *under DoS attacks (11.5) with* τ_D, b_{min}, l_{max}, *switched time delay system (11.8) is exponential stable with* H_∞ *performance level* $\bar{\gamma}$ *if Theorem 11.1 holds and there exist positive definite matrices* $P_i > 0, Q_i > 0, R_i > 0, Z_i > 0, i = 0, 1,$ *satisfying*

$$\begin{aligned} &P_1 \le \lambda_0 P_0,\ P_0 \le \lambda_1 e^{(\alpha_1+\alpha_0)d_m} P_1, \\ &Q_1 \le \lambda_0 Q_0,\ Q_0 \le \lambda_1 Q_1, \\ &R_1 \le \lambda_0 R_0,\ R_0 \le \lambda_1 R_1, \\ &Z_1 \le \lambda_0 Z_0,\ Z_0 \le \lambda_1 Z_1, \end{aligned} \tag{11.14}$$

$$(-ln(\lambda_1\lambda_0) - (\alpha_0 + \alpha_1)d_m - \alpha_0 l_{max} + \alpha_1 b_{min})/\tau_D > 0 \tag{11.15}$$

Proof See the Appendix B.

11.3 Resilient LFC Controller Design

Lemma 11.3 ([9]) *For real matrices Σ, Σ_0, and Σ_1, it holds*

$$\Sigma + \Sigma_0 \Lambda(t) \Sigma_1 + \Sigma_1^T \Lambda^T(t) \Sigma_0^T < 0,$$

for any $\Lambda(t)$ satisfying $\Lambda^T(t)\Lambda(t) \le I$, if and only if there exists a positive scalar $\varepsilon > 0$, such that

$$\Sigma + \varepsilon^{-1}\Sigma_0\Sigma_0^T + \varepsilon\Sigma_1^T\Sigma_1 < 0.$$

Based on the conclusion in Theorems 11.1 and 6.2, this section will propose the design method of the state feedback control gain by using LMI technique.

Theorem 11.3 *Given positive scalars $d_m > 0$, $\gamma > 0$, $\alpha_0 > 0$, $\alpha_1 > 0$, $\lambda_1 \ge 1$, $\lambda_0 \ge 1$, $\xi_i > 0$ $(i = 1, 2, 3, 4)$, $\delta > 0$, $\varepsilon > 0$, under DoS attacks (11.5) with τ_D, b_{min}, l_{max}, switched time delay system (11.8) is exponential stable with H_∞ performance level $\bar{\gamma}$ if (11.15) holds and there exist positive definite matrices $\tilde{P}_i, \tilde{Q}_i, \tilde{R}_i, \tilde{Z}_i$, $(i = 0, 1)$, and appropriate dimension matrices $\tilde{K}$, $\tilde{X}_0$, $\tilde{X}_1$ satisfying*

$$\hat{\Xi}_i < 0, (i = 0, 1), \tag{11.16}$$

$$\begin{bmatrix} \tilde{R}_1 & \tilde{X}_1 \\ \tilde{X}_1^T & \tilde{R}_1 \end{bmatrix} \ge 0, \begin{bmatrix} \tilde{R}_0 & \tilde{X}_0 \\ \tilde{X}_0^T & \tilde{R}_0 \end{bmatrix} \ge 0. \tag{11.17}$$

$$\begin{aligned} &\begin{bmatrix} -\lambda_0 \tilde{P}_0 & * \\ \tilde{P}_0 & -\tilde{P}_1 \end{bmatrix} < 0, \begin{bmatrix} -\lambda_1 e^{(\alpha_1+\alpha_0)d_m} \tilde{P}_1 & * \\ \tilde{P}_1 & -\tilde{P}_0 \end{bmatrix} < 0, \\ &\begin{bmatrix} -\lambda_0 \tilde{Q}_0 & * \\ \tilde{P}_0 & \hat{Q}_1 \end{bmatrix} < 0, \begin{bmatrix} -\lambda_1 \tilde{Q}_1 & * \\ \tilde{P}_1 & \hat{Q}_0 \end{bmatrix} < 0, \\ &\begin{bmatrix} -\lambda_0 \tilde{R}_0 & * \\ \tilde{P}_0 & \hat{R}_1 \end{bmatrix} < 0, \begin{bmatrix} -\lambda_1 \tilde{R}_1 & * \\ \tilde{P}_1 & \hat{R}_0 \end{bmatrix} < 0, \\ &\begin{bmatrix} -\lambda_0 \tilde{Z}_0 & * \\ \tilde{P}_0 & \hat{Z}_1 \end{bmatrix} < 0, \begin{bmatrix} -\lambda_1 \tilde{Z}_1 & * \\ \tilde{P}_1 & \hat{Z}_0 \end{bmatrix} < 0. \end{aligned} \tag{11.18}$$

where

$$\hat{\Xi}_1 = \begin{bmatrix} \hat{\Phi}_0 & * & * & * & * \\ d_m\hat{\Phi}_1 & \hat{\Xi}_1(2,2) & * & * & * \\ d_m\hat{\Phi}_1 & \varepsilon d_m^2 GG^T & \hat{\Xi}_1(3,3) & * & * \\ \hat{\Phi}_2 & 0 & 0 & -I & * \\ \hat{\Phi}_3 & 0 & 0 & 0 & -\varepsilon I \end{bmatrix},$$

$$\hat{\Xi}_1(2,2) = \hat{R}_1 + \varepsilon d_m^2 GG^T, \; \hat{\Xi}_1(3,3) = \hat{Z}_1 + \varepsilon d_m^2 GG^T,$$

$$\hat{\Phi}_0 = \begin{bmatrix} \hat{\Phi}_0(1,1) & \hat{\Phi}_0(1,2) & -\rho_1\tilde{X}_1 & F \\ * & \hat{\Phi}_0(2,2) & \rho_1(\tilde{X}_1 - \tilde{R}_1) & 0 \\ * & * & -e^{-\alpha_1 d_m}\tilde{Q}_1 + \rho_1\tilde{R}_1 & 0 \\ * & * & * & -\gamma^2 I \end{bmatrix},$$

$$\begin{aligned} \hat{\Phi}_0(1,1) =& He(A\tilde{P}_1 + B\tilde{K}) + \alpha_1\tilde{P}_1 + \tilde{Q}_1 \\ &+ \rho_1\tilde{R}_1 - \frac{\pi^2}{4}\tilde{Z}_1 + \varepsilon GG^T, \end{aligned}$$

$$\hat{\Phi}_0(1,2) = A_d\tilde{P}_1 - \rho_1\tilde{R}_1 + \rho_1\tilde{X}_1 + \frac{\pi^2}{4}\tilde{Z}_1,$$

$$\hat{\Phi}_0(2,2) = \rho_1(2\tilde{R}_1 - \tilde{X}_1 - \tilde{X}_1^T) - \frac{\pi^2}{4}\tilde{Z}_1,$$

$$\hat{\Phi}_1 = [A\tilde{P}_1 + B\tilde{K} + \varepsilon GG^T, A_d\tilde{P}_1, 0, F],$$

$$\hat{\Phi}_2 = [D\tilde{P}_1 + E\tilde{K}, 0, 0, 0], \quad \hat{\Phi}_3 = [H\tilde{P}_1, 0, 0, 0],$$

$$\hat{\Xi}_0 = \begin{bmatrix} \hat{\Gamma}_0 & * & * & * & * \\ d_m\hat{\Gamma}_1 & \hat{\Xi}_0(2,2) & * & * & * \\ d_m\hat{\Gamma}_1 & \varepsilon d_m^2 GG^T & \hat{\Xi}_0(3,3) & * & * \\ \hat{\Gamma}_2 & 0 & 0 & -I & * \\ \hat{\Gamma}_3 & 0 & 0 & 0 & -\varepsilon I \end{bmatrix},$$

$$\hat{\Xi}_0(2,2) = \hat{R}_0 + \varepsilon d_m^2 GG^T, \; \hat{\Xi}_0(3,3) = \hat{Z}_0 + \varepsilon d_m^2 GG^T,$$

$$\hat{\Gamma}_0 = \begin{bmatrix} \hat{\Gamma}_0(1,1) & \hat{\Gamma}_0(1,2) & -\rho_0\tilde{X}_0 & F \\ * & \hat{\Gamma}_0(2,2) & \rho_0(\tilde{X}_0 - \tilde{R}_0) & 0 \\ * & * & -e^{\alpha_0 d_m}\tilde{Q}_0 + \rho_0\tilde{R}_0 & 0 \\ * & * & * & -\gamma^2 I \end{bmatrix},$$

$$\begin{aligned} \hat{\Gamma}_0(1,1) =& He(A_0\tilde{P}_0) - \alpha_0\tilde{P}_0 + \tilde{Q}_0 \\ &+ \rho_0\tilde{R}_0 - \frac{\pi^2}{4}\tilde{Z}_0 + \varepsilon GG^T, \end{aligned}$$

$$\hat{\Gamma}_0(1,2) = A_d\tilde{P}_0 - \rho_0\tilde{R}_0 + \rho_0\tilde{X}_0 + \frac{\pi^2}{4}\tilde{Z}_0,$$

$$\hat{\Gamma}_0(2,2) = \rho_0(2\tilde{R}_0 - \tilde{X}_0 - \tilde{X}_0^T) - \frac{\pi^2}{4}\tilde{Z}_0,$$

$$\hat{\Gamma}_1 = [A\tilde{P}_0 + \varepsilon GG^T, A_d\tilde{P}_0, 0, F],$$

$$\hat{\Gamma}_2 = [D\tilde{P}_0, 0, 0, 0], \quad \hat{\Gamma}_3 = [H\tilde{P}_0, 0, 0, 0].$$

The state feedback control gain can be obtained by

$$K = \tilde{K}\tilde{P}_1^{-1}.$$

Proof See the Appendix C.

Note that the design of feedback control gain K depends on the selection of many parameters in Theorem 11.3. It is necessary to develop an algorithm to select parameters for achieving desired exponential exponential H_∞ performance of multi-area power systems.

11.4 Simulation Example

Simulation of a two-area LFC system under DoS attacks : The following simulation is carried out to show the influence of the attack beginning time. The parameters of a two-area power system borrowed from [10] are listed in Table 11.1.

The beginning time of DoS attacks are $\alpha = 0$(no attacks), $T_{DoS} = 0.1 * 30s = 3s$, $T_{DoS} = 0.2 * 30s = 6s$ and $T_{DoS} = 0.5 * 30s = 15s$, respectively. According to Theorem 11.3, the frequency derivations and tie-line power deviations are depicted in Figs. 11.3 and 11.4, respectively. It can be observed that the dynamic trajectories illustrated by red lines ($\alpha = 0.1$) have the largest settling time and overshoot by comparison. Because DoS attacks happen in the early time $T_{DoS} = 3s$ when power system have poor transient performance to be regulated by LFC scheme. This point has been discussed by [10]. By comparing with [10], the frequency derivations and tie-line power deviations have the less settling time and overshoot under our design resilient additional control scheme, as can be seen in Tables 11.2 and 11.3.

Simulation of a three-area LFC system under DoS attacks and time delay: The parameters of a three-area power system [11] are given in Table 11.4 with coefficients $T_{12} = 0.2$, $T_{13} = 0.12$, $T_{23} = 0.25$ (pu/rad). To solve the state feedback control gain, we set parameters as follows: $\gamma = 220$, $d_m = 0.1$, $k = 0.1$, $\alpha_0 = \alpha_1 = 0.1$, $\delta = 5$, $\lambda_0 = \lambda_1 = 1.01$. The load changes of the three-area power system is depicted in Fig. 11.5. According to Theorem 3, the frequency derivations and the tie-line power changes of the three-area power system are depicted in Figs. 11.6 and 11.7. The performance degradation of power system is obvious under the influence of DoS attacks and time delay. Even though there still exist DoS attacks and time delay, the dynamic trajectories are finally converging to zeros. Moreover, we get the H_∞ per-

Table 11.1 Parameters of two-area power system with LFC

Parameters	M	D	R	$T_g(s)$	$T_{ch}(s)$	β	T_{ij}
Area 1	10	1	0.05	0.1	0.3	21	0.1986
Area 2	12	1.5	0.05	0.17	0.4	21.5	0.1986

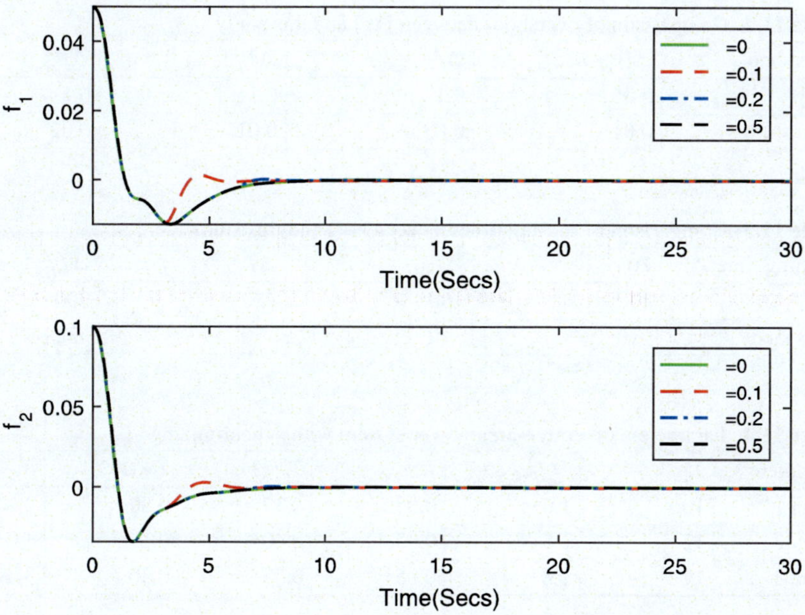

Fig. 11.3 Frequency derivations of the two-area power system under the influence of DoS attacks

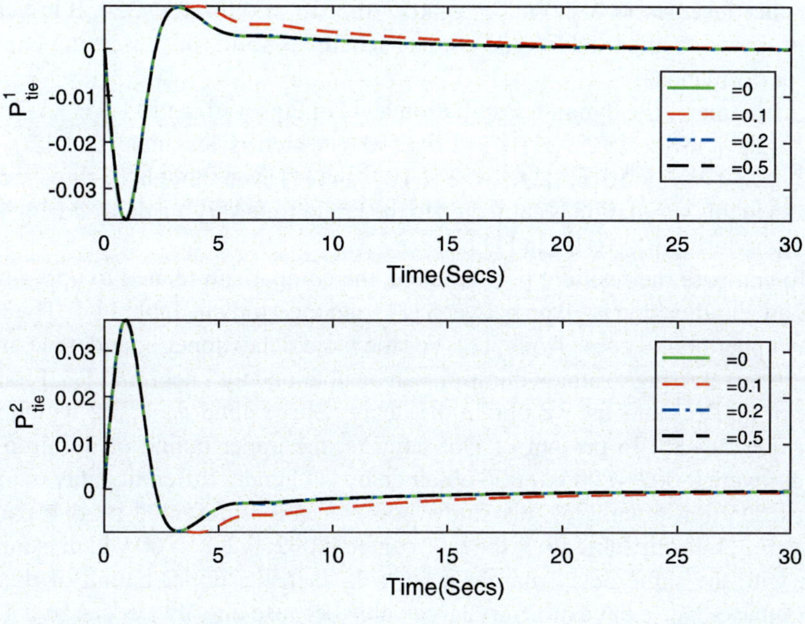

Fig. 11.4 Tie-line power deviations of the two-area power system under the influence of DoS attacks

Table 11.2 Comparison of overshoot between [10] and this study

Overshoot	Δf_1	Δf_2	ΔP_{tie}^1	ΔP_{tie}^2
[10]	0.05	0.1	0.1	0.1
This study	0.01	0.03	0.01	0.02

Table 11.3 Comparison of settling time between [10] and this study

Settling time(s)	Δf_1	Δf_2	ΔP_{tie}^1	ΔP_{tie}^2
[10]	100	100	200	200
This study	10	10	30	30

Table 11.4 Parameters of a three-area power system with synchronizing

Parameters	M	D	R	$T_g(s)$	$T_{ch}(s)$	β
Area 1	10	1.0	0.05	0.37	0.30	$2/R_1 + D_1$
Area 2	12	1.5	0.05	0.40	0.17	$4/R_2 + D_2$
Area 3	12	1.8	0.05	0.35	0.20	$3/R_3 + D_3$

formance lever $\gamma * = \|z\|/\|w\| = 0.0017/0.3513 = 0.0049 < 220$. It indicates that the proposed resilient additional control scheme is valid for preserving exponential H_∞ performance.

Furthermore, the dynamic oscillation level in Figs. 11.6 and 11.7 is $|\Delta f_1| < 10 \times 10^{-6}$, $|\Delta f_2| < 8 \times 10^{-6}$, $|\Delta f_1| < 1.5 \times 10^{-5}$ which is less than that $|\Delta f_1| < 2 \times 10^{-5}$, $|\Delta f_2| < 2 \times 10^{-5}$, $|\Delta f_1| < 2 \times 10^{-5}$ in [11] even though the three area power system in this simulation is subjected to the larger time varying delay during $[0, 0.1s]$ than that during $[0, 0.04s]$ in [11].

To compare the resilient performance, the comparison related to upper bound of DoS attacks duration is given between [11] and this study in Table 11.5. The values of delay upper bound come from [11]. For the fixed delay upper bound, the parameters l_{max}, b_{min} referring to attack duration can be obtained by Theorem 11.3. First, for the case of no DoS attacks, we obtain the delay upper bound $d_m = 3.2$ which is larger than $\bar{\eta} = 2.3885$. In present of DoS attacks, the upper bound of duration of DoS attacks were 0.002, 0.004, 0.006 obtained by [11] under different delay margin. The attack duration bounds of this study, however, are represented by two parameters (l_{max}, b_{min}). From Table 11.5, the l_{max} can be 0.002, 0.2, ..., 60 which is larger than $\tau_D h$ with the same delay margin $\bar{\eta}$ (d_m). In fact, the upper bound of duration of DoS attacks l_{max} can be arbitrary large value because one always can find a suitable large value b_{min} to satisfy the condition (11.15) in Theorem 11.3. Thus, our proposed method has less conservatism and more flexible than [11] from aspect of analysis of DoS attacks.

Simulation of a two-area LFC system under DoS attacks, time delay, uncertainty and GRC: This section will illustrate the validness of the proposed resilient

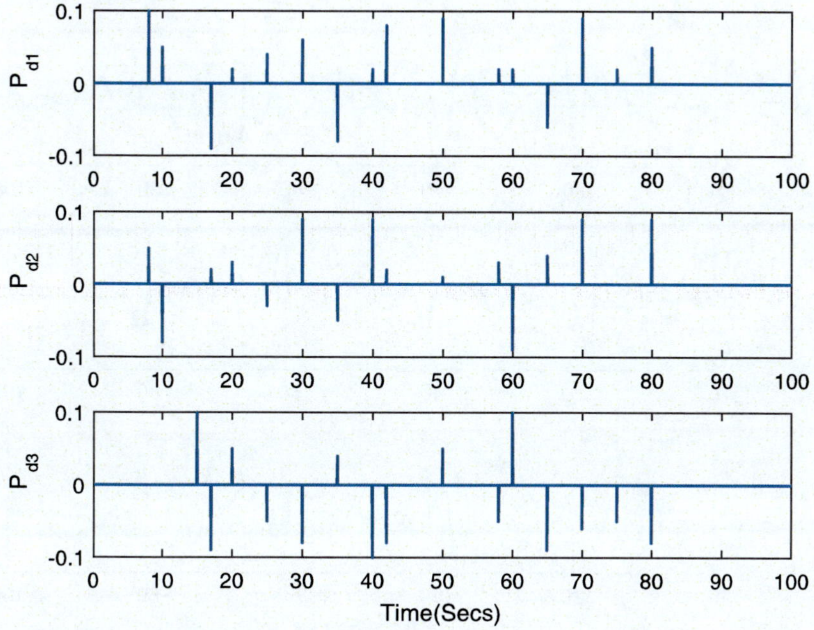

Fig. 11.5 Random impulsive load changes for the three-area power system

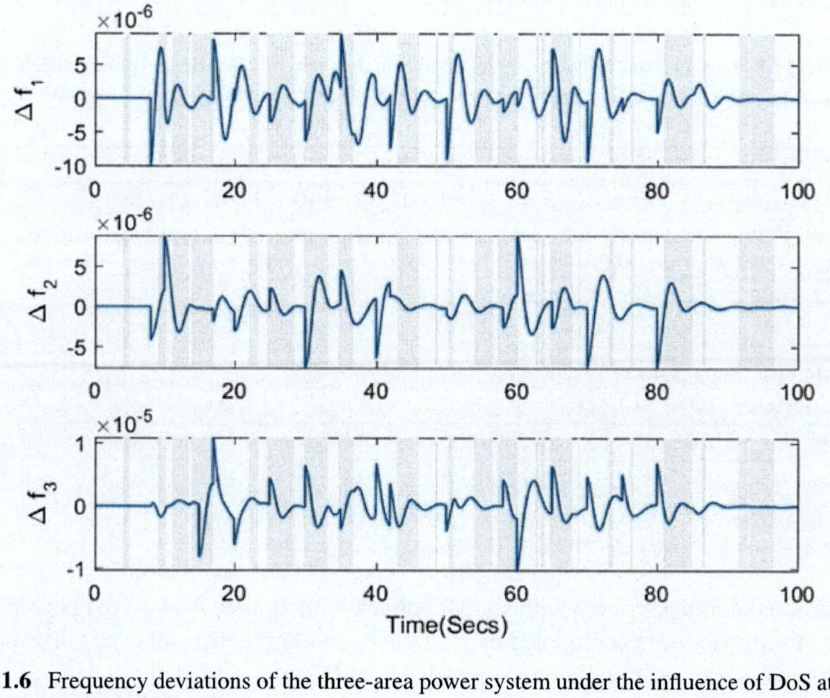

Fig. 11.6 Frequency deviations of the three-area power system under the influence of DoS attacks and time varying delay $[0, 0.1s]$

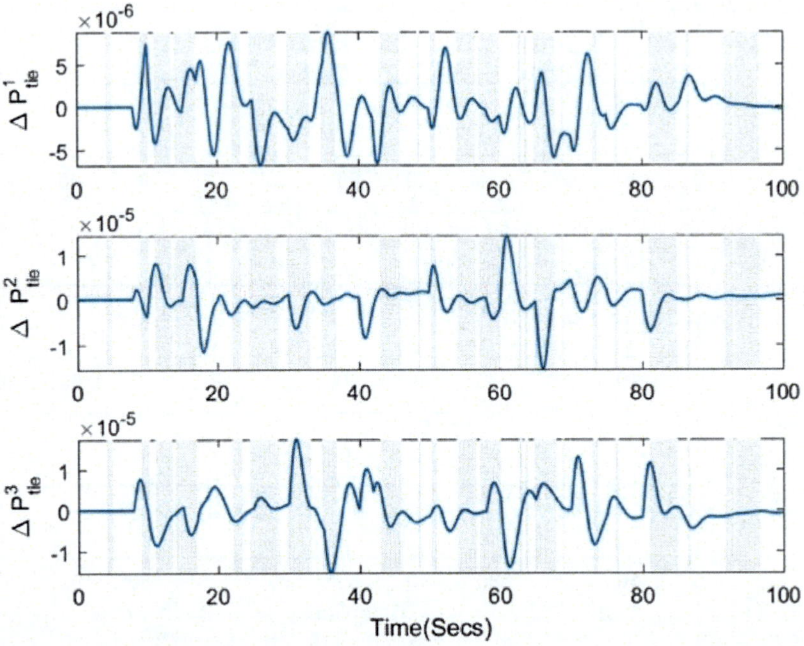

Fig. 11.7 Tie-line power deviations of the three-area power system under the influence of DoS attacks and time varying delay $[0, 0.1s]$

Table 11.5 Comparison between attack duration l_{max}, b_{min} of this study and attack duration $\tau_{dos-m} \leq \tau_D h$ ($h = 0.002s$) in [11] for various delay upper bound d_m ($\bar{\eta}$ in [11]) with $K_I = 0.1$

$\bar{\eta}$ (d_m)	$\tau_D h$ [11]	(l_{max}, b_{min}) this study
2.3885(3.2)	0	(0, inf)
1.8434(1.8434)	0.002	(0.002, 17) or (0.2, 19) or (20, 178)
1.700(1.700)	0.004	(0.004, 14) or (0.4, 17) or (40, 294)
1.2500(1.2500)	0.006	(0.006, 9) or (0.6, 13) or (60, 369)

Table 11.6 Parameters of two-area power system with LFC

Parameters	M	D	R	$T_g(s)$	$T_{ch}(s)$	β	k_i
Area 1	10	1	0.05	0.1	0.3	41	0.5
Area 2	12	1.5	0.05	0.4	0.17	81.5	0.5

additional control scheme with state feedback control law. A two-area power system with four generators is modeled by a single equivalent generator. The corresponding parameters are given in Table 11.6 with $T_{12} = 0.2(pu/rad)$ [2].

In this simulation, the uncertainty matrix $\varDelta A = GL(t)H$ (11.27) is specified by

$$G = diag\begin{bmatrix}G_1\ G_2 \ldots G_n\end{bmatrix},$$
$$L(t) = diag\begin{bmatrix}L_1(t)\ L_2(t) \ldots L_n(t)\end{bmatrix},$$
$$H = diag\begin{bmatrix}H_1\ H_2 \ldots H_n\end{bmatrix},$$

where $G_i = col\begin{bmatrix}0\ 0\ 0\ 1\ 0\end{bmatrix}$, $L_i(t) = \epsilon_i/1 + \epsilon_i$, $H_i = \begin{bmatrix}0.001/R_i T_{gi}\ 0\ 0\ 0\ 0\end{bmatrix}$.

To solve the state feedback control gain, we first set parameters as follows: $\gamma = 280$, $d_m = 0.5$, $\alpha_0 = 1$, $\alpha_1 = 5$, $\delta = 5$, $\lambda_0 = \lambda_1 = 1.01$, $\epsilon \in [0, 0.8]$, $\varepsilon = 10^{-3}$, $\bar{\upsilon} = 2$, $\underline{\upsilon} = -2$ and $\xi_1 = 0.1$, $\xi_2 = 0.1$, $\xi_3 = 0.1$, $\xi_4 = 90$.

Based on above parameters, the state feedback control gain K can be obtained by solving LMIs (11.16)–(11.18) in Theorem 11.3 by using MATLAB stools YALMIP with solver MOSEK.

$$K = [K1, K2],$$
$$K1 =$$
$$\begin{bmatrix}-11.9106 & 0.8927 & -0.4181 & -0.1415 & -0.2744\\ -0.4128 & 4.1687 & -0.0991 & -0.0390 & 4.5809\end{bmatrix},$$
$$K2 =$$
$$\begin{bmatrix}-1.8189 & 0.8761 & -0.0206 & -0.1796 & -0.2287\\ -102.4007 & 13.5772 & -1.4368 & -3.4687 & -4.6071\end{bmatrix}.$$

With the solved resilient control gain, under the load changes in Fig. 11.8, we will show the dynamic trajectories of the two-area power system. First, the frequency derivations of the two-area LFC system are depicted in Fig. 11.9. The grey area in Fig. 11.9 represent the intervals of DoS attacks. The frequency derivations are degraded under the load disturbances. Because the operation of power system is subjected to the comprehensive influence of DoS attacks, time delay, uncertainty and GRC. Besides, the tie-line power deviations are depicted in Fig. 11.10. One can observe that the serious oscillation happen during $[0, 10s]$ because of the high frequency DoS attacks. To preserve the exponential H_∞ performance, the resilient additional control inputs are generated as depicted in Fig. 11.11. The control inputs are zeros in grey area while the control inputs actions are performed during sleep intervals.

Under the resilient control inputs (11.7), however, the trajectories of frequency derivations are approaching zeros. It indicates that the closed-loop system (11.8) is exponential stable. Furthermore, by calculating the actual H_∞ performance level $\bar{\gamma}* = \parallel z(t) \parallel / \parallel w(t) \parallel = 20.5310$, one can observe that the $\bar{\gamma}*$ is far less than the respective one $\gamma = 280$. Thus, it concludes that the two-area power system is exponential stable with H_∞ performance level $\bar{\gamma}$ by using our designed resilient LFC scheme (11.7).

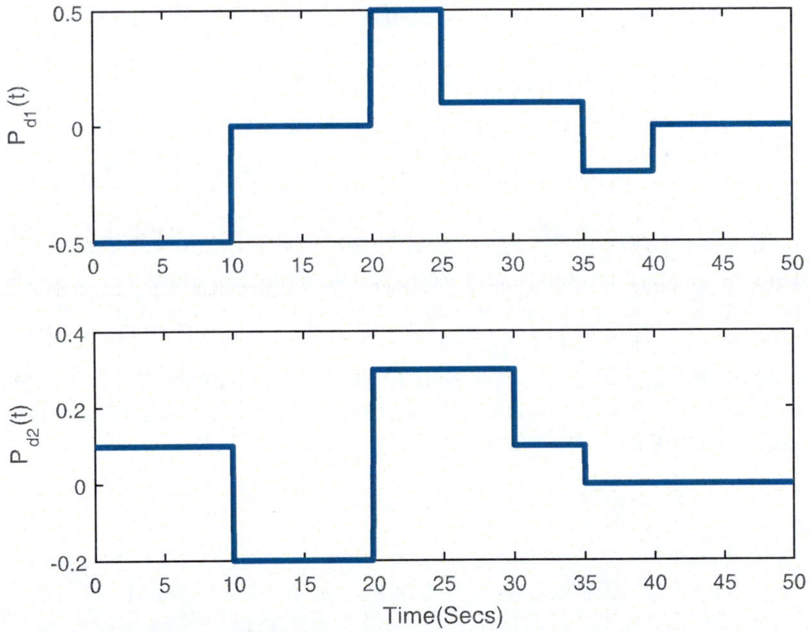

Fig. 11.8 Load changes in the two-area power system

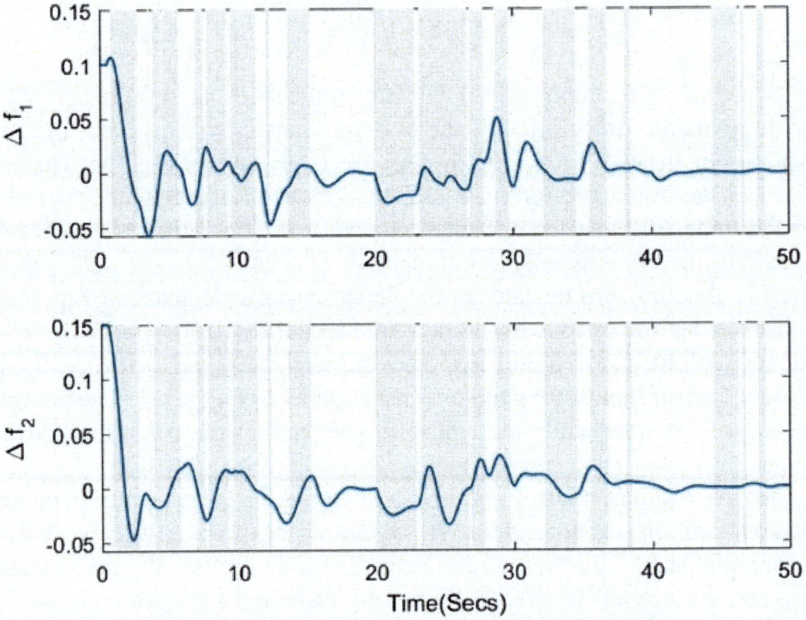

Fig. 11.9 Frequency derivations of the two-area power system under the influence of DoS attacks, time varying delay $[0, 0.5s]$, load changes ΔP_{di}, uncertainty ΔA and generator rate constraint $\bar{v}$, $\underline{v}$

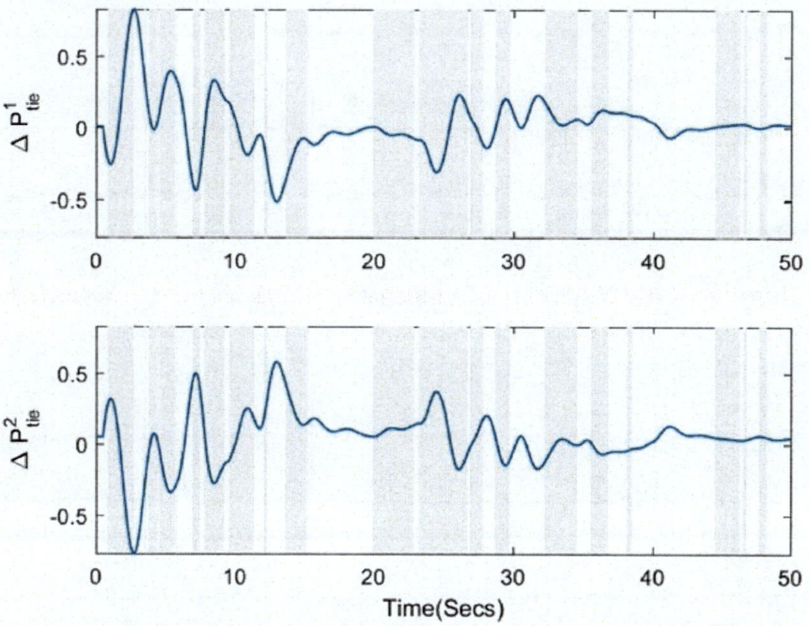

Fig. 11.10 Tie-line power deviations of the two-area power system under the influence of DoS attacks, time varying delay $[0, 0.5s]$, load changes ΔP_{di}, uncertainty ΔA and generator rate constraint $\bar{\upsilon}$, $\underline{\upsilon}$

With the above parameters, we investigate the relationship between b_{min} and l_{max}. For given l_{max}, the required b_{min} can be obtained according to (11.15) while other sufficient conditions are feasible in Theorem 11.3. The results are listed in Table 11.7. The b_{min} is increased with the increasing of l_{max}, which indicates that power system need more sleeping time to preserve the exponential H_∞ performance under the long time DoS attacks. Moreover, the decreasing trend of rate b_{min}/l_{max} indicates that the power system with longer sleep times has better resilient performance to tolerate longer DoS attacks even. The reason might be the decreasing of attack frequency estimated by $1/(l_{max} + b_{min})$.

According to Theorem 11.2, the exponential decay rates of the two-area LFC system are given under various uniform attack duration l_{max} Table 11.8. With the increasing of attack duration, the exponential decay rates β are decreasing. This result meets our common sense that the longer DoS attacks leads to the worse exponential stability of power system.

Beside of DoS attacks duration, DoS attacks frequency is another important factor to influence the resilience performance of a multi-area power system. As given in the process of system switching analysis in Theorem 2, we equivalently deal with each attack as periodic attack signals composed by uniform lower bound of sleep intervals

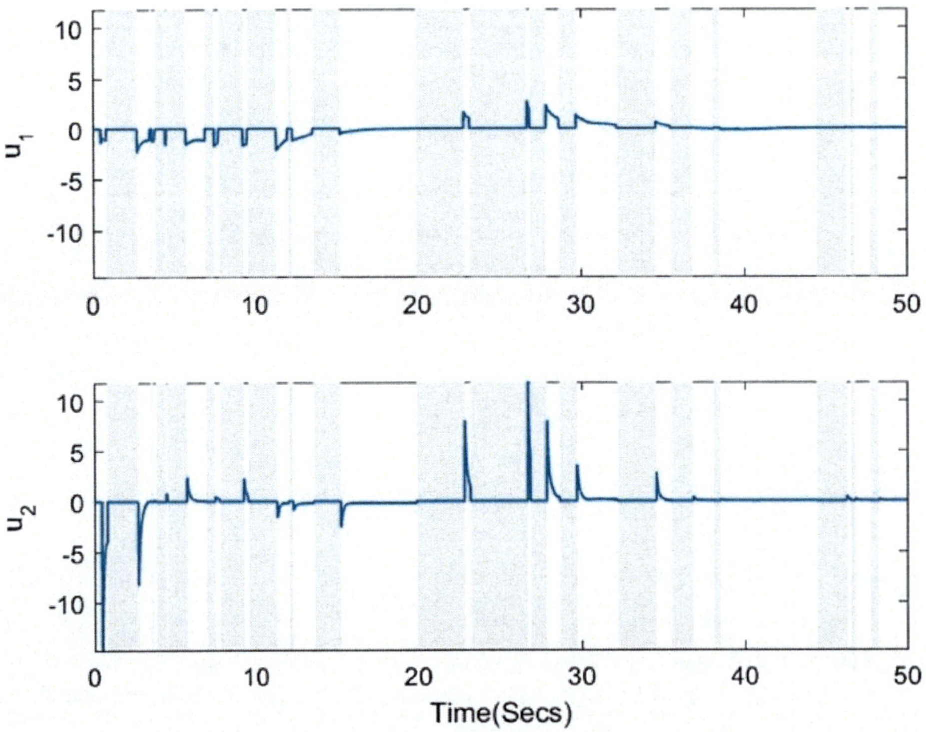

Fig. 11.11 Resilient additional control inputs for the two-area power system

Table 11.7 The required b_{min} for various DoS attacks duration l_{max}

l_{max}	1s	3s	5s	7s	9s
b_{min}	0.9s	1.3s	1.7s	2.1s	2.5s
b_{min}/l_{max} (%)	90.00	43.33	34.00	30.00	27.78

Table 11.8 The exponential decay rate β for various l_{max} with fixed $b_{min} = 2s$

l_{max}	1.5s	3.5s	4.5s	5.5s	6.5s
β	1.5657	0.6327	0.3816	0.1973	0.0565

b_{min} and uniform upper bound of attack intervals l_{max}. Thus, we estimate the attack frequency by $1/\tau_D = 1/(b_{min} + l_{max})$. Then, power system exponential decay rate β are given in Table 11.9 under different DoS attacks frequency $1/\tau_D$. Exponential decay rate β is calculated according to (11.15). From Table 11.9, the decay rate β is increased with the decreasing of attack frequency $1/\tau_D$. It indicates that the higher the attack frequency is, the worse the exponential stability of power system is.

Table 11.9 The exponential decay rate β for various attack frequency $1/\tau_D$

$1/\tau_D$	0.3226	0.1613	0.1075	0.0806	0.0645
β	0.1549	0.6420	0.8043	0.8855	0.9342

11.5 Conclusion

Resilient LFC design problem of a multi-area power system with uncertainty and physical constraint has been investigated under DoS attacks and time varying delay. Time delay system method and switched system method have been used in the resilient analysis. Sufficient conditions have been derived for preserving the exponential H_∞ performance. Under the tolerable DoS attacks, state feedback control law have been designed. In future study, the conservatism of the results can be improved by using advanced tight inequality technique [12]. Event through the influence of DoS attacks has been considered fully, the method need to be improved when the nonlinearities like GDB, boiler dynamics are considered. Further, based on the considered LFC scheme consisting of two control loops, we further study the active defense strategy by cooperating the two control loop, for example, redundancy switching control strategy.

Appendix A

Proof of Theorem 11.1: Choose a Lyapunov–Krasovskii functional for subsystem Σ_i (11.8), $i = 0, 1.$, respectively.

$$\begin{aligned}
V_i(t) = {} & x^T(t)P_i x(t) + \int_{t-d_m}^{t} x^T(s)exp(\star)Q_i x(s)ds \\
& + d_m \int_{-d_m}^{0}\int_{t+\theta}^{t} \dot{x}^T(s)exp(\star)R_i\dot{x}(s)dsd\theta, \\
& + d_m^2 \int_{t_k h}^{t} \dot{x}^T(s)exp(\star)Z_i\dot{x}(s)ds, \\
& - \frac{\pi^2}{4}\int_{t_k h}^{t} [x(s) - x(t_k h)]^T exp(\star)Z_i[x(s) - x(t_k h)]ds.
\end{aligned} \tag{11.19}$$

where $exp(\star) = e^{(-1)^i \alpha_i (t-s)}$ and $t_k h = t - d(t)$.

Taking derivation of LKF (11.19) with $i = 1$ along system Σ_1, it has

$$\begin{aligned}\dot{V}_1(t) \leq & -\alpha_1 V_1(t) + \alpha_1 x^T(t) P_1 x(t) \\ & + 2x^T(t) P_1[(\bar{A}_1 + \Delta A)x(t) + A_d x(t - d(t)) + Fw(t)] \\ & + x^T(t) Q_1 x(t) - e^{-\alpha_1 d_m} x^T(t - d_m) Q_1 x(t - d_m) \\ & + d_m^2 \xi^T(t) \Phi_1^T (R_1 + Z_1) \Phi_1 \xi(t) \\ & - d_m \int_{t-d_m}^{t} e^{\alpha_1 (s-t)} \dot{x}^T(s) R_1 \dot{x}(s) ds \\ & - \frac{\pi^2}{4} [x(t) - x(t - d(t))]^T Z_1 [x(t) - x(t - d(t))] \\ & + \xi^T(t) \Phi_2^T \Phi_2 \xi(t) - \gamma^2 w^T(t) w(t) \\ & - z^T(t) z(t) + \gamma^2 w^T(t) w(t),\end{aligned}$$

where $\xi(t) = col[x(t), x(t - d(t)), x(t - d_m), w(t)]$.

According to Lemmas 11.1 and 11.2, we have

$$\begin{aligned}-d_m \int_{t-d_m}^{t} & e^{\alpha_1 (s-t)} \dot{x}^T(s) R_1 \dot{x}(s) ds \\ & \leq \rho_1 \begin{bmatrix} \eta_1^T & \eta_2^T \end{bmatrix} \begin{bmatrix} R_1 & X_1 \\ X_1^T & R_1 \end{bmatrix} \begin{bmatrix} \eta_1 \\ \eta_2 \end{bmatrix}\end{aligned}$$

where $\rho_1 = \frac{\alpha_1 d_m}{1 - e^{\alpha_1 d_m}}$, $\eta_1 = x(t) - x(t - d(t))$, $\eta_2 = x(t - d(t)) - x(t - d_m))$.

By using Schur complement lemma, it is further obtained

$$\dot{V}_1(t) \leq -\alpha_1 V_1(t) + \xi^T(t) \Xi_1 \xi(t) - z^T(t) z(t) + \gamma^2 w^T(t) w(t).$$

According to $\Xi_1 \leq 0$ (11.12), the decay estimation of $V_1(t)$ (11.10) can be obtained.

Similarly, along system Σ_0, the decay estimation of LKF (11.19) with $i = 0$ can be derived by

$$\begin{aligned}\dot{V}_0(t) \leq & \alpha_0 V_0(t) - \alpha_0 x^T(t) P_0 x(t) \\ & + 2x^T(t) P_0[(\bar{A}_0 + \Delta A)x(t) + A_d x(t - d(t)) + Fw(t)] \\ & + x^T(t) Q_0 x(t) - e^{\alpha_0 d_m} x^T(t - d_m) Q_0 x(t - d_m) \\ & + d_m^2 \xi^T(t) \Gamma_1^T (R_0 + Z_0) \Gamma_1 \xi(t) \\ & - d_m \int_{t-d_m}^{t} e^{\alpha_0 (t-s)} \dot{x}^T(s) R_0 \dot{x}(s) ds \\ & - \frac{\pi^2}{4} [x(t) - x(t - d(t))]^T Z_0 [x(t) - x(t - d(t))] \\ & + \xi^T(t) \Gamma_2^T \Gamma_2 \xi(t) - \gamma^2 w^T(t) w(t) \\ & - z^T(t) z(t) + \gamma^2 w^T(t) w(t),\end{aligned}$$

where $\xi(t) = col[x(t), x(t-d(t)), x(t-d_m), w(t)]$.

According to Lemmas 11.1 and 11.2, we have

$$\begin{aligned} -d_m \int_{t-d_m}^{t} & e^{\alpha_0(t-s)}\dot{x}^T(s)R_0\dot{x}(s)ds \\ & \le \rho_0 \begin{bmatrix} \eta_1^T & \eta_2^T \end{bmatrix} \begin{bmatrix} R_0 & X_0 \\ X_0^T & R_0 \end{bmatrix} \begin{bmatrix} \eta_1 \\ \eta_2 \end{bmatrix} \end{aligned}$$

where $\rho_0 = \alpha_0 d_m/(e^{-\alpha_0 d_m} - 1)$.

By using Schur complement lemma, one can has

$$\dot{V}_0(t) \le \alpha_0 V_0(t) + \xi^T(t)\Xi_0\xi(t) - z^T(t)z(t) + \gamma^2 w^T(t)w(t).$$

According to $\Xi_0 \le 0$ (11.12), the decay estimation of $V_0(t)$ (11.11) can be obtained. This completes the proof.

Appendix B

Proof of Theorem 11.2: Choose a piecewise Lyapunov–Krasovskii functional as $V(t) = \begin{cases} V_1(t), & t \in I_{1,n} \\ V_0(t), & t \in I_{2,n} \end{cases}$ for the switched delay system (11.8).

According to (11.14), the inequality relationship of Lyapunov–Krasovskii functional (11.19) in switching points are derived

$$\begin{aligned} & V_1(g_n^+) \le \lambda_0 V_0(g_n^-), \\ & V_0((g_n+b_n)^+) \le \lambda_1 e^{(\alpha_1+\alpha_0)d_m} V_1((g_n+b_n)^-). \end{aligned} \tag{11.20}$$

First, we will analyze the exponential stability of the switched delay system (11.8) by assuming $w(t) = 0$. Case I: When $t \in I_{1,n} = [g_n, g_n + b_n)$, based on (11.10), (11.11) and (11.20), the decay estimation of $V(t)$ is developed by

$$\begin{aligned} V_1(t) & \le e^{-\alpha_1(t-g_n)} V_1(g_n^+) \\ & \le e^{-\alpha_1(t-g_n)} V_1(g_{n-1}^+)\lambda_1\lambda_0 e^{(\alpha_0+\alpha_1)d_m+\alpha_0 l_{n-1}-\alpha_1 b_{n-1}} \\ & \le e^{-\alpha_1(t-g_n)} V_1(g_{n-1}^+)\lambda_1\lambda_0 e^{(\alpha_0+\alpha_1)d_m+\alpha_0 l_{max}-\alpha_1 b_{min}} \\ & \vdots \\ & \le (\lambda_1\lambda_0 e^{(\alpha_0+\alpha_1)d_m+\alpha_0 l_{max}-\alpha_1 b_{min}})^n V_1(0). \end{aligned}$$

Due to $n = N(t, 0)$, based on Assumption 11.1, it has

$$V_1(t) \leq \varsigma_1 e^{-\beta t} V_1(0), \tag{11.21}$$

where $\varsigma_1 = e^{-\kappa\beta\tau_D}$ and $\beta = (-ln(\lambda_1\lambda_0) - (\alpha_0 + \alpha_1)d_m - \alpha_0 l_{max} + \alpha_1 b_{min})/\tau_D$.

Case II: When $t \in I_{2,n} = [g_n + b_n, g_{n+1})$, similarly, the decay estimation of $V(t)$ is given by

$$\begin{aligned}
V_0(t) \leq & e^{\alpha_0(t-g_n-b_n)} V_0((g_n + b_n)^+) \\
\leq & \lambda_0^{-1}\lambda_1\lambda_0 e^{(\alpha_1+\alpha_0)d_m+\alpha_0 l_{max}-\alpha_1 b_{min}} V_1(g_n^+) \\
\leq & \lambda_0{}^{-1}(\lambda_1\lambda_0 e^{(\alpha_1+\alpha_0)d_m+\alpha_0 l_{max}-\alpha_1 b_{min}})^{n+1} V_1(0).
\end{aligned}$$

Due to $n + 1 = N(t, 0)$, based on Assumption 11.2, it has

$$V_0(t) \leq \varsigma_0 e^{-\beta t} V_1(0), \tag{11.22}$$

where $\varsigma_0 = \lambda_0^{-1}\varsigma_1$.

Finally, by denoting $\varsigma = max(\varsigma_0, \varsigma_1)$, according to (11.21) and (11.22), we have

$$V(t) \leq \varsigma e^{-\beta t} V_1(0), \ \forall t > 0. \tag{11.23}$$

By denoting $\pi_1 = min\{\lambda(P_0), \lambda(P_1)\}, \pi_0 = \lambda_{max}(P_1) + d_m\lambda_{max}(Q_1) + d_m^3/2\lambda_{max}(R_1) + d_m^3\lambda_{max}(Z_1)$, $\|\chi_0\| = max\{\|x(t)\|, \|\dot{x}(t)\|\}, t \in [-d_m, 0]$, one has

$$\|x(t)\| \leq \sqrt{\varsigma\frac{\pi_0}{\pi_1}} e^{-\frac{\beta t}{2}} \|\chi_0\|$$

which indicates that the switched delay system (11.8) is exponential stable. Furthermore, we analyze the H_∞ performance of system (11.8) with $x(t_0) = 0$.

Based on (11.10) and (11.11), one has

$$\sum_{i=0}^{n} \int_{g_i}^{g_i+b_i} d[e^{\alpha_1(t-g_i)} V_1(t)] \leq \sum_{i=0}^{n} \int_{g_i}^{g_i+b_i} e^{\alpha_1(t-g_i)} \Delta(t)dt, \tag{11.24}$$

$$\sum_{i=0}^{n} \int_{g_i+b_i}^{g_{i+1}} d[e^{\alpha_0(g_{i+1}-t)} V_0(t)] \leq \sum_{i=0}^{n} \int_{g_i+b_i}^{g_{i+1}} e^{\alpha_0(g_{i+1}-t)} \Delta(t)dt, \tag{11.25}$$

where $\Delta(t) = -z^T(t)z(t) + \gamma^2 w^T(t)w(t)$.

Multiply both sides of (11.24) with λ_0^{-1} and add it with (11.25). According to (11.20), it is obtained under zero initial states.

$$\sum_{i=0}^{n}\{[\lambda_0^{-1}e^{\alpha_1 b_{min}} - \lambda_1 e^{\alpha_0 l_{max}+(\alpha_1+\alpha_0)d_m}]V_1(g_i+b_i)\} + \lambda_0^{-1}V_1(g_{n+1})$$

$$\leq \sum_{i=0}^{n}\{[\lambda_0^{-1}e^{\alpha_1 b_i} - \lambda_1 e^{\alpha_0 l_i+(\alpha_1+\alpha_0)d_m}]V_1(g_i+b_i) + \lambda_0^{-1}[V_1(g_{i+1}) - V_1(g_i)]\}$$

$$\leq \sum_{i=0}^{n}[\int_{g_i}^{g_i+b_i} \lambda_0^{-1}e^{\alpha_1(t-g_i)}\Delta(t)dt + \int_{g_i+b_i}^{g_{i+1}} e^{\alpha_0(g_{i+1}-t)}\Delta(t)dt].$$

According to (11.15), note that $\lambda_0^{-1}e^{\alpha_1 b_{min}} - \lambda_1 e^{\alpha_0 l_{max}} > 0$. Then, it has

$$\lambda_0^{-1}\sum_{i=0}^{n}\int_{g_i}^{g_i+b_i} e^{\alpha_1(t-g_i)}\Delta(t)dt + \sum_{i=0}^{n}\int_{g_i+b_i}^{g_{i+1}} e^{\alpha_0(g_{i+1}-t)}\Delta(t)dt \geq 0,$$

Further, with $n \to \infty$, it finally has

$$\int_0^\infty z^T(t)z(t)dt \leq \bar{\gamma}^2 \int_0^\infty w^T(t)w(t)dt, \tag{11.26}$$

where $\bar{\gamma} = \sqrt{\psi}\gamma$, $\psi = max(e^{\alpha_1 b_{max}}, \lambda_0 e^{\alpha_0 l_{max}})$ and $b_{max} = max\{b_i\}$. The proof is completed that system (11.8) is exponential stable with H_∞ disturbance attenuant level $\bar{\gamma}$.

Appendix C

Proof of Theorem 11.3: First, the uncertainty ΔA in (11.12) can be decomposed by

$$\Delta A = GL(t)H, \tag{11.27}$$

where G, H are known matrix and $L^T(t)L(t) \leq 1$.

Then, $\Xi_1 < 0$ and $\Xi_2 < 0$ in (11.12) are rewritten by

$$\Xi_1 + [\tilde{\Xi}_1, \mathbf{0}] + [\tilde{\Xi}_1, \mathbf{0}]^T < 0,$$
$$\Xi_0 + [\tilde{\Xi}_0, \mathbf{0}] + [\tilde{\Xi}_0, \mathbf{0}]^T < 0,$$

where

$$\tilde{\Xi}_1 = col[P_1\Delta A, 0, 0, 0, d_m\Delta A, d_m\Delta A, 0],$$
$$\tilde{\Xi}_0 = col[P_0\Delta A, 0, 0, 0, d_m\Delta A, d_m\Delta A, 0].$$

According to definition (11.27), one has

$$\Xi_1 + \Upsilon_1 L(t)\Psi + \Psi^T L^T(t)\Upsilon_1^T < 0,$$
$$\Xi_0 + \Upsilon_0 L(t)\Psi + \Psi^T L^T(t)\Upsilon_0^T < 0,$$

where

$$\Upsilon_1 = col[P_1G, 0, 0, 0, d_mG, d_mG, 0],$$
$$\Upsilon_0 = col[P_0G, 0, 0, 0, d_mG, d_mG, 0],$$
$$\Psi = [H, 0, 0, 0, 0, 0, 0].$$

According to Lemma 11.3, we obtain

$$\Xi_1 + \varepsilon\Upsilon_1\Upsilon_1^T + \varepsilon^{-1}\Psi^T\Psi < 0,$$
$$\Xi_0 + \varepsilon\Upsilon_0\Upsilon_0^T + \varepsilon^{-1}\Psi^T\Psi < 0,$$

By using Schur complement lemma, one has

$$\tilde{\Xi}_1 < 0,\ \tilde{\Xi}_0 < 0. \tag{11.28}$$

where

$$\tilde{\Xi}_1 = \begin{bmatrix} \tilde{\Phi}_0 & * & * & * & * \\ d_m\tilde{\Phi}_1 & \tilde{\Xi}_1(2,2) & * & * & * \\ d_m\tilde{\Phi}_1 & \varepsilon d_m^2GG^T & \tilde{\Xi}_1(3,3) & * & * \\ \Phi_2 & 0 & 0 & -I & * \\ \Phi_3 & 0 & 0 & 0 & -\varepsilon I \end{bmatrix},$$

$$\tilde{\Xi}_1(2,2) = -R_1^{-1} + \varepsilon d_m^2 GG^T,$$
$$\tilde{\Xi}_1(3,3) = -Z_1^{-1} + \varepsilon d_m^2 GG^T,$$

$$\tilde{\Phi}_0 = \begin{bmatrix} \tilde{\Phi}_0(1,1) & \Phi_0(1,2) & -\rho_1 X_1 & P_1F \\ * & \Phi_0(2,2) & \rho_1(X_1 - R_1) & 0 \\ * & * & -e^{-\alpha_1 d_m}Q_1 + \rho_1 R_1 & 0 \\ * & * & * & -\gamma^2 I \end{bmatrix},$$

$$\tilde{\Phi}_0(1,1) = He(P_1\bar{A}_1) + \alpha_1 P_1 + Q_1 + \rho_1 R_1 - \frac{\pi^2}{4}Z_1 + \varepsilon P_1GG^TP_1,$$

$$\Phi_0(1,2) = P_1A_d - \rho_1 R_1 + \rho_1 X_1 + \frac{\pi^2}{4}Z_1,$$

$$\Phi_0(2,2) = \rho_1(2R_1 - X_1 - X_1^T) - \frac{\pi^2}{4}Z_1,$$

$$\tilde{\Phi}_1 = [\bar{A}_1 + \varepsilon GG^TP_1, A_d, 0, F],\ \Phi_2 = [D + EK, 0, 0, 0],\ \Phi_3 = [H, 0, 0, 0],$$

$$\tilde{\Xi}_0 = \begin{bmatrix} \tilde{\Gamma}_0 & * & * & * & * \\ d_m\tilde{\Gamma}_1 & \tilde{\Xi}_0(2,2) & * & * & * \\ d_m\tilde{\Gamma}_1 & \varepsilon d_m^2 GG^T & \tilde{\Xi}_0(3,3) & * & * \\ \Gamma_2 & 0 & 0 & -I & * \\ \Gamma_3 & 0 & 0 & 0 & -\varepsilon I \end{bmatrix},$$

$$\tilde{\Xi}_0(2,2) = -R_0^{-1} + \varepsilon d_m^2 GG^T, \tilde{\Xi}_0(3,3) = -Z_0^{-1} + \varepsilon d_m^2 GG^T,$$

$$\tilde{\Gamma}_0 = \begin{bmatrix} \tilde{\Gamma}_0(1,1) & \Gamma_0(1,2) & -\rho_0 X_0 & P_0 F \\ * & \Gamma_0(2,2) & \rho_0(X_0 - R_0) & 0 \\ * & * & -e^{\alpha_0 d_m} Q_0 + \rho_0 R_0 & 0 \\ * & * & * & -\gamma^2 I \end{bmatrix},$$

$$\tilde{\Gamma}_0(1,1) = He(P_0\bar{A}_0) - \alpha_0 P_0 + Q_0 + \rho_0 R_0 - \frac{\pi^2}{4} Z_0 + \varepsilon P_0 GG^T P_0,$$

$$\Gamma_0(1,2) = P_0 A_d - \rho_0 R_0 + \rho_0 X_0 + \frac{\pi^2}{4} Z_0,$$

$$\Gamma_0(2,2) = \rho_0(2R_0 - X_0 - X_0^T) - \frac{\pi^2}{4} Z_0,$$

$$\tilde{\Gamma}_1 = [\bar{A}_0 + \varepsilon GG^T P_0, A_d, 0, F], \Gamma_2 = [D, 0, 0, 0], \Gamma_3 = [H, 0, 0, 0],$$

Further, denote $\tilde{P}_1 = P_1^{-1}, \tilde{K} = K\tilde{P}_1, \tilde{Q}_1 = \tilde{P}_1 Q_1 \tilde{P}_1, \tilde{R}_1 = \tilde{P}_1 R_1 \tilde{P}_1, \tilde{Z}_1 = \tilde{P}_1 Z_1 \tilde{P}_1$, $\tilde{X}_1 = \tilde{P}_1 X_1 \tilde{P}_1$, $\hat{Q}_1 = \delta^2 \tilde{Q}_1 - 2\delta \tilde{P}_1$, $\hat{R}_1 = \delta^2 \tilde{R}_1 - 2\delta \tilde{P}_1$ and $\hat{Z}_1 = \delta^2 \tilde{Z}_1 - 2\delta \tilde{P}_1$. Pre- and pro-multiply $diag\{\tilde{P}_1, \tilde{P}_1, \tilde{P}_1, I, I, I, I, I\}$ and its transpose on both sides of $\tilde{\Xi}_1 < 0$ in (11.28). By using inequality technique $-R_1^{-1} < \delta^2 \tilde{R}_1 - 2\delta \tilde{P}_1$, LMI $\hat{\Xi}_1 < 0$ (11.16) is derived.

Similarly, denoting $\tilde{P}_0 = P_0^{-1}$, $\tilde{Q}_0 = \tilde{P}_0 Q_0 \tilde{P}_0$, $\tilde{R}_0 = \tilde{P}_0 R_0 \tilde{P}_0$, $\tilde{Z}_0 = \tilde{P}_0 Z_0 \tilde{P}_0$, $\hat{Q}_0 = \delta^2 \tilde{Q}_0 - 2\delta \tilde{P}_0$, $\hat{R}_0 = \delta^2 \tilde{R}_0 - 2\delta \tilde{P}_0$ and $\hat{Z}_0 = \delta^2 \tilde{Z}_0 - 2\delta \tilde{P}_0$. Pre-and pro multiply $diag\{\tilde{P}_0, \tilde{P}_0, \tilde{P}_0, I, I, I, I, I\}$ and its transpose on both sides of $\tilde{\Xi}_0 < 0$ in (11.28), one has LMI $\hat{\Xi}_0 < 0$ in (11.16). Moreover, pre-and pro multiply $diag\{\tilde{P}_1, \tilde{P}_1\}, diag\{\tilde{P}_0, \tilde{P}_0\}$ and its transpose on both sides of (11.13), respectively, one has LMI in (11.17). By using Schur complement lemma, (11.14) are converted to LMIs (11.18) This completes the proof.

References

1. Z. Cheng, D. Yue, S. Hu, C. Huang, C. Dou, L. Chen, Resilient load frequency control design: Dos attacks against additional control loop. Int. J. Electr. Power & Energy Syst. **115**, 105496 (2020)
2. R. Dey, S. Ghosh, G. Ray, A. Rakshit, H_∞ load frequency control of interconnected power systems with communication delays. Int. J. Electr. Power & Energy Syst. **42**(1), 672–684 (2012)
3. A. Ahmadi, M. Aldeen, Robust overlapping load frequency output feedback control of multi-area interconnected power systems. Int. J. Electr. Power & Energy Syst. **89**, 156–172 (2017)
4. H. Bevrani, *Robust Power System Frequency Control* (Springer, New York, 2014)

5. S. Liu, W. Luo, L. Wu, Co-design of distributed model-based control and event-triggering scheme for load frequency regulation in smart grids. IEEE Trans. Syst. Man Cybern.: Syst. **50**(9), 3311–3319 (2020)
6. L. Jiang, W. Yao, Q.H. Wu, J.Y. Wen, S.J. Cheng, Delay-dependent stability for load frequency control with constant and time-varying delays. IEEE Trans. Power Syst. **27**(2), 932–941 (2012)
7. P. Park, J.W. Ko, C. Jeong, Reciprocally convex approach to stability of systems with time-varying delays. Automatica **47**(1), 235–238 (2011)
8. H. Trinh et al., Exponential stability of time-delay systems via new weighted integral inequalities. Appl. Math. Comput. **275**, 335–344 (2016)
9. Z. Wang, F. Yang, D.W. Ho, X. Liu, Robust H_∞ control for networked systems with random packet losses. IEEE Trans. Syst. Man Cybern. Part B (Cybernetics) **37**(4), 916–924 (2007)
10. S. Liu, X.P. Liu, A. El Saddik, Denial-of-service (DoS) attacks on load frequency control in smart grids, in *Innovative Smart Grid Technologies (ISGT), IEEE PES* (IEEE, 2013), pp. 1–6
11. C. Peng, J. Li, M. Fei, Resilient event-triggering H_∞ load frequency control for multi-area power systems with energy-limited DoS attacks. IEEE Trans. Power Syst. **32**(5), 4110–4118 (2017)
12. X.-M. Zhang, Q.-L. Han, A. Seuret, F. Gouaisbaut, An improved reciprocally convex inequality and an augmented Lyapunov-Krasovskii functional for stability of linear systems with time-varying delay. Automatica **84**, 221–226 (2017)

Chapter 12
Secure Distributed Optimal Frequency Regulation of Power Grid with Time-Varying Voltages Under False Data Injection Attacks

The distributed optimal frequency regulation of power grid containing plenty of distributed renewable energy resources is highly dependent on the communication network. However, the utilization of communication network makes the power grid vulnerable to false data injection attacks, which could deteriorate the control performance and even cause the failure of control task. To address this problem, this chapter investigates the secure distributed optimal frequency regulation of power gird under false data injection attacks. The stealthy false data injection attack model is constructed and its perniciousness on optimal frequency regulation is analyzed. A novel type of secure distributed optimal frequency regulation is developed by introducing secure-based communication network to improve the resilience against false data injection attacks. It is proved theoretically that the secure regulation can restore the frequency deviation of each area to zero and minimize the total generation cost to retain the economic efficiency even under false data injection attacks. Simulation result shows the effectiveness of the proposed method [1].

The chapter is organized as follows. The system model, control-based communication network and control purpose are presented in Sect. 12.1. Section 12.2 constructs the modality of false data injection attacks and analyzes its impact on power grid optimal frequency regulation. The secure distributed optimal frequency regulation against the false data injection attack is developed in Sect. 12.3. Section 12.4 illustrates the simulation results to demonstrate the effectiveness of our theory. Finally, the conclusion and future works are stated in Sect. 12.5.

D. Yue et al., *Secure Control of Networked Control Systems and Its Applications*,
https://doi.org/10.1007/978-981-33-6730-2_12

12.1 Problem Formulation

12.1.1 Power Grid Model

The considered power grid is partitioned into n smaller areas, where the dynamic behavior of each area can be regarded as an equivalent single generator as a result of coherency and aggregation techniques [2, 3]. Labelling the areas from 1 to n, the power network can be described by an undirected and connected graph $\mathcal{G} \triangleq (\mathcal{V}, \mathcal{E})$. $\mathcal{V} \triangleq \{1, 2, \ldots, n\}$ is the node set representing the areas, and $\mathcal{E} \triangleq \{(j, i)\} \subseteq \mathcal{V} \times \mathcal{V}$ is the edge set where (j, i) signifies the transmission line connecting the area j and i in power grid. The set of areas connected to area i by transmission lines is expressed as N_i. Assuming that there are a total of m transmission lines between the n areas, an incidence matrix $D \triangleq \{d_{ik}\} \in R^{n\times m}$ is used to represent the power network topology, where $d_{ik} = 1$ if node i is the positive end of edge k, and $d_{ik} = -1$ if node i is the negative end of edge k, otherwise $d_{ik} = 0$. Based on the partition technology of power grid, the individual generator and load buses are not distinguished [4] and the dynamics of area $i (i \in \mathcal{V})$ is described as follows [5].

$$\frac{d}{dt}\delta_i = \omega_i^b \tag{12.1}$$

$$M_i \frac{d}{dt}\omega_i^b = u_i - \sum_{j\in N_i} V_i V_j B_{ij} sin(\delta_i - \delta_j) - A_i(\omega_i^b - \omega^n) - P_i^l \tag{12.2}$$

$$\begin{aligned} \frac{T_{odi}}{X_{di} - X'_{di}} \frac{d}{dt} V_i = & \frac{E_{fi}}{X_{di} - X'_{di}} - \frac{1 - B_{ii}(X_{di} - X'_{di})}{X_{di} - X'_{di}} V_i \\ & + \sum_{j\in N_i} V_j B_{ij} cos(\delta_i - \delta_j) \end{aligned} \tag{12.3}$$

Mentioned that, the state variables δ_i, ω_i^b and V_i represent the voltage angle, frequency and voltage magnitude, the parameters ω^n, M_i, A_i, T_{doi}, X_{di} and X'_{di} denote the nominal frequency, moment of inertia, damping constant, direct axis transient open-circuit constant, direct synchronous reactance and direct synchronous transient reactance, B_{ij} means the susceptance of the transmission line, the power demand P_i^l is the uncontrollable inputs, and the controllable inputs E_{fi} and u_i describe the exciter voltage and controllable power generation respectively.

Remark 12.1 The system model (12.1)–(12.3) extends the classical second-order swing equations by introducing a differential equation (12.3) describing voltage dynamics. The conductance is neglected since the high voltage transmission network is considered in this chapter. Moreover, the exciter voltage E_{fi} is assumed to be constant and does not explicitly include exciter dynamics in order to keep the analysis concise.

The system model (12.1)–(12.3) can be equivalently re-written in the following compact form

$$\frac{d}{dt}\eta = D^T\omega \tag{12.4}$$

$$M\frac{d}{dt}\omega = u - D\Gamma(V)\mathbf{sin}(\eta) - A\omega - P^l \tag{12.5}$$

$$T\frac{d}{dt}V = -E(\eta)V + \overline{E}_{fd} \tag{12.6}$$

Noted that the vectors $\omega = (\omega_1^b - \omega^n, \ldots, \omega_n^b - \omega^n)^T$, $\delta = (\delta_1, \ldots, \delta_n)^T, u = (u_1, \ldots, u_n)^T, V = (V_1, \ldots, V_n)^T, P^l = (P_1^l, \ldots, P_n^l)^T, \overline{E}_{fd} = (\frac{E_{f1}}{X_{d1}-X'_{d1}}, \ldots, \frac{E_{fn}}{X_{dn}-X'_{dn}})^T$, D is the incidence matrix and $\eta = D^T\delta$, $\mathbf{sin}(\eta) = (sin(\eta_1), \ldots, sin(\eta_m))^T$, the matrix $M = diag\{M_1, \ldots, M_n\}$, $A = diag\{A_1, \ldots, A_n\}$ and $T = diag\{\frac{T_{od1}}{X_{d1}-X'_{d1}}, \ldots, \frac{T_{odn}}{X_{dn}-X'_{dn}}\}, \Gamma(V) = diag\{\gamma_1, \ldots, \gamma_m\}$ with $\gamma_k = V_iV_jB_{ij} = V_jV_i B_{ji}$, and the matrix $E(\eta)$ with $E_{ii} = \frac{1-B_{ii}(X_{di}-X'_{di})}{X_{di}-X'_{di}}$ and $E_{ij} = -B_{ij}cos(\eta_k)$ for $i \neq j$ where the index k denotes the edge (i, j). The matrix $E(\eta)$ is positive definite in realistic power grid [5].

12.1.2 Control-Based Communication Network

In order to implement the distributed optimal frequency regulation, the communication network should be deployed to transmit information among the areas in power network. This type of communication network is named as 'control-based communication network' since it is aimed at offering services for distributed optimal frequency regulation. The control-based communication network is represented by an undirected and connected graph $\mathcal{G}^c \triangleq (\mathcal{V}, \mathcal{E}^c)$. The node set $\mathcal{V}$ is defined in Sect. 12.1.1. $\mathcal{E}^c \triangleq \{(j, i)|\, if\ j \xrightarrow{\mathcal{G}^c} i\} \subseteq \mathcal{V} \times \mathcal{V}$ is the edge set where $j \xrightarrow{\mathcal{G}^c} i$ implies that area j can transmit its own information to area i through $\mathcal{G}^c$. The adjacency matrix of $\mathcal{G}^c$ is represented as $C^c \triangleq \{c_{ij}^c\} \in R^{n\times n}$ where $c_{ij}^c = 1$ if and only if $(i, j) \in \mathcal{E}^c$, otherwise $c_{ij}^c = 0$ and it is assumed that $c_{ii}^c = 0$. The neighbouring set of area i in control-based communication network is defined as $N_i^c \triangleq \{j \in \mathcal{V}|(j, i) \in \mathcal{E}^c\}$. Defining the matrix $D^c \triangleq diag\{d_1^c, \ldots, d_n^c\}$ where $d_i^c = \sum_{j\in N_i^c} c_{ij}^c$, the Laplacian matrix of $\mathcal{G}^c$ is given as $L^c = D^c - C^c$. It is known that the matrix L^c is semi-positive definite since $\mathcal{G}^c$ is undirected and connected [6]. Denote the eigenvalues of L^c as $0 = \lambda_1^c < \lambda_2^c \leq \cdots \leq \lambda_n^c$, and label the corresponding linearly independent orthonormal eigenvectors as $\xi_1^c, \ldots, \xi_n^c \in R^n$. Note that the control-based communication network $\mathcal{G}^c$ does not necessarily have to coincide with the power network $\mathcal{G}$.

12.1.3 Control Objective

When the load patterns in the power grid are changed subject to certain disturbances, the frequencies of the areas will deviate from their nominal value and the non-zero frequency deviation will occur. The main purpose of frequency regulation is to restore the non-zero frequency deviation ω in (12.4)–(12.6) to the steady state 0 by adjusting the power generation u. It means that the stability of steady state of system (12.4)–(12.6) should be considered.

For a given constant generation $u = \overline{u}$, the constant steady state $(\overline{\eta}, \overline{\omega}, \overline{V})$ of system (12.4)–(12.6) satisfies (12.7)–(12.9) and Lemma 12.1 [5].

$$0 = D^T\overline{\omega} \tag{12.7}$$

$$0 = \overline{u} - D\Gamma(\overline{V})\mathbf{sin}(\overline{\eta}) - A\overline{\omega} - P^l \tag{12.8}$$

$$0 = -E(\overline{\eta})\overline{V} + \overline{E}_{fd} \tag{12.9}$$

Lemma 12.1 *If there exist* $(\overline{\eta}, \overline{\omega}, \overline{V})$ *such that (12.7)–(12.9) hold, then* $\overline{\omega} = \omega_* \mathbf{1}$ *with*

$$\omega_* = \frac{(\overline{u} - P^l)^T \mathbf{1}}{\mathbf{1}^T A \mathbf{1}} = \frac{\sum_{i\in\mathcal{V}}(\overline{u}_i - P_i^l)}{\sum_{i\in\mathcal{V}} A_i}$$

and

$$\left(I - \frac{A\mathbf{1}\mathbf{1}^T}{\mathbf{1}^T A \mathbf{1}}\right)(\overline{u} - P^l) \in \mathcal{D}$$

where $\mathbf{1} = (1, \ldots, 1)^T$ *is the n dimensional vector, I is the identical matrix,* $\mathcal{D} = \{v \in \mathcal{R}(D) |\ v = D\Gamma(\overline{V})\mathbf{sin}(\overline{\eta}),\ \eta \in \mathcal{R}(D^T)\}$ *and* $\mathcal{R}(D)$ *denotes the range of D.*

It can be seen from Lemma 12.1 that the zero frequency deviation steady state, i.e. $\overline{\omega} = 0$, can be obtained when the total power demand and supply achieve a balance at the steady state, i.e. $(\overline{u} - P^l)^T\mathbf{1} = 0$. It means that the purpose of frequency regulation is to adjust the power generation u such that $(u - P^l)^T\mathbf{1} = 0$ at the steady state. Moreover, different generators have different associated cost functions generally, and there exists purpose in reducing generation cost by coordinating generators in an economically efficient way. As a consequence, the optimal frequency regulation problem is characterized as (12.10)–(12.11) by considering the frequency restoration and economic efficiency purposes simultaneously.

$$min_{\overline{u}} C(\overline{u}) = min_{\overline{u}} \sum_{i\in\mathcal{V}} C_i(\overline{u}_i) \tag{12.10}$$

$$s.t.\quad (\overline{u} - P^l)^T\mathbf{1} = 0 \tag{12.11}$$

where the cost function $C_i(\overline{u}_i) = \frac{1}{2}q_i\overline{u}_i^2$ with $q_i > 0$. Obviously, $C(\overline{u}) = \frac{1}{2}\overline{u}^T Q\overline{u}$ with the positive definite matrix $Q = diag\{q_1, \ldots, q_n\}$.

The following lemma [5] gives the constant generation $\overline{u}$ which is the solution of the optimal frequency regulation problem (12.10)–(12.11).

Lemma 12.2 *There exists a solution $\overline{u}$ for problem (12.10)–(12.11) if and only if*

$$\overline{u} = Q^{-1}\frac{\mathbf{1}\mathbf{1}^T P^l}{\mathbf{1}^T Q^{-1}\mathbf{1}} \tag{12.12}$$

Assumption 12.1 For a given P^l, there exist $\overline{\eta} \in \mathcal{R}(D^T)$, $\overline{V} \in R^n_{>0}$ and $\overline{E}_{fd} \in R^n$ such that $0 = -E(\overline{\eta})\overline{V} + \overline{E}_{fd}$ and $(Q^{-1}\frac{\mathbf{1}\mathbf{1}^T}{\mathbf{1}^T Q^{-1}\mathbf{1}} - I)P^l \in \mathcal{D}$, where $\mathcal{D}$ is defined in Lemma 12.1.

Assumption 12.1 shows that the constant steady state $(\overline{\eta}, \overline{\omega}, \overline{V})$ satisfying (12.7)–(12.9) exists under the constant optimal control (12.12). By an internal model approach, the distributed feedback controller (12.13)–(12.14) is designed by Trip et al. [5], which is able to generate the feedforward input (12.12) under Assumptions 12.1 and 12.2. It shows that the optimal frequency regulation problem (12.10)–(12.11) is solved by (12.13)–(12.14) under ideal information transmission.

$$u_i = \frac{\theta_i}{q_i} \tag{12.13}$$

$$\frac{d}{dt}\theta_i = \sum_{j \in N_i^c} (\theta_j - \theta_i) - \frac{\omega_i}{q_i} \tag{12.14}$$

Assumption 12.2 The steady states $\overline{\eta} \in (-\frac{\pi}{2}, \frac{\pi}{2})^m$ and $\overline{V} \in R^n_{>0}$ satisfy

$$E(\overline{\eta}) - diag(\overline{V})^{-1}|D|\Gamma(\overline{V})diag(\mathbf{sin}(\overline{\eta}))diag(\mathbf{cos}(\overline{\eta}))^{-1}diag(\mathbf{sin}(\overline{\eta}))|D|^T$$
$$diag(\overline{V})^{-1} > 0$$

where the elements in matrix $|D|$ are the absolute values of those in incidence matrix D.

It should be mentioned that, the introducing of communication network makes the power grid vulnerable to cyber-attack. Moreover, the distributed control architecture is more likely to suffer cyber-attack due to the lack of a central authority. The attack launched by adversary may deteriorate the control performance of (12.13)–(12.14) and even cause the failure of optimal frequency regulation, which signifies that the security issue should be a major concern. In next sections, the modality and impact of cyber-attack will be analyzed, and then the secure distributed optimal frequency regulation strategy will be constructed.

12.2 Modality and Impact of False Data Injection Attack

Considering the controlled-based communication network suffering from false data injection attacks, the ideal optimal frequency regulation control (12.13)–(12.14) is affected by an uncertainty injection as follows

$$u_i = \frac{\theta_i}{q_i} \tag{12.15}$$

$$\frac{d}{dt}\theta_i = \sum_{j \in N_i^c} \big((\theta_j - d_j) - (\theta_i - d_i)\big) - \frac{\omega_i}{q_i} \tag{12.16}$$

where the variable d_i denotes the uncertain injection caused by cyber-attack at area i. Denoting $u = (u_1, \ldots, u_n)^T$, $\theta = (\theta_1, \ldots, \theta_n)^T$ and $d = \{d_1, \ldots, d_n\}$, (12.15)–(12.16) can be re-written as

$$u = Q^{-1}\theta \tag{12.17}$$

$$\frac{d}{dt}\theta = -L^c\theta + L^c d - Q^{-1}\omega \tag{12.18}$$

The dynamic of injection d is constructed by adversary as follows

$$\frac{d}{dt}d = F_a d + B_a \theta \tag{12.19}$$

where the matrix F_a and B_a are chosen by adversary arbitrarily to destroy the frequency restoration and economic efficiency. Note that F_a should be chosen as a Hurwitz matrix.

Theorem 12.1 *Supposing the matrix $F_a = -\lambda_a I$ and $B_a = I$, the states of the closed-loop system constructed by (12.4)–(12.6) and (12.17)–(12.19) approach to infinite for all $\lambda_a \in (0, 1)$, which demonstrates the failure of optimal frequency regulation.*

Proof Equations (12.18) and (12.19) can be written as follows

$$\frac{d}{dt}\begin{pmatrix}\theta \\ d\end{pmatrix} = \begin{pmatrix}-L^c & L^c \\ I & -\lambda_a I\end{pmatrix}\begin{pmatrix}\theta \\ d\end{pmatrix} - \begin{pmatrix}Q^{-1} \\ \mathbf{0}\end{pmatrix}\omega$$

Denoting the matrix $A \triangleq \begin{pmatrix}-L^c & L^c \\ I & -\lambda_a I\end{pmatrix}$, the characteristic equation of A is

$$det\big(s^2 I + (\lambda_a I + L^c)s + (\lambda_a - 1)L^c\big) = 0 \tag{12.20}$$

where $det(\cdot)$ denotes the determinant of matrix. Equation (12.20) indicates that the eigenvalues of matrix A is positive for $\lambda_a \in (0, 1)$ [7]. This implies that θ converges to infinite, which means that the closed-loop system is disable and the optimal frequency regulation fails.

Theorem 12.1 shows that the conventional regulation under cyber-attack (12.17)–(12.18) can not accomplish the optimal frequency regulation purpose of power grid, which means that the distributed optimal frequency regulation designed under ideal information transmission (12.13)–(12.14) is not applicable in the case of the considered cyber-attack. It should be mentioned that, there are numerous choices of F_a and B_a that would render the failure of optimal frequency regulation except the choice in Theorem 12.1. Therefore, it is necessary to study the secure distributed optimal frequency regulation strategy which is robust against the uncertain dynamic injections caused by cyber-attack, and this will be elaborated in next section.

12.3 Secure Distributed Optimal Frequency Regulation

A novel type of communication network which is named as 'secure-based communication network' is introduced to ensure the construction of secure distributed optimal frequency regulation under cyber-attack. The secure-based communication network, which connects all the areas in power grid, is described by an undirected connected graph $\mathcal{G}^s \triangleq (\mathcal{V}, \mathcal{E}^s)$. The node set $\mathcal{V}$ is the same as that in Sect. 12.1.1 and $\mathcal{E}^s \triangleq \{(j,i)|\ if\ j \xrightarrow{\mathcal{G}^s} i\} \subseteq \mathcal{V} \times \mathcal{V}$ is the edge set where $j \xrightarrow{\mathcal{G}^s} i$ implies that area j can transmit its own information to area i through $\mathcal{G}^s$. Define the neighboring set $N_i^s \triangleq \{j \in \mathcal{V}|(j,i) \in \mathcal{E}^s\}$ of area i in secure-based communication network. The semi-positive Laplacian matrix of graph $\mathcal{G}^s$ is marked as L^s. Denote the eigenvalues of L^s as $0 = \lambda_1^s < \lambda_2^s \leq \cdots \leq \lambda_n^s$, and label the corresponding orthonormal linearly independent eigenvectors as $\xi_1^s, \ldots, \xi_n^s \in R^n$. It should be mentioned that, the edges in the secure-based communication network $\mathcal{G}^s$ does not necessarily have to coincide with those in the control-based communication network $\mathcal{G}^c$ and power network $\mathcal{G}$. The three types of networks are illustrated in Fig. 12.1.

The secure distributed optimal frequency regulation as shown in (12.21)–(12.23) is proposed to resist the influences caused by the considered cyber-attack.

$$u_i = \frac{\theta_i}{q_i} \tag{12.21}$$

$$\frac{d}{dt}\theta_i = \sum_{j \in N_i^c} (\theta_j - \theta_i) - \beta \sum_{j \in N_i^s} (z_j - z_i) - \frac{\omega_i}{q_i} \tag{12.22}$$

$$\frac{d}{dt}z_i = \sum_{j \in N_i^s} (z_j - z_i) + \beta \sum_{j \in N_i^s} (\theta_j - \theta_i) \tag{12.23}$$

where the additional variable z_i is introduced into the regulation and β is a positive constant.

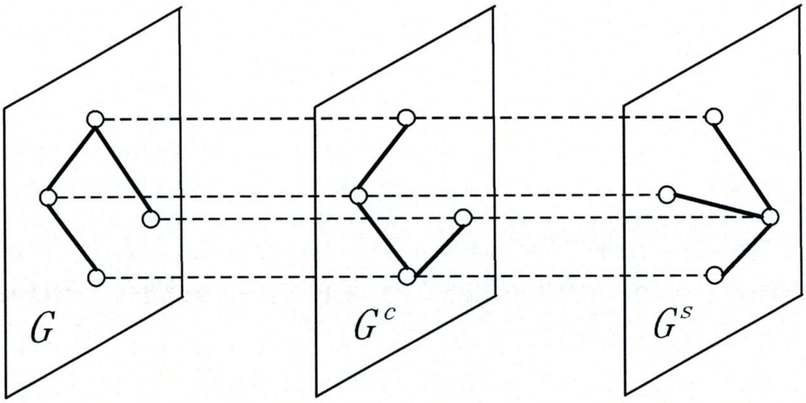

Fig. 12.1 An example of the relationship of power network $\mathcal{G}$, control-based communication network $\mathcal{G}^c$ and secure-based communication network $\mathcal{G}^s$. The nodes are the same in the three types of networks, but the edges can be different

Remark 12.2 Compared with the conventional regulation (12.13)–(12.14), the secure one introduces the additional variable z_i whose dynamic is represented in (12.23) to resist the cyber-attack. It can be seen from (12.21)–(12.23) that the variables θ_i and z_i is transmitted among the areas through secure-based communication network in order to implement the secure regulation, although it may lead to the additional expense of a redundant communication network. The redundancy means that the other communication network $\mathcal{G}^s$, whose topology is undirected and connected, is needed except the conventional control-based communication network $\mathcal{G}^c$. From the perspective of practice, the redundancy implies that the additional information flows are required, and this can be implemented by using wireless or Internet. The redundant network increases the resilience of power grid in the information layer while thwarting cyber-attack.

Mentioned that the secure distributed optimal frequency regulation (12.21)–(12.23) is in the case of non-attack. We will prove that the optimal frequency regulation problem (12.10)–(12.11) can be addressed by the regulation (12.21)–(12.23) in the absence of cyber-attack.

Denoting $z = (z_1, \ldots, z_n)^T$, re-write (12.21)–(12.23) in the following compact form

$$u = Q^{-1}\theta \tag{12.24}$$

$$\frac{d}{dt}\theta = -L^c\theta + \beta L^s z - Q^{-1}\omega \tag{12.25}$$

$$\frac{d}{dt}z = -L^s z - \beta L^s \theta \tag{12.26}$$

The closed-loop system under the regulation (12.21)–(12.23) is given by (12.4)–(12.6) and (12.24)–(12.26). For the constant optimal generation $\overline{u}$ given in (12.12), the steady state $(\overline{\eta}, \overline{\omega}, \overline{V}, \overline{\theta}, \overline{z})$ of this closed-loop system exists under Assumption 12.1, where $\overline{\omega} = 0$, $\overline{\theta} = \frac{\mathbf{1}\mathbf{1}^T P^l}{\mathbf{1}^T Q^{-1}\mathbf{1}}$, $\overline{\eta}$, $\overline{V}$ and $\overline{z} = a\mathbf{1}$ are constant for some constant $a \in R$. Moreover, the steady state satisfies (12.7)–(12.9) and the following (12.27)–(12.28).

$$\frac{d}{dt}\overline{\theta} = -L^c\overline{\theta} + \beta L^s\overline{z} \tag{12.27}$$

$$\frac{d}{dt}\overline{z} = -L^s\overline{z} - \beta L^s\overline{\theta} \tag{12.28}$$

Theorem 12.2 *Under Assumptions 12.1 and 12.2, the secure distributed optimal frequency regulation (12.21)–(12.23) (or (12.24)–(12.26) equivalently) guarantees the solution of the closed-loop system (12.4)–(12.6) and (12.24)–(12.26), which start in a neighborhood of the steady state* $(\overline{\eta}, \overline{\omega}, \overline{V}, \overline{\theta}, \overline{z})$, *to converge asymptotically to the new steady state* $(\widetilde{\eta}, \overline{\omega}, \widetilde{V}, \overline{\theta}, \widetilde{z})$ *where* $E(\widetilde{\eta})\widetilde{V} - \overline{E}_{fd} = 0$ *for constant* $\widetilde{V}$ *and* $\widetilde{\eta}$, *and* $\widetilde{z} = b\mathbf{1}$ *with some constant* $b \in R$.

Proof Defining the following functions

$$W_1(\omega) \triangleq \frac{1}{2}(\omega - \overline{\omega})^T M(\omega - \overline{\omega}) \tag{12.29}$$

and

$$W_2(\eta, V) \triangleq -\mathbf{1}^T \Gamma(V)\mathbf{cos}(\eta) + \mathbf{1}^T \Gamma(\overline{V})\mathbf{cos}(\overline{\eta}) - \left(\Gamma(\overline{V})\mathbf{sin}(\overline{\eta})\right)^T(\eta - \overline{\eta})$$
$$-\overline{E}_{fd}(V - \overline{V}) + \frac{1}{2}V^T F V - \frac{1}{2}\overline{V}^T F\overline{V} \tag{12.30}$$

where F is a diagonal matrix with $F_{ii} = \frac{1 - B_{ii}(X_{di} - X'_{di})}{X_{di} - X'_{di}}$, the Lyapunov function is constructed as

$$U(\eta, \omega, V, \theta, z) = W_1 + W_2 + \frac{1}{2}(\theta - \overline{\theta})^T(\theta - \overline{\theta}) + \frac{1}{2}(z - \overline{z})^T(z - \overline{z})$$

Mentioned that the function W_2 has a strict local minimum at $(\overline{\eta}, \overline{V})$ under Assumption 12.2 [5]. This implies that the Lyapunov function U has the local minimum point $(\overline{\eta}, \overline{\omega}, \overline{V}, \overline{\theta}, \overline{z})$, which is the steady state.

The derivation of W_1 with respect to t along the solution of closed-loop system is

$$\begin{aligned}\frac{d}{dt}W_1 &= (\omega-\overline{\omega})^T M\frac{d}{dt}\omega\\ &= (\omega-\overline{\omega})^T(u - D\Gamma(V)\mathbf{sin}(\eta) - A\omega - P^l)\\ &= (\omega-\overline{\omega})^T\big(u - D\Gamma(V)\mathbf{sin}(\eta) - A\omega - \overline{u} + D\Gamma(\overline{V})\mathbf{sin}(\overline{\eta}) + A\overline{\omega}\big)\end{aligned} \tag{12.31}$$

where (12.5) and (12.8) are used here, and the derivation of W_2 is

$$\begin{aligned}\frac{d}{dt}W_2 &= \left(\frac{\partial W_2}{\partial\eta}\right)^T\frac{d\eta}{dt} + \left(\frac{\partial W_2}{\partial V}\right)^T\frac{dV}{dt}\\ &= \omega^T D\big(\Gamma(V)\mathbf{sin}(\eta) - \Gamma(\overline{V})\mathbf{sin}(\overline{\eta})\big) - \|\nabla_V W_2\|^2_{T^{-1}}\end{aligned} \tag{12.32}$$

where (12.4) and (12.6) are used, and the notation $\|\nabla_V W_2\|_{T^{-1}} \triangleq (\frac{\partial W_2}{\partial V})^T T^{-1}(\frac{\partial W_2}{\partial V}) = \big(E(\eta)V - \overline{E}_{fd}\big)^T T^{-1}\big(E(\eta)V - \overline{E}_{fd}\big)$.

According to (12.25)–(12.28), we have

$$\frac{d}{dt}(\theta-\overline{\theta}) = -L^c(\theta-\overline{\theta}) + \beta L^s(z-\overline{z}) - Q^{-1}\omega \tag{12.33}$$

and

$$\frac{d}{dt}(z-\overline{z}) = -L^s(z-\overline{z}) - \beta L^s(\theta-\overline{\theta}) \tag{12.34}$$

Based on (12.31)–(12.34), the derivation of U with respect to t satisfies

$$\begin{aligned}\frac{d}{dt}U =& -(\omega-\overline{\omega})^T A(\omega-\overline{\omega}) - \|\nabla_V W_2\|^2_{T^{-1}} + (\omega-\overline{\omega})^T(u-\overline{u})\\ &-(\theta-\overline{\theta})^T L^c(\theta-\overline{\theta}) - (\theta-\overline{\theta})^T Q^{-1}\omega - (z-\overline{z})^T L^s(z-\overline{z})\\ =& -(\omega-\overline{\omega})^T A(\omega-\overline{\omega}) - \|\nabla_V W_2\|^2_{T^{-1}} - (\theta-\overline{\theta})^T L^c(\theta-\overline{\theta})\\ &-(z-\overline{z})^T L^s(z-\overline{z})\end{aligned} \tag{12.35}$$

where the last equation can be obtained since $u = Q^{-1}\theta$ and $\overline{\omega} = 0$.

According to (12.35) and Lasalle's invariance principle, the solution of closed-loop system converges into the largest invariant set contained in the following set

$$S \triangleq \{(\eta,\omega,V,\theta,z)|\omega = \overline{\omega} = 0, \|\nabla_V W_2\|_{T^{-1}} = 0, \theta = \overline{\theta} + c_1\mathbf{1}, z = \overline{z} + c_2\mathbf{1}\}$$

where c_1 and c_2 are some constants.

Equations (12.5) and (12.8) yield that $M\frac{d}{dt}\omega = u - \overline{u} - A(\omega-\overline{\omega}) - D\big(\Gamma(V)$ $\mathbf{sin}(\eta) - \Gamma(\overline{V})\mathbf{sin}(\overline{\eta})\big)$, and thus $0 = c_1 Q^{-1}\mathbf{1} - D\big(\Gamma(V)\mathbf{sin}(\eta) - \Gamma(\overline{V})\mathbf{sin}(\overline{\eta})\big)$ in S since $u = Q^{-1}\theta$. Multiplying both sides of the above equation by $\mathbf{1}^T$ yields $0 = c_1\mathbf{1}^T Q^{-1}\mathbf{1}$ since $\mathbf{1}^T D = 0$, and then it is obtained that $c_1 = 0$. Moreover,

$\|\nabla_V W_2\|_{T^{-1}} = 0$ implies $E(\eta)V - \overline{E}_{fd} = 0$. It indicates that V is constant in set S since $\frac{d}{dt}V = 0$ according to (12.6) and the constant V is denoted as $\widetilde{V}$. This further demonstrates that η is constant in set S and is denoted as $\widetilde{\eta}$. It can be seen that the constant $b = a + c_2$. This concludes the proof.

Remark 12.3 The dynamical scenario of optimal frequency regulation for power grid can be depicted as follows. The power grid is operated at an acceptable steady state $(\overline{\eta}, \overline{\omega}, \overline{V}, \overline{\theta^0}, \overline{z})$ initially for the initial power demand $P^{l,0}$. Mentioned that we have $(Q^{-1}\overline{\theta^0} - P^{l,0})^T\mathbf{1} = 0$ since the balance between total power demand and supply should be maintained at the steady state. When the power demand changes from $P^{l,0}$ to P^l due to certain disturbances of load patterns, the initial power flow is mismatched and can not satisfy the optimal frequency regulation requirement since $(Q^{-1}\overline{\theta^0} - P^l)^T\mathbf{1} \neq 0$ usually. Suddenly, the regulation actions (12.21)–(12.23) will govern the closed-loop system trajectory to another new steady state $(\widetilde{\eta}, \overline{\omega}, \widetilde{V}, \overline{\theta}, \widetilde{z})$ under the new power demand P^l, and the new steady state is also acceptable for the power grid stable operation [5, 8]. Mentioned that the action (12.21)–(12.23) addressed the optimal frequency regulation problem (12.10)–(12.11) since the power generation u achieves $Q^{-1}\frac{\mathbf{1}\mathbf{1}^T P^l}{\mathbf{1}^T Q^{-1}\mathbf{1}}$, which is the optimal solution of (12.10)–(12.11) for the new power demand P^l as illustrated in Lemma 12.2, at the new steady state $(\widetilde{\eta}, \overline{\omega}, \widetilde{V}, \overline{\theta}, \widetilde{z})$.

Remark 12.4 The solution of closed-loop system (12.4)–(12.6) and (12.24)–(12.26) converges to the new steady state $(\widetilde{\eta}, \overline{\omega}, \widetilde{V}, \overline{\theta}, \widetilde{z})$ instead of the considered one $(\overline{\eta}, \overline{\omega}, \overline{V}, \overline{\theta}, \overline{z})$ as shown in Theorem 12.2. It should be mentioned that both of $(\widetilde{\eta}, \overline{\omega}, \widetilde{V}, \overline{\theta}, \widetilde{z})$ and $(\overline{\eta}, \overline{\omega}, \overline{V}, \overline{\theta}, \overline{z})$ can address the optimal frequency regulation problem (12.10)–(12.11) since the same power generation $\overline{u} = Q^{-1}\overline{\theta}$, which is the optimal solution of problem (12.10)–(12.11), is achieved under these two types of steady state. Based on the dynamical scenario explained in Remark 12.3, the initial value of the solution is $(\overline{\eta}, \overline{\omega}, \overline{V}, \overline{\theta^0}, \overline{z})$ when the power demand changes from $P^{l,0}$ to P^l, where $\mathbf{1}^T(Q^{-1}\overline{\theta^0} - P^{l,0}) = 0$ and $\overline{\theta^0} = c^0\mathbf{1}$ for some constant $c^0 \in R$. On the other hand, we have $\mathbf{1}^T(Q^{-1}\overline{\theta} - P^l) = 0$ and $\overline{\theta} = \overline{c}\mathbf{1}$ for some constant $\overline{c} \in R$. This implies that $\mathbf{1}^T Q^{-1}(\overline{\theta} - \overline{\theta^0}) = \mathbf{1}^T(P^l - P^{l,0})$, and thus we get $\overline{c} - c^0 = \frac{\mathbf{1}^T(P^l - P^{l,0})}{\mathbf{1}^T Q^{-1}\mathbf{1}}$. Based on the Lasalle's invariance principle and the fact that $\frac{d}{dt}U \leq 0$ shown in the proof of Theorem 12.2, the new steady state $(\widetilde{\eta}, \overline{\omega}, \widetilde{V}, \overline{\theta}, \widetilde{z})$ belongs to the compact Lyapunov level set $\Gamma = \{(\eta, \omega, V, \theta, z)|U(\eta, \omega, V, \theta, z) \leq U(\overline{\eta}, \overline{\omega}, \overline{V}, \overline{\theta^0}, \overline{z})\}$. Mentioned that $U(\overline{\eta}, \overline{\omega}, \overline{V}, \overline{\theta^0}, \overline{z}) = \frac{1}{2}(\overline{\theta} - \overline{\theta^0})^T(\overline{\theta} - \overline{\theta^0}) = \frac{1}{2}(\overline{c} - c^0)^2\mathbf{1}^T\mathbf{1} = \frac{n}{2}(\overline{c} - c^0)^2 = \frac{n}{2}(\frac{\mathbf{1}^T(P^l - P^{l,0})}{\mathbf{1}^T Q^{-1}\mathbf{1}})^2$. It demonstrates that $U(\widetilde{\eta}, \overline{\omega}, \widetilde{V}, \overline{\theta}, \widetilde{z}) \leq \frac{n}{2}(\frac{\mathbf{1}^T(P^l - P^{l,0})}{\mathbf{1}^T Q^{-1}\mathbf{1}})^2$, which further implies

$$W_2(\widetilde{\eta}, \widetilde{V}) + \frac{1}{2}(\widetilde{z} - \overline{z})^T(\widetilde{z} - \overline{z}) \leq \frac{n}{2}\left(\frac{\mathbf{1}^T(P^l - P^{l,0})}{\mathbf{1}^T Q^{-1}\mathbf{1}}\right)^2 \tag{12.36}$$

according to the definition of function U. It should be seen that, formula (12.36) exhibits the distance relation between the steady states $(\widetilde{\eta}, \overline{\omega}, \widetilde{V}, \overline{\theta}, \widetilde{z})$ and $(\overline{\eta},$

$\overline{\omega}, \overline{V}, \overline{\theta}, \overline{z}$) since the function $W_2(\eta, V)$ reaches the strict minimum at $(\overline{\eta}, \overline{V})$. Moreover, formula (12.36) shows that the distances about voltage magnitude and power flow are dependent on the change of total power demand $\mathbf{1}^T(P^l - P^{l,0})$, which is convincing in practical power grid regulation.

Theorem 12.2 demonstrates that the secure distributed frequency regulation (12.21)–(12.23) can accomplish the optimal frequency regulation problem in the case of non-attack. When the considered cyber-attack carries out over the control-based communication network as illustrated in Sect. 12.2, the uncertain injection $d_i (i = 1, \dots, n)$ launched by the adversary affects the secure regulation as follows.

$$u_i = \frac{\theta_i}{q_i} \tag{12.37}$$

$$\frac{d}{dt}\theta_i = \sum_{j \in N_i^c}((\theta_j - d_j) - (\theta_i - d_i)) - \beta \sum_{j \in N_i^s}(z_j - z_i) - \frac{\omega_i}{q_i} \tag{12.38}$$

$$\frac{d}{dt}z_i = \sum_{j \in N_i^s}(z_j - z_i) + \beta \sum_{j \in N_i^s}(\theta_j - \theta_i) \tag{12.39}$$

which can be written as

$$u = Q^{-1}\theta \tag{12.40}$$

$$\frac{d}{dt}\theta = -L^c\theta + L^c d + \beta L^s z - Q^{-1}\omega \tag{12.41}$$

$$\frac{d}{dt}z = -L^s z - \beta L^s \theta \tag{12.42}$$

where the dynamic of injection d is given in (12.19).

The following theorem interprets the effectiveness of secure distributed frequency regulation (12.37)–(12.39) against cyber-attack.

Theorem 12.3 *Under Assumptions 12.1 and 12.2, the secure distributed optimal frequency regulation (12.37)–(12.39) (or (12.40)–(12.42) equivalently) with sufficiently large β guarantees the solution of the closed-loop system (12.4)–(12.6) and (12.40)–(12.42), which start in a neighborhood of the power grid steady state $(\overline{\eta}, \overline{\omega}, \overline{V}, \overline{\theta}, \overline{z})$, to converge asymptotically into an arbitrary small neighborhood of the new steady state $(\widetilde{\eta}, \overline{\omega}, \widetilde{V}, \overline{\theta}, \widetilde{z})$ where $E(\widetilde{\eta})\widetilde{V} - \overline{E}_{fd} = 0$ for constant $\widetilde{V}$ and $\widetilde{\eta}$, and $\widetilde{z} = b\mathbf{1}$ with some constant $b \in R$.*

Proof Construct the function $\widetilde{W}_1(\omega) = \beta W_1$ and $\widetilde{W}_2(\eta, V) = \beta W_2$ where W_1 and W_1 are defined as (12.29) and (12.30) respectively. Define the following function

$$\widetilde{W}_3(\theta, z, d) = \frac{\beta}{2}(\theta - \overline{\theta})^T(\theta - \overline{\theta}) + \frac{1}{2}\begin{pmatrix} z - \overline{z} \\ d \end{pmatrix}^T \begin{pmatrix} \beta I & I \\ I & P_a \end{pmatrix} \begin{pmatrix} z - \overline{z} \\ d \end{pmatrix}$$

where the positive definite matrix P_a is selected such that $F_a^T P_a + P_a F_a = -Q_a$ for positive definite matrix Q_a. Mentioned that $\widetilde{W}_3$ has the minima $(\overline{\theta}, \overline{z}, 0)$ when $\beta > \frac{1}{\lambda_{min}(P_a)}$ where $\lambda_{min}(P_a)$ denotes the minimum eigenvalue of matrix P_a.

Consider the Lyapunov function of the closed-loop system as

$$\widetilde{U}(\eta, \omega, V, \theta, z, d) = \widetilde{W}_1(\omega) + \widetilde{W}_2(\eta, V) + \widetilde{W}_3(\theta, z, d)$$

According to the above analysis, the function $\widetilde{U}$ has the local minimum point $(\overline{\eta}, \overline{\omega}, \overline{V}, \overline{\theta}, \overline{z}, 0)$ under Assumption 12.2.

Similar with the proof of Theorem 12.2, we have

$$\frac{d}{dt}(\widetilde{W}_1 + \widetilde{W}_2) = -\beta(\omega - \overline{\omega})^T A(\omega - \overline{\omega}) - \beta\|\nabla_V W_2\|^2_{T^{-1}} + \beta(\omega - \overline{\omega})^T (u - \overline{u}) \tag{12.43}$$

Based on (12.41) and (12.42), it yields

$$\frac{d}{dt}(\theta - \overline{\theta}) = -L^c(\theta - \overline{\theta}) + L^c d + \beta L^s(z - \overline{z}) - Q^{-1}\omega \tag{12.44}$$

$$\frac{d}{dt}(z - \overline{z}) = -L^s(z - \overline{z}) - \beta L^s(\theta - \overline{\theta}) \tag{12.45}$$

The derivation of $\widetilde{W}_3$ with respect to t along the solution of closed-loop system satisfies

$$\frac{d}{dt}\widetilde{W}_3 = \beta(\theta - \overline{\theta})^T \frac{d}{dt}(\theta - \overline{\theta}) + \begin{pmatrix} z - \overline{z} \\ d \end{pmatrix}^T \begin{pmatrix} \beta I & I \\ I & P_a \end{pmatrix} \frac{d}{dt} \begin{pmatrix} z - \overline{z} \\ d \end{pmatrix} \tag{12.46}$$

Combining (12.44)–(12.46) yields

$$\begin{aligned} \frac{d}{dt}\widetilde{W}_3 = & -\beta(\theta - \overline{\theta})^T L^c(\theta - \overline{\theta}) - \beta(z - \overline{z})^T L^s(z - \overline{z}) - \frac{1}{2}d^T Q_a d \\ & + (z - \overline{z})^T (F_a - L^s)d + d^T P_a B_a \theta - \beta(\omega - \overline{\omega})^T (u - \overline{u}) \\ & + (z - \overline{z})^T B_a \theta \end{aligned} \tag{12.47}$$

According to (12.43) and (12.47), we can get

$$\begin{aligned} \frac{d}{dt}\widetilde{U} = & -\beta(\omega - \overline{\omega})^T A(\omega - \overline{\omega}) - \beta\|\nabla_V W_2\|^2_{T^{-1}} - \beta(\theta - \overline{\theta})^T L^c(\theta - \overline{\theta}) \\ & - \beta(z - \overline{z})^T L^s(z - \overline{z}) - \frac{1}{2}d^T Q_a d \\ & + (z - \overline{z})^T (F_a - L^s)d + d^T P_a B_a \theta + (z - \overline{z})^T B_a \theta \end{aligned}$$

Defining the orthogonal matrix $T^c \triangleq (\xi_1^c, \ldots, \xi_n^c)^T$ and $T^s \triangleq (\xi_1^s, \ldots, \xi_n^s)^T$ where the orthonormal eigenvectors ξ_i^c and ξ_i^s are defined in Sects. 12.1.2 and 12.3 before, it can be obtained that $T^c L^c (T^c)^T = \Lambda^c \triangleq diag\{0, \lambda_2^c, \ldots, \lambda_n^c\}$ and $T^s L^s (T^s)^T = \Lambda^s \triangleq diag\{0, \lambda_2^s, \ldots, \lambda_n^s\}$. Denoting $\widetilde{\theta} \triangleq T^c(\theta - \overline{\theta})$, $\widetilde{z} \triangleq T^s(z - \overline{z})$ and $W = (0, I_{n-1})^T \in R^{n\times(n-1)}$, we have

$$\begin{aligned}\frac{d}{dt}\widetilde{U} =& -\beta(\omega-\overline{\omega})^T A(\omega-\overline{\omega}) - \beta\|\nabla_V W_2\|_{T^{-1}}^2 - \beta\widetilde{\theta}^T\Lambda^c\widetilde{\theta} - \beta\widetilde{z}^T\Lambda^s\widetilde{z}\\&-\frac{1}{2}d^T Q_a d + \widetilde{z}^T T^s(F_a - L^s)d + d^T P_a B_a (T^c)^T\widetilde{\theta} + d^T P_a B_a\overline{\theta}\\&+\widetilde{z}^T T^s B_a (T^c)^T\widetilde{\theta} + \widetilde{z}^T T^s B_a\overline{\theta}\\=& -\beta(\omega-\overline{\omega})^T A(\omega-\overline{\omega}) - \beta\|\nabla_V W_2\|_{T^{-1}}^2 - \beta\widetilde{\theta}_*^T\Lambda_*^c\widetilde{\theta}_* - \beta\widetilde{z}_*^T\Lambda_*^s\widetilde{z}_*\\&-\frac{1}{2}d^T Q_a d + \widetilde{z}_*^T W^T T^s(F_a - L^s)d + d^T P_a B_a (T^c)^T W\widetilde{\theta}_* + d^T P_a B_a\overline{\theta}\\&+\widetilde{z}_*^T W^T T^s B_a (T^c)^T W\widetilde{\theta}_* + \widetilde{z}_*^T W^T T^s B_a\overline{\theta}\end{aligned}\tag{12.48}$$

Note that $\Lambda_*^c = diag\{\lambda_2^c, \ldots, \lambda_n^c\}$ and $\Lambda_*^s = diag\{\lambda_2^s, \ldots, \lambda_n^s\}$ are positive definite matrix, and $\widetilde{\theta}_*, \widetilde{z}_* \in R^{n-1}$ are the vectors which are composed of the last $(n-1)$ components of vectors $\widetilde{\theta}$ and $\widetilde{z}$ respectively.

Since the matrix $W, L^s, T^s, T^c, F_a, P_a$ and B_a are all constant, by the help of Young's inequality, it can be get from (12.48) that

$$\begin{aligned}\frac{d}{dt}\widetilde{U} \le& -\beta(\omega-\overline{\omega})^T A(\omega-\overline{\omega}) - \beta\|\nabla_V W_2\|_{T^{-1}}^2 - \beta\lambda_2^c \parallel \widetilde{\theta}_* \parallel^2 - \beta\lambda_2^s \parallel \widetilde{z}_* \parallel^2\\&-\frac{1}{2}\lambda_{Q,min} \parallel d \parallel^2 + \frac{1}{2}(a_1 \parallel W^T T^s(F_a - L^s) \parallel^2 + a_3 \parallel W^T T^s B_a \parallel^2\\&+a_2 \parallel W^T T^s B_a (T^c)^T W \parallel^2) \parallel \widetilde{z}_* \parallel^2 + \frac{1}{2}(\frac{1}{a_1} + \frac{1}{a_4} + \frac{1}{a_5}) \parallel d \parallel^2\\&+\frac{1}{2}(\frac{1}{a_2} + a_4 \parallel P_a B_a (T^c)^T W \parallel^2) \parallel \widetilde{\theta}_* \parallel^2 + \frac{a_5}{2} \parallel P_a B_a\overline{\theta} \parallel^2 + \frac{\parallel \overline{\theta} \parallel^2}{2a_3}\end{aligned}\tag{12.49}$$

where $\lambda_{Q,min}$ is the minimum eigenvalue of Q, the positive constants a_1, a_2, a_3, a_4 and a_5 are selected to satisfy $\frac{1}{a_1} + \frac{1}{a_4} + \frac{1}{a_5} < \lambda_{Q,min}$.

Equation (12.49) indicates that the solution of closed-loop system converges into the following set $\widetilde{S}$ by selecting sufficiently large β,

$$\begin{aligned}\widetilde{S} \triangleq& \{(\eta, \omega, V, \theta, z, d) | \beta(\omega-\overline{\omega})^T A(\omega-\overline{\omega}) + \beta\|\nabla_V W_2\|_{T^{-1}}^2\\&+(\beta\lambda_2^c - \beta_1) \parallel \widetilde{\theta}_* \parallel^2 + (\beta\lambda_2^s - \beta_2) \parallel \widetilde{z}_* \parallel^2 + \gamma \parallel d \parallel^2 \le \epsilon\}\end{aligned}$$

where the positive constants $\beta_1 = \frac{1}{2}(\frac{1}{a_2} + a_4 \parallel P_a B_a (T^c)^T W \parallel^2)$, $\beta_2 = \frac{1}{2}(a_1 \parallel W^T T^s (F_a - L^s) \parallel^2 + a_3 \parallel W^T T^s B_a \parallel^2 + a_2 \parallel W^T T^s B_a (T^c)^T W \parallel^2)$, $\gamma = \frac{1}{2}(\lambda_{Q,min} - \frac{1}{a_1}$

$-\frac{1}{a_4}-\frac{1}{a_5})$ and $\epsilon = \frac{a_5}{2} \parallel P_a B_a \overline{\theta} \parallel^2 + \frac{\|\overline{\theta}\|^2}{2a_3}$. The parameter β can be selected sufficiently large such that $\beta\lambda_2^c - \beta_1 > 0$ and $\beta\lambda_2^s - \beta_2 > 0$.

Furthermore, we have the following formula since $\parallel \widetilde{\theta}_* \parallel = \parallel (T^c)^{-1} \parallel \cdot \parallel \theta - \overline{\theta} \parallel = \parallel T^c \parallel \cdot \parallel \theta - \overline{\theta} \parallel$ and $\parallel \widetilde{z}_* \parallel = \parallel (T^s)^{-1} \parallel \cdot \parallel z - \overline{z} \parallel = \parallel T^s \parallel \cdot \parallel z - \overline{z} \parallel$.

$$\widetilde{S} \subseteq \left\{ (\eta, \omega, V, \theta, z, d) | \parallel \omega - \overline{\omega} \parallel^2 \leq \frac{\epsilon}{\beta\lambda_{A,min}}, \|\nabla_V W_2\|^2_{T^{-1}} \leq \frac{\epsilon}{\beta}, \right.$$
$$\left. \parallel \theta - \overline{\theta} \parallel^2 \leq \frac{\epsilon}{\parallel T^c \parallel^2 (\beta\lambda_2^c - \beta_1)}, \parallel z - \overline{z} \parallel^2 \leq \frac{\epsilon}{\parallel T^s \parallel^2 (\beta\lambda_2^s - \beta_2)}, \parallel d \parallel^2 \leq \frac{\epsilon}{\gamma} \right\}$$

where $\lambda_{A,min}$ is the minimum eigenvalue of matrix A. By the help of the proof in Theorem 12.2, the trajectory of $(\eta, \omega, V, \theta, z)$ in set $\widetilde{S}$ is in an arbitrary small neighborhood of the new steady state $(\widetilde{\eta}, \overline{\omega}, \widetilde{V}, \overline{\theta}, \widetilde{z})$ which is demonstrated in Theorem 12.2 for sufficiently large β. This concludes the proof.

Remark 12.5 Theorem 12.3 indicates that the secure regulation (12.37)–(12.39) drives the system state of power grid into an arbitrary small neighborhood of the new steady state $(\widetilde{\eta}, \overline{\omega}, \widetilde{V}, \overline{\theta}, \widetilde{z})$, under the considered cyber-attack launched by adversary. Mentioned that the states of power grid, such as frequency, voltage magnitude, power generation and so on, are not necessary to asymptotically converge to the steady state strictly in application. Instead, these states are allowed to fluctuate in a certain range around the steady state in practical power grid. This certifies that the uniform boundedness result represented in Theorem 12.3 can still fulfill the optimal frequency regulation requirement of power grid.

12.4 Simulation Example

A test power grid containing four areas, which is equivalent to IEEE New England 39-bus system or the South Eastern Australian 59-bus system [9], is used to verify the effectiveness of the secure distributed optimal frequency regulation. The network topology of test system is given in Fig. 12.2. The susceptance of transmission lines is given as $B_{14} = 21.0$, $B_{12} = 25.6$, $B_{23} = 33.1$ and $B_{34} = 16.6$, and the other parameters are shown in Table 12.1. The power demand changed from $P^l = (2.00, 1.00, 1.50, 1.00)^T$ to $P^l = (2.20, 1.05, 1.55, 1.10)^T$ at 20 s.

12.4.1 *Response Under Attack-Free Case*

The first scenario is that there is no cyber-attack in power grid and the simulation result is shown in Figs. 12.3 and 12.4. Figure 12.3 illustrates the effectiveness of the conventional distributed optimal frequency regulation (12.13)–(12.14), while Fig. 12.4 shows that of the secure distributed optimal frequency regulation (12.21)–

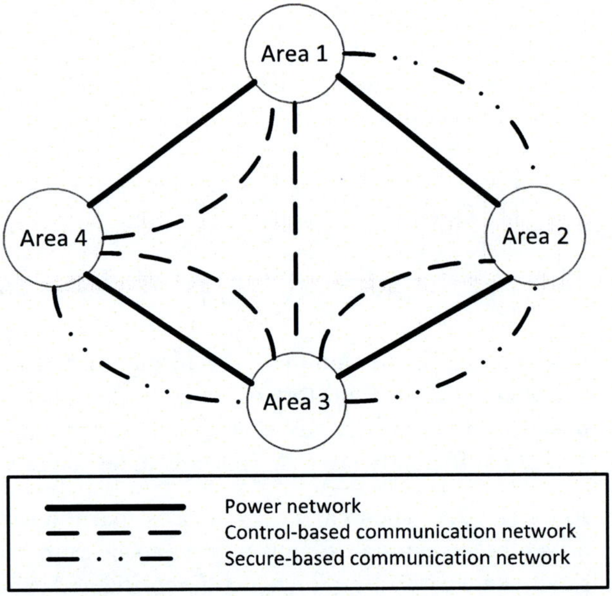

Fig. 12.2 The test power grid

Table 12.1 The value of parameters in simulation

	Area 1	Area 2	Area 3	Area 4
M_i (p.u.)	5.22	3.98	4.49	4.22
A_i (p.u.)	1.60	1.22	1.38	1.42
T_{doi} (s)	5.54	7.41	6.11	6.22
X_{di} (p.u.)	1.84	1.62	1.80	1.94
X'_{di} (p.u.)	0.25	0.17	0.36	0.44
E_{fdi} (p.u.)	4.41	4.20	4.37	4.45
B_{ii} (p.u.)	−49.61	−61.66	−52.17	−40.18
q_i ($\$10^4/h$)	1.00	0.75	1.50	0.50

(12.23) with the parameter $\beta = 14$. It can be seen that the action results of these two types of regulations are almost the same. Both of them can drive the frequency derivation to 0 when the power demand changes. The power generations of the four areas take a different share of the power demand according to the cost function in order to minimize the total cost. Moreover, the voltage magnitudes do not deviate much from their nominal value of 1 p.u. under the two types of regulations, which is acceptable in practical applications. Therefore, the optimal frequency regulation problem is addressed, and the effectiveness of the constructed secure distributed optimal frequency regulation in the case of no cyber-attack is verified here.

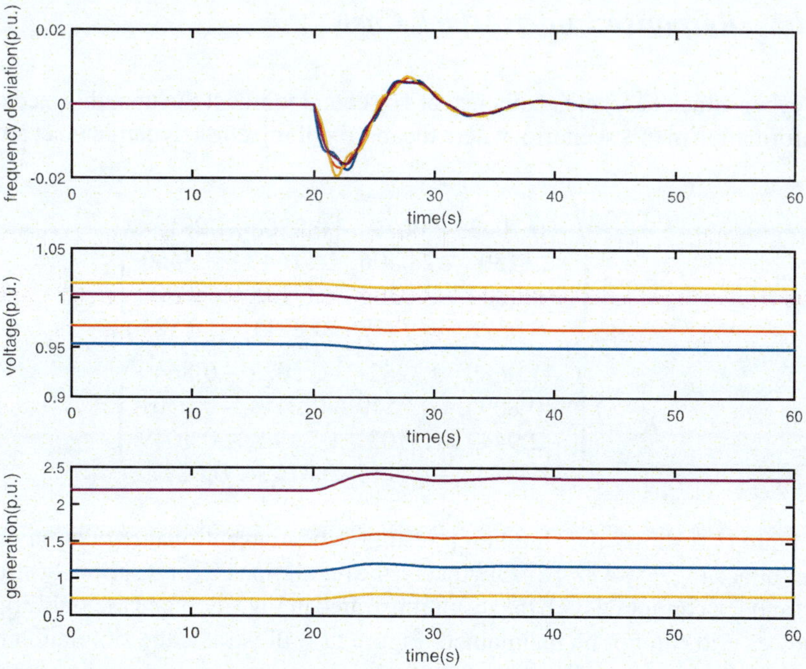

Fig. 12.3 Response of conventional regulation without cyber-attack

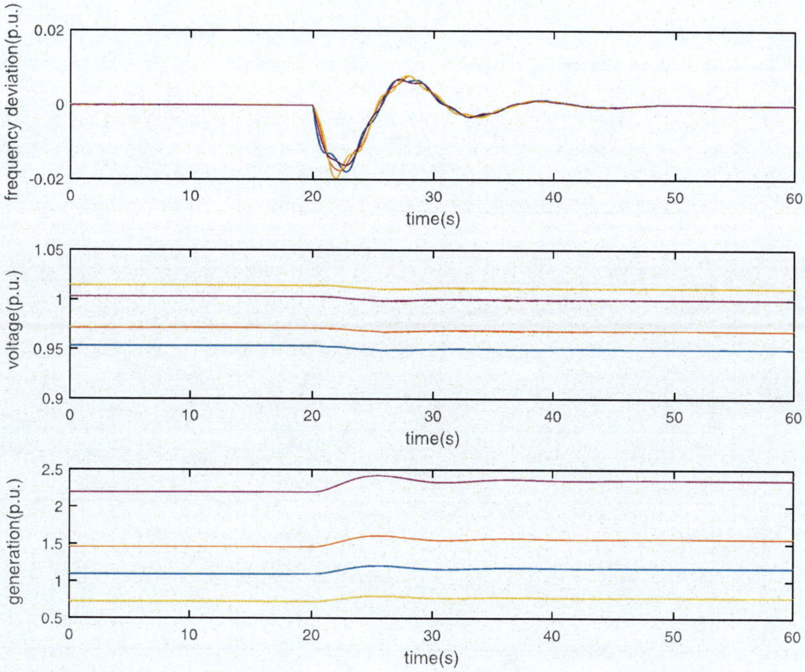

Fig. 12.4 Response of secure regulation without cyber-attack

12.4.2 Response Under Attack Case

The cyber-attack (12.19) is launched at 10 second to inject the unauthorized changes of information in this scenario, where the matrix of injection dynamic is set arbitrarily as

$$F_a = \begin{pmatrix} -0.7043 & 2.0679 & -0.2568 & 4.2815 \\ 2.1130 & -2.0406 & 1.6164 & -0.3450 \\ -6.0804 & -3.0383 & -2.7128 & 5.8092 \\ -0.5326 & -0.7417 & -0.8711 & -1.1329 \end{pmatrix},$$

$$B_a = \begin{pmatrix} -1.0689 & 0.3252 & -0.1022 & -0.8649 \\ -0.8095 & -0.7549 & -0.2414 & -0.0301 \\ -2.9443 & 1.3703 & 0.3192 & -0.1649 \\ 1.4384 & -1.7115 & 0.3129 & 0.6277 \end{pmatrix}$$

Figure 12.5 shows the simulation result of the conventional regulation under the cyber-attack (12.15)–(12.16). Obviously, the regulation effect is seriously affected by cyber-attack compared with the result illustrated in Fig. 12.3, and the stable operation of power grid can not be maintained. Figure 12.6 illustrates the simulation result of the constructed secure distributed optimal frequency regulation under cyber-attack (12.37)–(12.39). Similar to the result shown in Fig. 12.4, the frequency deviation is

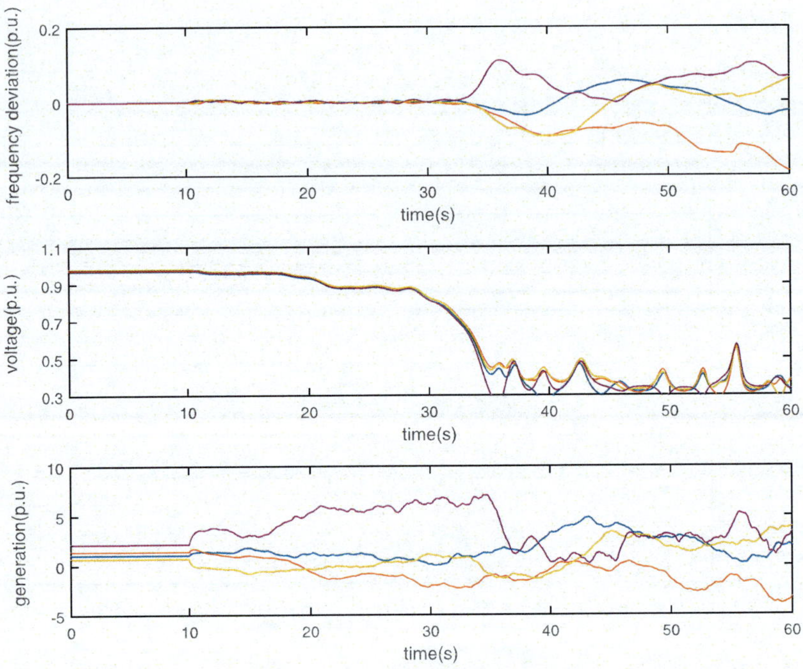

Fig. 12.5 Response of conventional regulation under cyber-attack

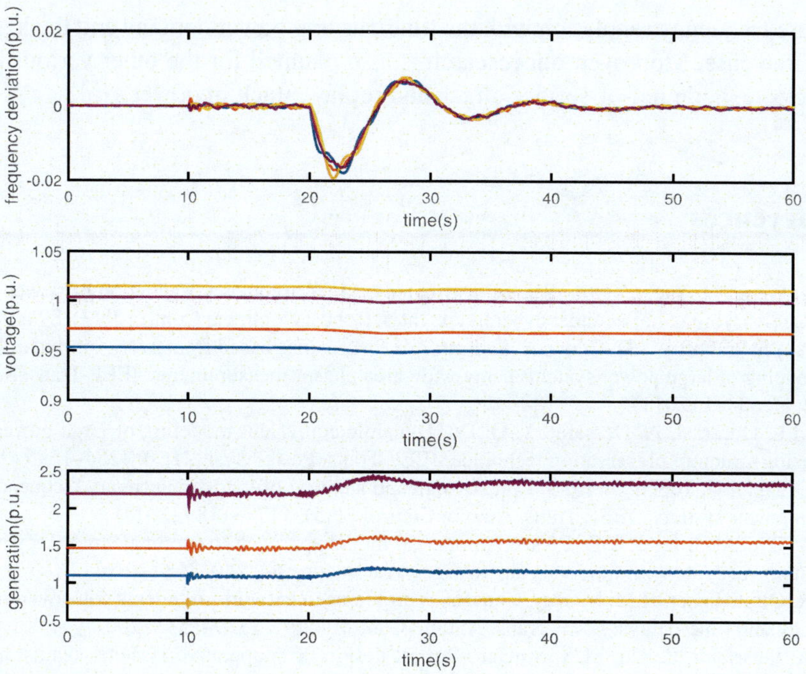

Fig. 12.6 Response of secure regulation under cyber-attack

restored and the power generation is allocated to minimize the cost function. This implies that the optimal frequency regulation task is performed under the secure regulation even in the case of cyber-attack. Mentioned that although the states of power grid fluctuate in a small range around the steady state as shown in Fig. 12.6, it still fulfills the requirement of optimal frequency regulation in practical power grid.

12.5 Conclusion

The problem of secure distributed optimal frequency regulation of power grid under false data injection attack has been investigated in this chapter. The false data injection attacks launched by adversary injects unauthorized changes into the transmitted information among areas to destroy the optimal frequency regulation task with stealthy characteristic. The model of cyber-attack is constructed and its perniciousness on optimal frequency regulation is analyzed. A novel type of secure distributed optimal frequency regulation is developed by introducing secure-based communication network to improve the resilience against cyber-attack. It is proved theoretically that the secure regulation can restore the frequency deviation and retain the economic efficiency simultaneously even under the considered cyber-attack, and the simulation result confirms the effectiveness of theoretical method. Our future work concerns the secure frequency regulation with the time-varying power demand and the lossy electric line case. Moreover, our research is also planned for the other various kinds of attacks, e.g.. denial-of-service attack and replay attack in power grid.

References

1. S. Weng, D. Yue, C. Dou, Secure distributed optimal frequency regulation of power grid with time-varying voltages under cyberattack. Int. J. Robust Nonlinear Control **30**(3), 894–909 (2020)
2. A. Chakrabortty, J.H. Chow, A. Salazar, A measurement-based framework for dynamic equivalencing of large power systems using wide-area phasor measurements. IEEE Trans. Smart Grid **2**(1), 68–81 (2011)
3. M.L. Ourari, L.A. Dessaint, V.-Q. Do, Dynamic equivalent modeling of large power systems using structure preservation technique. IEEE Trans. Power Syst. **21**(3), 1284–1295 (2006)
4. C. Zhao, U. Topcu, N. Li, S. Low, Design and stability of load-side primary frequency control in power systems. IEEE Trans. Autom. Control **59**(5), 1177–1189 (2014)
5. S. Trip, M. Burger, C.D. Persis, An internal model approach to (optimal) frequency regulation in power grids with time-varying voltages. Automatica **64**, 240–253 (2016)
6. R. Olfati-Saber, R.M. Murray, Consensus problems in networks of agents with switching topology and time-delays. IEEE Trans. Autom. Control **49**(9), 1520–1533 (2004)
7. A. Gusrialdi, Z. Qu, M.A. Simaan, Robust design of cooperative systems against attacks, in *2014 American Control Conference* (2014), pp. 1456–1462
8. L.-Y. Lu, C.-C. Chu, Consensus-based droop control synthesis for multiple dics in isolated micro-grids. IEEE Trans. Power Syst. **30**(5), 2243–2256 (2015)
9. S. Nabavi, A. Chakrabortty, Topology identification for dynamic equivalent models of large power system networks, in *2013 American Control Conference* (2013), pp. 1138–1143

Index

D. Yue et al., *Secure Control of Networked Control Systems and Its Applications*,
https://doi.org/10.1007/978-981-33-6730-2